Gertz Likhtenshtein

Solar Energy Conversion

Related Titles

Strehmel, B., Strehmel, V., Malpert, J. H.
Applied and Industrial Photochemistry
2012
ISBN: 978-3-527-32668-6

Scheer, R., Schock, H.-W.
Chalcogenide Photovoltaics
Physics, Technologies, and Thin Film Devices
2011
ISBN: 978-3-527-31459-1

Alkire, R. C., Kolb, D. M., Lipkowski, J., Ross, P. (Eds.)
Photoelectrochemical Materials and Energy Conversion Processes
2010
ISBN: 978-3-527-32859-8

Würfel, P.
Physics of Solar Cells
From Basic Principles to Advanced Concepts
2009
ISBN: 978-3-527-40857-3

Pagliaro, M., Palmisano, G., Ciriminna, R.
Flexible Solar Cells
2008
ISBN: 978-3-527-32375-3

De Vos, A.
Thermodynamics of Solar Energy Conversion
2008
ISBN: 978-3-527-40841-2

Brabec, C., Scherf, U., Dyakonov, V. (Eds.)
Organic Photovoltaics
Materials, Device Physics, and Manufacturing Technologies
2008
ISBN: 978-3-527-31675-5

Gertz Likhtenshtein

Solar Energy Conversion

Chemical Aspects

WILEY-VCH Verlag GmbH & Co. KGaA

The Author

Prof. Gertz Likhtenshtein
Department of Chemistry
Ben Gurion University
P.O. Box 653
84105 Beer-Sheva
Israel

All books published by **Wiley-VCH** are carefully produced. Nevertheless, authors, editors, and publisher do not warrant the information contained in these books, including this book, to be free of errors. Readers are advised to keep in mind that statements, data, illustrations, procedural details or other items may inadvertently be inaccurate.

Library of Congress Card No.: applied for

British Library Cataloguing-in-Publication Data
A catalogue record for this book is available from the British Library.

Bibliographic information published by the Deutsche Nationalbibliothek
The Deutsche Nationalbibliothek lists this publication in the Deutsche Nationalbibliografie; detailed bibliographic data are available on the Internet at http://dnb.d-nb.de.

© 2012 Wiley-VCH Verlag & Co. KGaA, Boschstr. 12, 69469 Weinheim, Germany

All rights reserved (including those of translation into other languages). No part of this book may be reproduced in any form – by photoprinting, microfilm, or any other means – nor transmitted or translated into a machine language without written permission from the publishers. Registered names, trademarks, etc. used in this book, even when not specifically marked as such, are not to be considered unprotected by law.

Composition Thomson Digital, Noida, India
Printing and Binding Markono Print Media Pte Ltd, Singapore
Cover Design Schulz Grafik-Design, Fußgönheim

Printed in Singapore
Printed on acid-free paper

Print ISBN: 978-3-527-32874-1

Contents

Preface IX

1 **Electron Transfer Theories** 1
1.1 Introduction 1
1.2 Theoretical Models 1
1.2.1 Basic Two States Models 1
1.2.1.1 Landau–Zener Model 1
1.2.1.2 Marcus Model 3
1.2.1.3 Electronic and Nuclear Quantum Mechanical Effects 5
1.2.2 Further Developments in the Marcus Model 7
1.2.2.1 Electron Coupling 7
1.2.2.2 Driving Force and Reorganization Energy 9
1.2.3 Zusman Model and its Development 17
1.2.4 Effect of Nonequilibrity on Driving Force and Reorganization 21
1.2.5 Long-Range Electron Transfer 24
1.2.6 Spin Effects on Charge Separation 28
1.2.7 Electron–Proton Transfer Coupling 29
1.2.8 Specificity of Electrochemical Electron Transfer 33
1.3 Concerted and Multielectron Processes 38
References 40

2 **Principal Stages of Photosynthetic Light Energy Conversion** 45
2.1 Introduction 45
2.2 Light-Harvesting Antennas 46
2.2.1 General 46
2.2.2 Bacterial Antenna Complex Proteins 47
2.2.2.1 The Structure of the Light-Harvesting Complex 47
2.2.2.2 Dynamic Processes in LHC 48
2.2.3 Photosystems I and II Harvesting Antennas 49
2.3 Reaction Center of Photosynthetic Bacteria 53
2.3.1 Introduction 53
2.3.2 Structure of RCPB 56
2.3.3 Kinetics and Mechanism of Electron Transfer in RCPB 58

2.3.4	Electron Transfer and Molecular Dynamics in RCPB	64
2.4	Reaction Centers of Photosystems I and II	68
2.4.1	Reaction Centers of PS I	69
2.4.2	Reaction Center of Photosystem II	72
2.5	Water Oxidation System	76
	References	84
3	**Photochemical Systems of the Light Energy Conversion**	**91**
3.1	Introduction	91
3.2	Charge Separation in Donor–Acceptor Pairs	92
3.2.1	Introduction	92
3.2.2	Cyclic Tetrapyrroles	93
3.2.3	Miscellaneous Donor–Acceptor Systems	101
3.2.4	Photophysical and Photochemical Processes in Dual Flourophore–Nitroxide Molecules (FNO)	106
3.2.4.1	System 1	107
3.2.4.2	Systems 2	109
3.3	Electron Flow through Proteins	113
3.3.1	Factors Affecting Light Energy Conversion in Dual Fluorophore–Nitroxide Molecules in a Protein	116
3.3.2	Photoinduced Interlayer Electron Transfer in Lipid Films	118
	References	121
4	**Redox Processes on Surface of Semiconductors and Metals**	**127**
4.1	Redox Processes on Semiconductors	127
4.1.1	Introduction	127
4.1.2	Interfacial Electron Transfer Dynamics in Sensitized TiO_2	127
4.1.3	Electron Transfer in Miscellaneous Semiconductors	130
4.1.3.1	Single-Molecule Interfacial Electron Transfer in Donor–Bridge–Nanoparticle Acceptor Complexes	130
4.1.4	Redox Processes on Carbon Materials	133
4.2	Redox Processes on Metal Surfaces	136
4.3	Electron Transfer in Miscellaneous Systems	144
	References	147
5	**Dye-Sensitized Solar Cells I**	**151**
5.1	General Information on Solar Cells	151
5.2	Dye-Sensitized Solar Cells	153
5.2.1	General	153
5.2.2	Primary Grätzel DSSC	155
5.3	DSSC Components	158
5.3.1	Sensitizers	158
5.3.1.1	Ruthenium Complexes	158
5.3.1.2	Metalloporphyrins	159
5.3.1.3	Organic Dyes	161

5.3.1.4	Semiconductor Sensitizes *167*	
5.3.2	Photoanode *168*	
5.3.3	Injection and Recombination *171*	
5.3.4	Charge Carrier Systems *175*	
5.3.5	Cathode *182*	
5.3.6	Solid-State DSSC *185*	
	References *189*	

6 Dye-Sensitized Solar Cells II *199*
6.1 Optical Fiber DSSC *199*
6.2 Tandem DSSC *202*
6.3 Quantum Dot Solar Cells *208*
6.4 Polymers in Solar Cells *211*
6.5 Fabrication of Solar Cell Components *219*
6.6 Fullerene-Based Solar Cells *223*
References *229*

7 Photocatalytic Reduction and Oxidation of Water *235*
7.1 Introduction *235*
7.2 Photocatalytic Dihydrogen Production *236*
7.2.1 Photocatalytic H_2 Evolution over TiO_2 *236*
7.2.2 Miscellaneous Semiconductor Photocatalysts for H_2 Evolution *237*
7.2.3 Photocatalytic H_2 Evolution from Water Based on Platinum and Palladium Complexes *240*
7.3 Water Splitting into O_2 and H_2 *241*
7.3.1 Thermodynamics and Feasable Mechanism of the Water Splitting *241*
7.3.2 Mn Clusters as Water Oxidizing Photocatalysts *242*
7.3.2.1 Structure and Catalytic Activity of Cubane Manganese Clusters *242*
7.3.2.2 Catalytic Activity and Mechanism of WOS in Manganese Clusters *244*
7.3.3 Heterogeneous Catalysts for WOS *248*
7.3.3.1 General *248*
7.3.3.2 Photocatalysts Based on Titanium Oxides *250*
7.3.3.3 Miscellaneous Semiconductors for the WOS Catalysis *251*
References *256*

Conclusions *261*
References *263*

Index *265*

Preface

World energy consumption is about 4.7 10^{20} J and is expected to grow at the rate of 2% each year for the next 25 years. Since the emergence of the apparition of the impending energy crises, various avenues are being explored to replace fossil fuels with renewable energy from solar power. In the past decades, the development of advanced molecular materials and nanotechnology, solar cells, and dye-sensitized solar cells, first of all, has initiated a new set of ideas that can dramatically improve energy conversion efficiency and reduce prices of alternative energy sources. Nevertheless, there are many fundamental problems to be solved in this area. Commercial competition of the new materials with existing fossil energy sources remains one of the most challenging problems for mankind.

This book embraces all principal aspects of structure and physicochemical action mechanisms of dye-sensitized collar cells (DSSCs) and photochemical systems of light energy conversion and related areas. A large body of literature exists on this subject and many scientists have made important contributions to this the field. The Internet program SkiFinder shows 44979 references for "dye sensitized" and 8493 references for "dye-sensitized collar cells." It is impossible in the space allowed in this book to give a representative set of references. The author apologizes to those he has not been able to include. More than 1000 references are given in the book, which should provide a key to essential relevant literature.

Chapter 1 of the monograph is a brief outline of the contemporary theories of electron transfer in donor–acceptor pairs and between a dye and a semiconductor. Principal stages of light energy conversion in biological photosystems, in which the Nature demonsrates excellent examples for solving problems of conversion of light energy to energy of chemical compounds, are described in Chapter 2. The light energy conversion in donor–acceptor pairs in solution and on templates is the subject of Chapter 3. Chapter 4 describes redox processes on the surface of semiconductors and metals. Chapters 1–4 form the theoretical and experimental background for the central Chapters 5 and 6. In Chapter 5, a general survey is made of fundamentals of the primary Gertzel dye-sensitized solar cell and its rapid development. Advantages in design of new type of dye-sensitized solar cells such as optical fiber, tandem, and solid-state DSSC and fabrication of its components are reviewed in Chapter 6. Chapter 7 gathers information on recent progress made in photocatalytic reduction and oxidation of water.

The monograph is intended for scientists and engineers working on dye-sensitized collar cells and other molecular systems of light energy conversion and related areas such as photochemistry and photosynthesis and its chemical mimicking. The book can be used as a subsidiary manual for instruction for graduate and undergraduate students of university chemistry, physics, and biophysics departments.

Gertz Likhtenshtein

1
Electron Transfer Theories

1.1
Introduction

Electron transfer (ET) is one of the most ubiquitous and fundamental phenomena in chemistry, physics, and biology [1–34]. Nonradiative and radiative ET are found to be a key elementary step in many important processes involving isolated molecules and supermolecules, ions and excess electrons in solution, condensed phase, surfaces and interfaces, electrochemical systems and biology, and in solar cells, in particular.

As a light microscopic particle, an electron easily tunnels through a potential barrier. Therefore, the process is governed by the general tunneling law formulated by Gamov [35]. The principal theoretical cornerstone for condensed phase ET was laid by Franck and Libby (1949–1952) who asserted that the Franck–Condon principle is applicable not only to the vertical radiative processes but also to nonradiative horizontal electron transfer. The next decisive step in the field was taken by Marcus and his colleagues [2, 17, 36] and Hash [37]. These authors articulated the need for readjustment of the coordination shells of reactants in self-exchange reactions and of the surrounding solvent to the electron transfer. They also showed that the electronic interaction of the reactants gives rise to the splitting at the intersection of the potential surfaces, which leads to a decrease in the energy barrier.

1.2
Theoretical Models

1.2.1
Basic Two States Models

1.2.1.1 Landau–Zener Model
The nonadiabatic electron transfer between donor (D) and acceptor (A) centers is treated by the Fermi's golden rule (FGR) [38]

1 Electron Transfer Theories

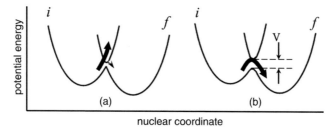

Figure 1.1 Variation in the energy of the system along the reaction coordinate: (a) diabatic terms of the reactant (i) and products (f); (b) adiabatic terms of the ground state (f) and excited state (f). V is the resonance integral [9].

$$k_{ET} = \frac{2\pi V^2 FC}{h} \tag{1.1}$$

where FC is the Franck–Condon factor related to the probability of reaching the terms crossing area for account of nuclear motion and V is an electronic coupling term (resonance integral) depending on the overlap of electronic wave functions in initial and final states of the process (see Figure 1.1).

At the transition of a system from one state to another, with a certain value of the coordinate Q_{tr}, the energy of the initial (i) and final (f) states of energy terms is the same and the law of energy conservation permits the term–term transition

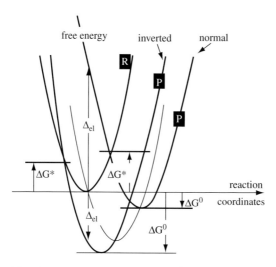

Figure 1.2 Energy versus reaction coordinates for the reactants and the products for normal and inverted reactions, and the inversion curve. The electron in the initial state requires a positive excitation energy Δel for the normal reaction and a negative excitation energy - Δel for the inverted reaction (which could be directly emitted as light). There is a positive energy barrier ΔF^* in both cases between the reactants and the products that requires thermal activation for the reaction to occur. This energy barrier as well as the energy for a direct electron excitation vanishes for the inversion curve and then the electron transfer becomes ultrafast. Reproduced from Ref. [61].

(Figure 1.2). Generally, the rate constant of the transition in the crossing area is dependent on the height of the energetic barrier (activation energy, E_a), the frequency of reaching of the crossing area (ψ), and the transition coefficient (κ):

$$k_{tr} = \kappa \nu \exp(-E_a) \tag{1.2}$$

The transition coefficient κ is related to the probability of the transition in the crossing area (P) and is described by the Landau–Zener equation [39, 40]:

$$\kappa = \frac{2P}{(1+P)} \tag{1.3}$$

where

$$P = 1 - \exp\left[\frac{-4\pi^2 V^2}{h\nu(S_i - S_f)}\right] \tag{1.4}$$

V is the electronic coupling factor (the resonance integral), ν is the velocity of nuclear motion, and S_i and S_f are the slopes of the initial and final terms in the Q_{tr} region. If the exponent of the exponential function is small, then

$$P = \frac{4\pi^2 V^2}{h\nu(S_i - S_f)} \tag{1.5}$$

and the process is nonadiabatic. Thus, the smaller the magnitude of the resonance integral V, the smaller is the probability of nonadiabatic transfer. The lower the velocity of nuclear motion and smaller the difference in the curvature of the terms, the smaller is the probability of nonadiabatic transfer. At $P = 1$, the process is adiabatic and treated by classical Arrhenius or Eyring equations.

The theory predicts a key role by electronic interaction, which is quantitatively characterized by the value of resonance integral V in forming energetic barrier. If this value is sufficiently high, the terms are split with a decreasing activation barrier and the process occurs adiabatically. In another nonadiabatic extreme, where the interaction in the region of the coordinate Q_{tr} is close to zero, the terms practically do not split, and the probability of transition i → f is very low.

1.2.1.2 Marcus Model

According to the Marcus model [2, 3, 5, 17, 36], the distortion of the reactants, products, and solvent from their equilibrium configuration is described by identical parabolas, shifted relative to each other according to the driving force of the value of the process, standard Gibbs free energy ΔG_0 (Figure 1.2). Within the adiabatic regime (strong electronic coupling, the resonance integral $V > 200\,\text{cm}^{-1}$), in the frame of the Eyring theory of the transition state, the value of the electron transfer rate constant is

$$k_{ET} = \left(\frac{h\nu}{k_B T}\right) \exp-\left[\frac{(\lambda + \Delta G_0)^2}{4\lambda k_B T}\right] \tag{1.6}$$

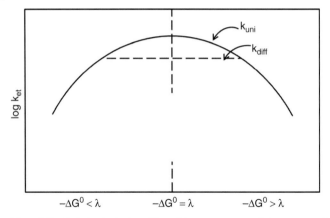

Figure 1.3 Variation in the logarithm of the rate constant of electron transfer with the driving force for the reaction after Marcus [50].

where λ is the reorganization energy defined as energy for the vertical electron transfer without replacement of the nuclear frame. Equation 1.6 predicts the log $k_{ET} - \Delta G_0$ relationships depending on the relative magnitudes of λ and ΔG_0 (Figure 1.3): (1) $\lambda > \Delta G_0$, when log k increases if ΔG_0 decreases (normal Marcus region); (2) $\lambda = \Delta G_0$, the reaction becomes barrierless; and (3) $\lambda < \Delta G_0$, when log k decreases with increasing driving force.

The basic Marcus equation is valid in following conditions:

1) All reactive nuclear modes, that is, local nuclear modes, solvent inertial polarization modes, and some other kinds of collective modes, are purely classical. The electronic transition in the ET process is via the minimum energy at the crossing of the initial and final state potential surfaces.
2) The potential surfaces are essentially diabatic surfaces with insignificant splitting at the crossing and of parabolic shape. The latter reflects harmonic molecular motion with equilibrium nuclear coordinate displacement and a linear environmental medium response.
3) The vibrational frequencies and the normal modes are the same in the initial and final states.

The Marcus theory also predicts the Brönsted slope magnitude in the normal Marcus region:

$$\alpha_B = \frac{d\Delta G^{\#}}{d\Delta G_0} = \frac{1}{2}\left(1 + \frac{\Delta G_0}{\lambda}\right) \tag{1.7}$$

The processes driving force (ΔG_0) can be measured experimentally or calculated theoretically. For example, when solvation after the process of producing photoinitiated charge pairing is rapid, ΔG_0 can be approximately estimated by the following equation:

$$\Delta G_0 = E_{D/D^+} - (E_{A/A^+} + E_{D^*}) - \frac{e^2}{\varepsilon}(r_{D^+} + r_{A^-}) \tag{1.8}$$

where E_{D/D^+} and $E_{A^+/A}$ are the standard redox potential of the donor and acceptor, respectively, E_{D^*} is the energy of the donor exited state, r_{D^+} and r_{A^-} are the radii of the donor and acceptor, respectively, and ε is the medium dielectric constant.

The values of λ can be roughly estimated within the framework of a simplified model suggesting electrostatic interactions of oxidized donor (D^+) and reduced acceptor (A^-) of radii r_{D^+} and r_{A^-} separated by the distance R_{DA} with media of dielectric constant e_0 and refraction index n:

$$\lambda = \frac{e^2}{2}\left(\frac{1}{n^2} - \frac{1}{e_0}\right)\left(\frac{1}{r_{D^+}} + \frac{1}{r_{A^-}} - \frac{2}{R_{DA}}\right) \tag{1.9}$$

1.2.1.3 Electronic and Nuclear Quantum Mechanical Effects

The nonadiabatic electron transfer between donor (D) and acceptor (A) centers is treated by the FGR (Equation 1.1). The theory of nonadiabatic electron transfer was developed by Levich, Dogonadze, and Kuznetsov [41–43]. These authors, utilizing the Landau–Zener theory for the intersection area crossing suggesting harmonic one-dimensional potential surface, proposed a formula for nonadiabatic ET energy:

$$k_{ET} = \frac{2\pi V^2}{h\sqrt{4\pi\lambda k_B T}} \exp\left[-\frac{(\lambda + \Delta G_0)^2}{4\lambda k_B T}\right] \tag{1.10}$$

Therefore, the maximum rate of ET at $\lambda = \Delta G_0$ is given by

$$k_{ET\,(max)} = \frac{2\pi V^2}{h\sqrt{4\pi\lambda k_B T}} \tag{1.11}$$

Involvement of intramolecular high-frequency vibrational modes in electron transfer was considered [44–49]. For example, when the high-frequency mode ($h\nu$) is in the low-temperature limit and solvent dynamic behavior can be treated classically [1], the rate constant for nonadiabatic ET in the case of parabolic terms is given by

$$k_{ET} = \frac{\sum_j 2\pi F_j V^2}{h\lambda k_B T} \exp\left[-\frac{(jh\nu + \lambda_s + \Delta G_0)^2}{4\lambda k_B T}\right] \tag{1.12}$$

where j is the number of high-frequency modes, $F_j = e^{-S}/j!$, $S = \lambda_v/h\nu$, and λ_v and λ_s denote the reorganization inside the molecule and solvent, respectively.

In the case of thermal excitation of the local molecular and medium high-frequency modes, theories mentioned before predicted the classical Marcus relation in the normal Marcus region. While in the inverted region, significant deviation on the parabolic energy gap dependence is expected (Figure 1.4). The inverted Marcus region cannot be experimentally observed if the stabilization of the first electron transfer product for the accounting of the high-frequency vibrational mode occurs faster than the equilibrium of the solvent polarization with the momentary charge distribution can be established. Another source of the deviation is the nonparabolic shape of the activation barrier [1].

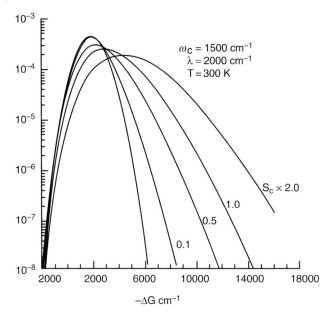

Figure 1.4 The energy gap dependence of the nuclear Franck–Condon factor, which incorporates the role of the high-frequency intramolecular modes. $S_c = \Delta/2$ is the dimensionless electron–vibration coupling, given in terms that reduce replacement (Δ) between the minimum of the nuclear potential surfaces of the initial and final electronic states [1]. Reproduced with permission.

A nonthermal electron transfer assisted by an intramolecular high-frequency vibrational mode has been theoretically investigated [18]. An analytical expression for the nonthermal transition probability in the framework of the stochastic point transition approach has been derived. For the strong electron transfer, the decay of the product state can vastly enhance the nonthermal transition probability in the whole range of parameters except for the areas where the probability is already close to unity. If the initial ion state is formed either by forward electron transfer or by photoexcitation, it may be visualized as a wave packet placed on the ion free energy term above the ion and the ground-state terms intersection (see Figure 6.1).

The Marcus inverted region cannot be observed experimentally when term-to-term transition in the crossing region is not a limiting step of the process as a whole (Figure 1.3) [50]. When ET reaction is very fast in the region of maximum rate, the process can be controlled by diffusion and, therefore, is not dependent on λ, V^2, and ΔG_0. The integral encounter theory (IET) has been extended to the reactions limited by diffusion along the reaction coordinate to the level crossing points where either thermal or hot electron transfer occurs [18]. IET described the bimolecular ionization of the instantaneously excited electron donor D^* followed by the hot geminate backward transfer that precedes the ion pair equilibration

$$D^* + A \underset{w_{21}}{\overset{w_{12}}{\longleftrightarrow}} [D^+ \underset{DA}{\overset{}{\cdots}} A^-] \overset{\bar{\varphi}}{\longrightarrow} D^+ + A^-$$
$$\phantom{D^* + A \underset{w_{21}}{\overset{w_{12}}{\longleftrightarrow}} [D^+} w_{32} \updownarrow w_{23}$$

and its subsequent thermal recombination tunneling is strong. It was demonstrated that the fraction of ion pairs that avoids the hot recombination is much smaller than their initial number when the electron tunneling is strong. The kinetics of recombination/dissociation of photogenerated radical pairs (RPs) was described with a generalized model (GM), which combines exponential models (EMs) and contact models (CMs) of cage effect dynamics [31]. Kinetics of nonthermal electron transfer controlled by the dynamical solvent effect was discussed in Ref. [11]. Recombination of ion pairs created by photoexcitation of viologen complexes is studied by a theory accounting for diffusion along the reaction coordinate to the crossing points of the electronic terms. The kinetics of recombination convoluted with the instrument response function were shown to differ qualitatively from the simplest exponential decay in both the normal and the inverted Marcus regions. The deviations of the exponentiality are minimal only in the case of activationless recombination and are reduced even more by taking into consideration a single quantum mode assisting the electron transfer

1.2.2
Further Developments in the Marcus Model

1.2.2.1 Electron Coupling
Variational transition-state theory was used to compute the rate of nonadiabatic electron transfer for a model of two sets of shifted harmonic oscillators [51]. The relationship to the standard generalized Langevin equation model of electron transfer was established and provided a framework for the application of variational transition-state theory in simulation of electron transfer in a microscopic (nonlinear) bath. A self-consistent interpretation based on a hybrid theoretical analysis that includes *ab initio* quantum calculations of electronic couplings, molecular dynamics simulations of molecular geometries, and Poisson–Boltzmann computations of reorganization energies was offered [52]. The analysis allowed to estimate the following parameters of systems under investigation: (1) reorganization energies, (2) electronic couplings, (3) access to multiple conformations differing both in reorganization energy and in electronic coupling, and (4) donor–acceptor coupling dependence on tunneling energy, associated with destructively interfering electron and hole-mediated coupling pathways. Fundamental arguments and detailed computations show that the influence of donor spin state on long-range electronic interactions is relatively weak.

The capability of multilevel Redfield theory to describe ultrafast photoinduced electron transfer reactions and the self-consistent hybrid method was investigated [53]. Adopting a standard model of photoinduced electron transfer in a condensed phase environment, the authors considered electron transfer reactions in the normal and inverted regimes, as well as for different values of the electron transfer parameters, such as reorganization energy, electronic coupling, and temperature.

A semiclassical theory of electron transfer reactions in Condon approximation and beyond was developed in [54]. The effect of the modulation of the electronic wave functions by configurational fluctuations of the molecular environment on the kinetic parameters of electron transfer reactions was discussed. A new formula for the transition probability of nonadiabatic electron transfer reactions was obtained and regular method for the calculation of non-Condon corrections was suggested. Quantum Kramers-like theory of the electron transfer rate from weak-to-strong electronic coupling regions using Zhu–Nakamura nonadiabatic transition formulas was developed to treat the coupled electronic and nuclear quantum tunneling probability [55]. The quantum Kramers theory to electron transfer rate constants was generalized. The application in the strongly condensed phase manifested that the approach correctly bridges the gap between the nonadiabatic (Fermi's golden rule) and adiabatic (Kramers theory) limits in a unified way, and leads to good agreement with the quantum path integral data at low temperature.

In work [56], electron transfer coupling elements were extracted from constrained density functional theory (CDFT). This method made use of the CDFT energies and the Kohn–Sham wave functions for the diabatic states. A method of calculation of transfer integrals between molecular sites, which exploits few quantities derived from density functional theory electronic structure computations and does not require the knowledge of the exact transition state coordinate, was conceived and implemented [57]. The method used a complete multielectron scheme, thus including electronic relaxation effects. The computed electronic couplings can then be combined with estimations of the reorganization energy to evaluate electron transfer rates. On the basis of the generalized nonadiabatic transition-state theory [58], the authors of the work [59, 60] presented a new formula for electron transfer rate, which can cover the whole range from adiabatic to nonadiabatic regime in the absence of solvent dynamics control. The rate was expressed as a product of the Marcus theory and a coefficient that represents the effects of nonadiabatic transition at the crossing seam surface. The numerical comparisons were performed with different approaches and the present approach showed an agreement with the quantum mechanical numerical solutions from weak to strong electronic coupling.

A nonadiabatic theory for electron transfer and application to ultrafast catalytic reactions has been discussed in Ref. [61]. The author proposed a general formalism that not only extends those used for the standard theory of electron transfer but also becomes equivalent to it far from the inversion point. In the vicinity of the inversion point when the energy barrier for ET is small, the electronic frequencies become of the order of the phonon frequencies and the process of electron tunneling is nonadiabatic because it is strongly coupled to the phonons. It was found that when the model parameters are fine-tuned, ET between donor and acceptor becomes reversible and this system is a coherent electron–phonon oscillator (CEPO). The acceptor that does not capture the electron may play the role of a catalyst (Figure 1.5). Thus, when the catalyst is fine-tuned with the donor in order to form a CEPO, it may trigger an irreversible and ultrafast electron transfer (UFET) at low temperature between the donor and an extra acceptor. Such a trimer system may be regulated by small perturbations and behaves as a molecular transistor.

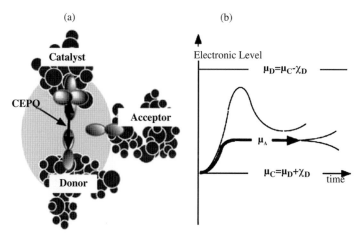

Figure 1.5 Principle of ET with a coherent electron–phonon oscillator [61].

Two weakly coupled molecular units of donor and catalyst generate a CEPO. This system is weakly coupled to a third unit, the acceptor (Figure 1.5a). An electron initially on the donor generates an oscillation of the electronic level of the CEPO. If the bare electronic level of a third molecular unit (acceptor) is included in the interval of variation, as soon as resonance between the CEPO and the acceptor is reached, ET is triggered irreversibly to the acceptor (Figure 1.5b). The authors suggested that because of their ability to produce UFET, the concept of CEPOs could be an essential paradigm for understanding the physics of the complex machinery of living systems.

Perturbation molecular orbital (PMO) theory was used to estimate the electronic matrix element in the semiclassical expression for the rate of nonadiabatic electron transfer at ion–molecular collisions [62]. It was shown that the electron transfer efficiency comes from the calculated ET rate divided by the maximum calculated ET rate and by dividing the observed reaction rate by the collision rate, calculated by the PMO treatment of ion–molecular collision rates.

1.2.2.2 Driving Force and Reorganization Energy

Several works were devoted to models for medium reorganization and donor–acceptor coupling [63–78]. The density functional theory based on *ab initio* molecular dynamics method combines electronic structure calculation and statistical mechanics and was used for first-principles computation of redox free energies at one-electron energy [66]. The authors showed that this is implemented in the framework of the Marcus theory of electron transfer, exploiting the separation in vertical ionization and reorganization contributions inherent in Marcus theory. Direct calculation of electron transfer parameters through constrained density functional theory was a subject of the work by Wu and Van Voorhis [67]. It was shown that constrained density functional theory can be used to access diabatic potential energy surfaces in the Marcus theory of electron transfer, thus providing a means to directly calculate the driving force and the inner sphere reorganization energy. The influence

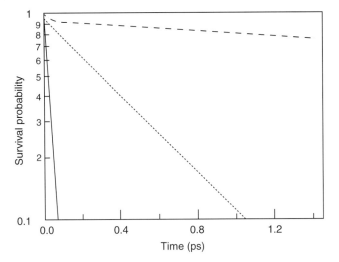

Figure 1.6 Survival probability as a function of time for D-B-A systems containing three (solid line), four (dotted line), and five (dashed line) subunits. All curves were calculated for the D-B-A system with the energy gap $\Delta\varepsilon$ between the donor and the equienergetic bridge equal to 1.2 eV. The value of the charge transfer integral V was taken to be equal to 0.3 eV [68].

of static and dynamic torsional disorder on the kinetics of charge transfer (CT) in donor–bridge–acceptor (D-B-A) systems has been investigated theoretically using a simple tight binding model [68]. Modeling of CT beyond the Condon approximation revealed two types of non-Condon (NC) effects. It was found that if τ_{rot} is much less than the characteristic time, τ_{CT}, of CT in the absence of disorder, the NC effect is static and can be characterized by rate constant for the charge arrival on the acceptor. For larger τ_{rot}, the NC effects become purely kinetic and the process of CT in the tunneling regime exhibits timescale invariance, the corresponding decay curves become dispersive, and the rate constant turns out to be time dependent. In the limit of very slow dynamic fluctuations, the NC effects in kinetics of CT were found to be very similar to the effects revealed for bridges with the static torsional disorder. The authors argued that experimental data reported in the literature for several D-B-A systems must be attributed to the multistep hopping mechanism of charge motion rather than to the mechanism of single-step tunneling. Survival probability as a function of time for D-B-A systems is shown in Figure 1.6.

The theory developed by Fletcher in Refs [71, 72] took into account the fact that charge fluctuations contribute to the activation of electron transfer, besides dielectric fluctuations. It was found that highly polar environments are able to catalyze the rates of thermally activated electron transfer processes because under certain well-defined conditions, they are able to stabilize the transient charges that develop on transition states. Plots of rate constant for electron transfer versus driving force are shown in Figure 1.7, which is drawn on the assumption that electron transfer is nonadiabatic and proceeds according to Dirac's time-dependent perturbation theory. On the

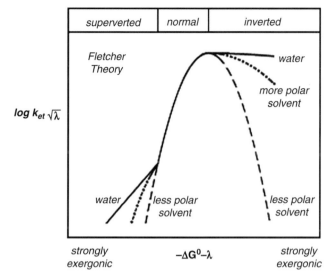

Figure 1.7 The rate constant for electron transfer (k_{ET}) as a function of the driving force ($-\Delta G^0$) and reorganization energy (λ) on the Fletcher theory [71, 72]. Note the powerful catalytic effect of polar solvents (such as water) on strongly exergonic reactions.

theory, the relative permittivity of the environment exerts a powerful influence on the reaction rate in the highly exergonic region (the "inverted" region) and in the highly endergonic region (the "superverted" region).

According to authors, nonadiabatic electron transfer is expected to be observed whenever there is small orbital overlap (weak coupling) between donor and acceptor states, so that overall electron transfer rates are slow compared to the media dynamics. For strongly exergonic electron transfer reactions that are activated by charge fluctuations in the environment, the activation energy was determined by the intersection point of thermodynamic potentials (Gibbs energies) of the reactants and products. The following equations for $G_{reactants}$ and $G_{products}$, which are the total Gibbs energies of the reactants and products (including their ionic atmospheres), respectively, were suggested:

$$G_{reactants} = \frac{1}{2} Q_1^2 \left(\frac{1}{4\pi\varepsilon_0}\right)\left(\frac{1}{\varepsilon(0)}\right)\left(\frac{1}{a_D} + \frac{1}{a_A} - \frac{2}{d}\right) \tag{1.13}$$

$$G_{products} = \frac{1}{2} Q_2^2 \left(\frac{1}{4\pi\varepsilon_0}\right)\left(\frac{1}{\varepsilon(\infty) + f_1\left[\varepsilon(0) - \varepsilon(\infty)\right]}\right)\left(\frac{1}{a_D} + \frac{1}{a_A} - \frac{2}{d}\right) \tag{1.14}$$

Q_1 and Q_2 are the charge fluctuations that build up on them, $\varepsilon(0)$ is the relative permittivity of the environment in the low-frequency limit (static dielectric constant), $\varepsilon(\infty)$ is the relative permittivity of the environment in the high-frequency limit ($\varepsilon \approx 2$), a_A is the radius of the acceptor in the transition state (including its ionic atmosphere), a_D is the radius of the electron donor in the transition state (including its ionic

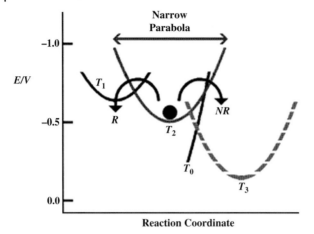

Figure 1.8 Superimposed Gibbs energy profiles in the vicinity of the electron trap T_2. Trapping is thermodynamically reversible, so the electron can return to T_0 radiatively (R) via T_1 or nonradiatively (NR) via the inverted region. Both routes are kinetically hindered by the extreme narrowness of the Gibbs energy parabola, however. This narrowness is conferred by the extremely nonpolar environment surrounding T_2. Trapping state T_3 is the final acceptor [72].

atmosphere), and f_1 is a constant ($0 < f_1 < 1$) that quantifies the extent of polar screening by the environment, d is the distance between the electron donor and acceptor. Figure 1.8 shows the Gibbs energy for electron transfer through an intermediate.

Authors of paper [73] focused on the microscopic theory of intramolecular electron transfer rate. They examined whether or not and/or under what conditions the widely used Marcus-type equations are applicable to displaced–distorted (D-D) and displaced–distorted–rotated (D-D-R) harmonic oscillator (HO) cases. For this purpose, the cumulant expansion (CE) method was applied to derive the ET rate constants for these cases. In the CE method, the analytical condition was derived upon which the Marcus-type equation of the Gaussian form was obtained for the D-D HO case. In the frame of theory, the following equation for the ET rate constant was derived:

$$W_{b \to a} = \frac{|J_{ab}|^2}{\hbar^2} \sqrt{\frac{\pi \hbar^2}{\lambda k_B T}} \exp\left(-\frac{[\hbar \omega_{ab} + \langle V_{ab}(0)\rangle - \lambda + \lambda]^2}{4\lambda k_B T}\right) \tag{1.15}$$

where $\Delta G_{ab} = \hbar \omega_{ab} + \langle V_{ab}(0)\rangle - \lambda$ and $\hbar \omega_{ab} = E_a - E_b$.

The quantity $\hbar \omega_{ab} + \langle V_{ab}(0)\rangle$ has the following physical meaning. The quantity $\langle V_{ab}(0)\rangle$ is the vibrational energy acquired in the final state through vertical or FC transition from the initial state, averaged over the initial vibrational states under condition of vibrational thermal equilibrium in the initial potential energy surface.

It was found that the reorganization energy and the free energy change for the D-D HO depend on the temperature. As a consequence, the preexponential factor of the ET rate shows a temperature dependence different from the usual Arrhenius

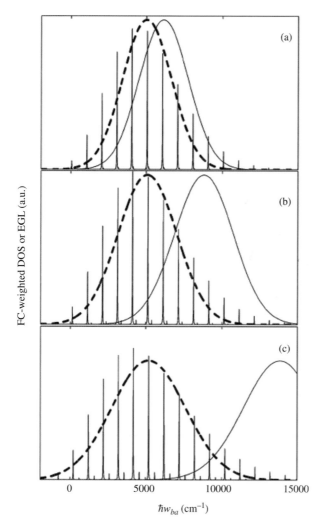

Figure 1.9 The dependent of the Franck–Condon weighted density 1 (FC DOS) (y-axis) on the degree of distortion for a model one-mode D-D HO system described in the text. The degrees of distortion are 0.9 in (a), 0.75 in (b), and 0.6 in (c). The peaks are the EGL calculated with the exact TCF method. The thick dashed lines denote the energy gap law (EGL) of the ET rate EGL calculated with the CE method using the same parameters. The thin smooth lines are the EGL calculated with the conventional Marcus theory [73].

behavior. The dependence of the Franck–Condon weighted density 1 on the degree of distortion for a model one-mode D-D HO system is presented in Figure 1.9.

The temperature dependence of the ET rate at different degrees of mixing for two modes whose frequencies are 100 and 30 cm^{-1} is shown in Figure 1.10.

The influence of spatial charge redistribution modeled by a change in the dipole moment of the reagent that experiences excitation on the dynamics of ultrafast

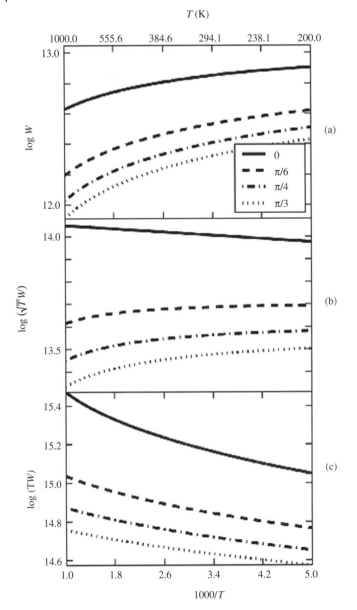

Figure 1.10 The temperature dependence of the ET rate at different degrees of mixing for two modes whose frequencies are 100 and 30 cm^{-1} in the lower level, without distortion. The abscissa is the inverse temperature, labeled in 1000/T. On the top, T (K) is also shown. (a) The ET rate on logarithmic scale versus the inverse temperature, for four different angles of rotation (labeled in the legend in radian). (b) The ET rate multiplied by the square root of the temperature, on logarithmic scale, versus the inverse temperature. (c) The ET rate multiplied by the temperature, on logarithmic scale, versus the inverse temperature [73].

photoinduced electron transfer was studied [22]. A two-center model based on the geometry of real molecules was suggested. The model described photoexcitation and subsequent electron transfer in a donor–acceptor pair. The rate of electron transfer was shown to depend substantially on the dipole moment of the donor at the photoexcitation stage and the direction of subsequent electron transfer. The authors of the work [74] have shown that the polarization fluctuation and the energy gap formulations of the reaction coordinate follow naturally from the Marcus theory of outer electron transfer. The Marcus formula modification or extension led to a quadratic dependence of the free energies of the reactant and product intermediates on the respective reaction coordinates. Both reaction coordinates are linearly related to the Lagrangian multiplier m in the Marcus theory of outer sphere electron transfer, so that m also plays the role of a natural reaction coordinate. When $m = 0$, $F^*(m = 0)$ is the equilibrium free energy of the reactant intermediate X^* at the bottom of its well, and when $m = -1$, $F(m = -1)$ is the corresponding equilibrium free energy of the product intermediate X. At $m = -1$ the free energy of reorganization of the solvent from its equilibrium configuration at the bottom of the reactant. A theory of electron transfer with torsionally induced non-Condon (NC) effects was developed by Jang and Newton [69]. The starting point of the theory was a generalized spin-boson Hamiltonian, where an additional torsional oscillator bilinearly coupled to other bath modes causes a sinusoidal non-Condon modulation. Closed form time-dependent nonadiabatic rate expressions for both sudden and relaxed initial conditions, which are applicable for general spectral densities and energetic condition, were derived. Under the assumption that the torsional motion is not correlated with the polaronic shift of the bath, simple stationary limit rate expression was obtained. Model calculation of ET illustrated the effects of torsional quantization and gating on the driving force and temperature dependence of the electron transfer rate. The Born–Oppenheimer (BO) formulation of polar solvation is developed and implemented at the semiempirical (PM3) CI level, yielding estimation of ET coupling elements (V_0) for intramolecular ET in several families of radical ion systems [70]. The treatment yielded a self-consistent characterization of kinetic parameters in a two-dimensional solvent framework that includes an exchange coordinate. The dependence of V_0 on inertial solvent contributions and on donor–acceptor separation was discussed (see Figure 1.11).

In the work [76], it was demonstrated that constrained density functional theory allowed to compute the three key parameters entering the rate constant expression: the driving force (ΔG^0), the reorganization energy (λ), and the electronic coupling H_{DA}. The results confirm the intrinsic exponential behavior of the electronic coupling with the distance separating D and A or, within the pathway paradigm, between two bridging atoms along the pathway. Concerning the "through space" decay factors, the CDFT results suggested that a systematic parameterization of the various kinds of weak interactions encountered in biomolecules should be undertaken in order to refine the global "through space" decay factors. Such a work has been initiated in this paper. The hydrogen bond term has also been adjusted. Besides the refinement of the distance component, the authors underlined the appearance of an angular dependence and the correlation factor between R and ϕ.

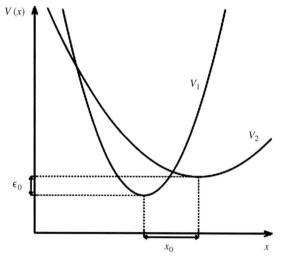

Figure 1.11 ET dynamics in a low temperature case for different reorganization energies (see details in Ref. [87, 88]).

Taking into account the volume of reagents, the theory gives the following Equation 1.16 [77]:

$$\lambda = \frac{e^2}{2}\left(\frac{1}{n^2} - \frac{1}{e_0}\right)\left(\frac{1}{r_{D^+}} + \frac{1}{r_{A^-}} - \frac{2}{R_{DA}} + \frac{[r_{D^+}^3 + r_{A^-}^3]}{2R_{DA}^4}\right) \tag{1.16}$$

Further development of theory of reorganization energy consisted in taking into consideration the properties of medium and manner in which it interfaces with the solute [65]. These properties must include both size and shape of the solute and solvent molecules, distribution of electron density in reagents and products, and the frequency domain appropriate to medium reorganization. When the symmetry of donor and acceptor is equivalent, reorganization energy can be generalized as

$$\lambda = C \Delta g_{\text{eff}}(e^2)\left(\frac{1}{r_{\text{eff}}} - \frac{1}{R_{DA}}\right) \tag{1.17}$$

where $C = 0.5$ is a coefficient, Δg_{eff} is the effective charge, and r_{eff} is the effective radius of charge separated centers. More general theory of the reorganization energy takes the difference between energies of the reactant state and product state, U_R and U_P, with the same nuclear coordinates q, as the reaction coordinate [78]:

$$\Delta e(q) = U_P(q) - U_R(q) \tag{1.18}$$

In this theory, the reorganization energy is related to the equilibrium mean square fluctuation of the reaction coordinate as

$$L = \frac{1}{2}(\beta < \Delta e - \langle \Delta e \rangle)^2 \tag{1.19}$$

The atoms in the systems are divided into four groups: donor (D) and acceptor (A) sites of a reaction complex (as in protein), nonredox site atoms, and water atoms as the environment. The following calculation determined each component's contribution to Δe and, therefore, to the reorganization energy. In the case of thermal excitation of the local molecular and medium high-frequency modes, before theories mentioned predicted the classical Marcus relation in the normal Marcus region. While in the inverted region, significant deviation in the parabolic energy gap dependence is expected. The inverted Marcus region cannot be experimentally observed if the stabilization of the first electron transfer product for the accounting of the high-frequency vibrational mode occurs faster than the equilibrium of the solvent polarization with the momentary charge distribution can be established. Another source of the deviation is the nonparabolic shape of the activation barrier.

The effect of solvent fluctuations on the rate of electron transfer reactions was considered using linear expansion theory and a second-order cumulant expansion [79]. An expression was obtained for the rate constant in terms of the dielectric response function of the solvent and was proved to be valid not only for approximately harmonic systems such as solids but also for strongly molecularly anharmonic systems such as polar solvents.

Microscopic generalizations of the Marcus nonequilibrium free energy surfaces for the reactant and the product, constructed as functions of the charging parameter, were presented [55]. Their relation to surfaces constructed as functions of the energy gap is also established. The Marcus relation was derived in a way that clearly shows that it is a good approximation in the normal region even when the solvent response is significantly nonlinear. The hybrid molecular continuum model for polar solvation, combining the dielectric continuum approximation for treating fast electronic (inertialess) polarization effects, was considered [32]. The slow (inertial) polarization component, including orientational and translational solvent modes, was treated by a combination of the dielectric continuum approximation and a molecular dynamics simulation, respectively. This approach yielded an ensemble of equilibrium solvent configurations adjusted to the electric field created by a charged or strongly polar solute. Both equilibrium and nonequilibrium solvation effects were studied by means of this model, and their inertial and inertialess contributions were separated. Three types of charge transfer reactions were analyzed. It was shown that the standard density linear response approach yields high accuracy for each particular reaction, but proves to be significantly in error when reorganization energies of different reactions were compared.

1.2.3
Zusman Model and its Development

The Zusman equation (ZE) has been widely used to study the effect of solvent dynamics on electron transfer reactions [80–90]. In this equation, dynamics of the electronic degrees of freedom is coupled to a collective nuclear coordinate. Application of this equation is limited by the classical treatment of the nuclear degrees of freedom. The Zusman theory is based on description of the solvent complex permittivity in the Debey approach. According to the theory, the Gibbs energy activation

$$\Delta G^{\neq} = (\Delta G_0 + E_p)^2/4E_p \tag{1.20}$$

where

$$E_p = (8\pi)^{-1}(1/\varepsilon_m - 1/\varepsilon_s) \, \text{Int} \, \Delta D^2(r) dr \tag{1.21}$$

where ε_m is the dielectric constant of the solvent at intermediate frequency, ε_s is the dielectric constant of the solvent, and $\Delta D(r)$ is the difference of inductions in the first and second dynamics state. The difference of the Equations 1.20 and 1.21 from the corresponding classic Marcus equation (1.5) is that the "equilibrium" reorganization energy λ was replaced for the dynamic reorganization energy (E_p) of the slow degree of freedom of the solvent, while preexponential factor for nonadiabatic reaction including the coupling factor (V) and effective frequency v_{ef} of the solvent $1/v_{ef} = \varepsilon_s/\varepsilon_{op}\tau_D$, where ε_{op} is the optical dielectric constant and τ_D is the average time for orientation of dipoles of solvent. The Zusman theory is limited because of the neglection of quantum effects in describing the dynamics of the nuclear degrees of freedom.

The authors of paper [84] derived the ZE as a high-temperature approximation to the exact theory. In this work, the authors applied the recently developed hierarchical equations of motion (HEOM) method. A multistate displaced oscillator system strongly coupled to a heat bath was considered a model of an ET reaction system [85]. By performing canonical transformation, the model was reduced to the multistate system coupled to the Brownian heat bath defined by a nonohmic spectral distribution. For this system, the hierarchy equations of motion for a reduced density operator was derived. To demonstrate the formalism, the time-dependent ET reaction rates for a three-state system were calculated for different energy gaps. An analytic study of the density matrix and Wigner representation equations for dissipative electron transfer was presented [86]. Obtained expression showed a very fast relaxation in time if the barrier to reaction is greater than the thermal energy. The fast off-diagonal relaxation disallows an adiabatic elimination of the momentum even in the large friction limit. These equations are a generalization to phase space of the large friction Zusman equations [80]. Taking into account the quantum effect of nuclear dynamics, formalism provided an exact solution to the ET dynamics.

By taking into account both the quantum fluctuations of the collective bath coordinate and its non-Markovian dynamics, an exact solution to the ET dynamics was provided [88]. The forward ET reaction rates were calculated using the HEOM, the ZE, and also the Fermi' golden rule, resulting in

$$k^{FGR} = 2V^2 \, \text{Re} \int_0^\infty dt e^{-iE_0 t} \exp\left[-\int_0^\infty d\omega \frac{4J(\omega)}{\pi\omega^2}\right.$$
$$\left. \times \{\coth(\beta\omega/2)[1-\cos(\omega t)] - i\sin(\omega t)\}\right] \tag{1.22}$$

The interacting spectral density of the harmonic bath was defined as

$$J(\omega) = \frac{\pi}{2} \sum_i \frac{c_i^2}{\omega_i} \delta(\omega - \omega_i) \tag{1.23}$$

The solvent reorganization energy is determined as

$$\lambda = \frac{4}{\pi} \int_0^\infty d\omega \, \frac{J(\omega)}{\omega} \tag{1.24}$$

Using solvent-controlled charge transfer dynamics on diabatic surfaces with different curvatures, the framework of Zusman was theoretically analyzed [88]. A generalization of the nonadiabatic Marcus–Levich–Dogonadze rate expression was obtained for the case of different forward and backward reorganization energies and a corresponding generalization of the Zusman rate expression that bridges between nonadiabatic and solvent-controlled adiabatic electron transfer was provided. The derived analytical rate expressions were compared with the precise numerics. The proposed mechanism consisted of spontaneously arising inhomogeneous in space fluctuation in polarization relaxations to equilibrium, not only via solvent dipoles rotation but also via solvent molecule self-diffusion. This process of polarization diffusion leads to the modulation of electronic energy levels that are now fluctuating faster if only the rotational motions of the solvent dipoles are accounted for. It was found that if the rate of tunneling is large enough, the rate constant of the reaction is controlled by the solvent dynamics and the polarization diffusion also contributes. It is also shown that the contribution of the polarization diffusion can become a dominating one in the solvents with large enough diffusion coefficients.

The theory of electron transfer reactions accounting for the influence of the solvent polarization diffusion on the rate of reaction was developed [82]. It was shown that in the limit when reaction is controlled by the solvent dynamics, the effective frequency of the medium fluctuations consists of the sum of two contributions – the contribution of the solvent dipole rotation and the contribution of the solvent polarization diffusion. The model of charge distribution of the products and reagents of the reaction is suggested. Starting from the Zusman equations to the case of parabolic diabatic curves with different curvatures, a generalized master equation for the populations and formal expressions for their long-time limit was derived [89]. In the limit of very small tunnel splitting, a novel rate formula for the nonadiabatic transitions was obtained. For larger values of the tunnel splitting, the consecutive step approximation leading to a rate formula was used that bridges between the nonadiabatic and the solvent-controlled adiabatic regimes.

In Ref. [84], the Zusman equation in the framework of the exact hierarchical equations of motion formalism was revisited. It was shown that a high-temperature approximation of the hierarchical theory is equivalent to the Zusman equation in describing electron transfer dynamics and the exact hierarchical formalism naturally extends the Zusman equation to include quantum nuclear dynamics at low temperatures. Numerical exact results are also presented for the electron transfer reaction dynamics and rate constant calculation. The authors derived the ZE as a

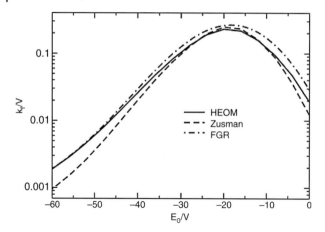

Figure 1.12 The temperature effect on ET rates derived from different theories (see details in Ref. [84]).

high-temperature approximation to the exact theory, the HEOM method. The latter provides an exact solution to the ET dynamics, by taking into account both the quantum fluctuations of the collective bath coordinate and its non-Markovian dynamics. Since the HEOM method is based on nonperturbative propagation of the reduced density matrix, it can be used to study both nonequilibrium dynamics and dissipative dynamics under a driving laser field interaction. Figure 1.12 shows the temperature effect on ET rates derived from different theories.

It can be seen that the rate from the ZE agrees well with the exact result at high temperatures and deviates from the exact result at low temperatures.

The comprehensive theory of charge transfer in polar media was used for the treatment of experimental data for complex systems, with due account of large-amplitude strongly anharmonic intramolecular reorganization [90]. Equations for the activation barrier and free energy relationships taking into account vibrational frequency changes, local mode anharmonicity, and rotational reorganization, in both diabatic and adiabatic limits were provided. Possible modifications of the Zusman stochastic equations aimed to account for quantum interference of basis states of the system have been investigated. A set of equations that includes nonequilibrium distribution in the momentum space at short timescale in a strong friction limit was obtained. The authors stressed the following features of ET of complex molecules:

1) "Large" intramolecular reorganization, that is, coordinate displacements in excess of, say, ≈ 0.2 Å;
2) As a consequence of (1), significant vibrational frequency changes (potential surface distortion) and/or large anharmonicity;
3) Intramolecular rotational reorganization;
4) Ion pair reorganization in weakly polar solvents;
5) Multi-ET and long-range ET reactions;
6) All correlations are smooth and appear to resemble one another;

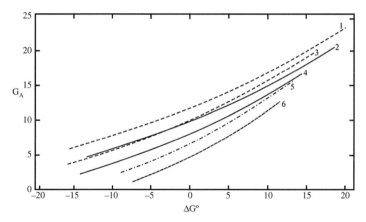

Figure 1.13 Activation Gibbs free energy dependence on reaction free energy for charge transfer with coupling to a local Morse potential and a harmonic solvent continuum, calculated with different values of the parameter d. The harmonic local mode $d = D_f/D_i X_0^2$. Energies in units of $k_B T$, $T = 298$ K. $E_r^{sol} = E_r^Q = 20$. Resonance splitting $\Delta E = 4$. The following parametric variations: (1) $d=2$, diabatic limit; (2) $d=2$, adiabatic limit; (3) $d=1$, diabatic limit; (4) $d=1$, adiabatic limit; (5) $d=0.5$, diabatic limit; (6) $d=0.5$, adiabatic limit. Large d corresponds to strong repulsion in the repulsive branch of the Morse potential, small d to weak repulsion. Large d therefore gives the higher activation free energy. $d = D_f/D_i X_0^2$ Morse potential ($d=0.5$), adiabatic limit [90].

7) The activation free energy increases with increasing d, which characterizes the repulsive branches of the Morse potentials in the initial and final states. Small d represents a shallower potential in the final than in the initial state, and vice versa. Smaller values of d therefore give a smaller activation Gibbs free energy (Figure 1.13).

Harmonic and anharmonic potentials were compared. The harmonic local mode corresponds to $d=1$. The activation free energy for a shallow final-state anharmonic mode ($d=0.5$) is lower than for the harmonic mode ($d=1$) but higher for a steep repulsive final state branch ($d=2$). In addition, the correlation was notably asymmetric around zero driving force for the anharmonic mode, with larger curvature for negative than for positive ΔG^0. Resonance splitting ($\Delta E = 0.1$ and 0.2 eV) lowers the activation free energy by approximately $\Delta E/2$ over most of the driving force range.

$$\alpha = \frac{dG_A}{d\Delta G^0} = \frac{1}{2}\left[1 + \frac{\Delta G^0}{E_r^{sol}}\right] \quad (1.25)$$

1.2.4
Effect of Nonequilibrity on Driving Force and Reorganization

When the initial state distribution remains in thermal equilibrium throughout the ET process, the process driving force is related to the standard Gibbs energy (ΔG_0).

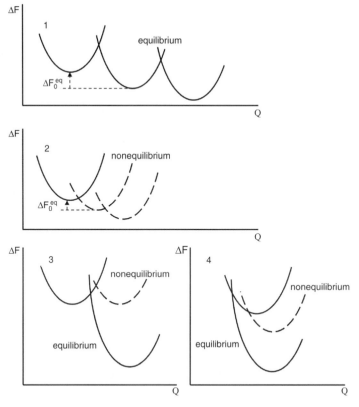

Figure 1.14 Schematic representation of electronic potential energy surfaces: (1) consecutive conformational and solvational equilibrium processes with the essential change in the nuclear coordinates Q and the standard Gibbs energy ΔG_0; (2) consecutive nonequilibrium processes with small changes in Q and ΔG_0; (3, 4) equilibrium (full line) and nonequilibrium (broken line) processes in the normal and inverted Marcus regions, respectively. Reproduced with permission from Ref. [91].

A different situation takes place if the elementary act of ET occurs before the formation of conformational and solvational states of the medium. In fact, two consecutive stages take place: ET for the accounting of fast vibration translation modes of the system and the media relaxation. In such a case, the thermodynamic standard energy for the elementary act (ΔG_0^{neq}) appears to be less than that involved in the case of the equilibrium dielectric stabilization of redox centers ΔG_0 [9, 19, 91]. It can be concluded that the initial and final energy terms in the nonequilibrium case will be positioned closer to each other in space and energy than in equilibrium (Figure 1.14). Consequently, in the inverted Marcus region, the value of the reorganization, Gibbs and activation energies are expected to be markedly lower than that in the equilibrium case. In the normal Marcus region, we predict a larger activation energy and slower ET rate for nonequilibrium processes than for

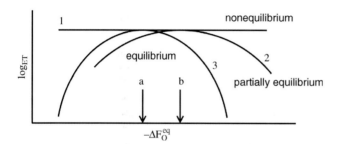

Figure 1.15 Schematic representation of the dependence of the ET constants logarithm on the equilibrium Gibbs energy ΔG_0: (1) nonequilibrium conformational and solvational processes; (2) partial nonequilibrium processes, λ^{neq} and ΔG_0^{neq}, are slightly dependent on ΔG_0; (3) equilibrium processes. Arrows a and b are conditions for the maximum $\lambda = \Delta G_0$ and $\lambda^{neq} = \Delta G_0^{neq}$, respectively. Reproduced with permission from Ref. [91]

equilibrium processes when differences in their standard Gibbs energy would be larger than that in the reorganization energy. In general, the situation would be dependent on the interplay of both parameters of the Marcus model. The second property expected for nonequilibrium processes is the lack of dependence (Figure 1.15, curve 1) or weak dependence (curve 2) of the experimental rate constant of ET in both Marcus regions (inverted and noninverted), compared to that predicted by the classic Marcus expression (curve 3).

During the ET processes, the nuclear degrees of freedom need to reorganize in responding to the change in charge distributions upon the electronic state transitions. The interplay of electronic coupling and nuclear reorganization dynamics could thus result in rich dynamical behaviors in ET reactions. An example of nuclear dynamical effects on ET reaction is that, when increasing the nuclear relaxation time, ET reactions change from the nonadiabatic regime to the solvent-controlled adiabatic regime.

A microscopic theory for the rate of nonadiabatic electron transfer, including generalizations of the Marcus nonequilibrium free energy surfaces for the reactant and the product, was developed and its relation to classical Marcus theory was analyzed [92]. A simple algorithm was proposed for calculating free energy changes from computer simulations on just three states: the reactant, the product, and an "anti"-product formed by transferring a positive unit charge from the donor to the acceptor. The activation energy as a parabolic function of the free energy change of reaction was derived when the solvent response is significantly nonlinear. The electron transfer theory for the high-frequency intramolecular mode and low-frequency medium mode for a single-mode case when the reactant surface is not in a thermal equilibrium has been rederived [93]. In the limit of very low and very high temperatures, the expressions were analyzed and compared with the case of thermal distribution and a Franck–Condon factor for a multimode displaced, distorted, and Duschinsky rotated adiabatic potential surfaces has been derived to obtain the ET rate.

1.2.5
Long-Range Electron Transfer

Long-range Electron Transfer (LRET) between donor (D) and acceptor (A) centers can occur by three mechanisms: (1) direct transfer that involves direct overlap between electron orbitals of the donor and acceptor, (2) consecutive electron jumps via chemical intermediates with a fixed structure, and (3) superexchange via intermediate orbitals. In direct LRET the direct electronic coupling between D and A is negligible and this mechanism is not practically realized in condensed media being noncompetitive with the consecutive and superexchange processes. In theoretical consideration of the consecutive LRET, a relevant theory of ET in two-term systems can be applied.

Of considerable interest is the superexchange process [9, 91, 94–110]. According to the Fermi's golden rule (Equation 1.5), the nonadiabatic ET rate constant is strongly dependent on electronic coupling between the donor state D and the acceptor state A connected by a bridge (V_{AB}) that is given by an expression derived from the weak perturbation theory

$$V_{AB} = \frac{\sum V_{A\alpha} V_{\alpha B}}{\Delta E_\alpha} \tag{1.26}$$

where $V_{A\alpha}$ and $V_{\alpha B}$ are the couplings between bridge orbitals and acceptor and donor orbitals, respectively, and ΔE_α is the energy of the bridge orbitals relative to the energy of the donor orbital. The summation over α includes both occupied and unoccupied orbitals of the bridge.

Using static perturbation theory, time-dependent perturbation theory, and direct time-dependent dynamics within generalized tight binding models, the authors of work [99] examined the role of energy gaps, relative energetics of donor and acceptor orbitals with hole-type and electron-type superexchange sites, damping and dephasing, and overall energetics in electron transfer. The dynamic studies indicated some important phenomena, which include quantum interferences between different pathways, recurrences, and oscillations and competitive effects of hole-type and electron-type superexchange. An integral equation approach to nonlinear effects in the free energy profile of electron transfer reaction has been developed [103]. Electronic and dynamical aspects of superexchange-assisted through-bridge electron transfer were considered. This approach was extended to a more general case, where D is connected to A by a number of atomic orbitals. A special, the so-called "artificial intelligence," search procedure was devised to select the most important amino acid residues, which mediate long-range transfer [104–106].

According to the approach of Onuchic and coworkers [107], for a pathway between bridged donor and acceptor groups, the coupling element can be written as

$$V_{AB} = V_0 \prod_i^N \varepsilon_i \tag{1.27}$$

where V_0 is the coupling between the donor and the first bond of the pathway and ε_i is a decay factor associated with the decay of electron density from one bond to another.

The ε_B, ε_H, and ε_s values are related to superexchange through two covalent bonds sharing a common atom, an H bond, and space, respectively. The decay factor is approximated by equation

$$\varepsilon_i = \varepsilon_i^0 \exp\left[\beta_i(R - R_i^0)\right] \tag{1.28}$$

where R_i^0 is the equilibrium length bond or van der Waals distance, β_i is some factor, specific to the distance R, which depends on the orbital interactions and ε_i^0 is the value of ε_i for $R = R_i^0$, which is proportional to factor σ related to the interaction orientation. The values of $\varepsilon_B = 0.4$–0.6, $\sigma_H = \sigma_B = 1.0$, and $\beta_s = 1.7\,\text{Å}$ were taken for the calculation of V_{AB}. According to this theory, the increase in connectivity for the electron transfer is about 0.24 per atom.

A semiempirical approach for the quantitative estimation of the effect bridging the group on LRET was developed by Likhtenshtein [9, 91].

The basic idea underlying this approach is an analogy between superexchange in electron transfer and such electron exchange processes as triplet–triplet energy transfer (TTET) and spin exchange (SE). The ET rate constant is proportional to the square of the resonance integral V_{ET}. The rate constant of TTET is

$$k_{TT} = \frac{2\pi}{h} J_{TT} FC \tag{1.29}$$

where J_{TT} is the TT exchange integral. The Hamiltonian of the exchange interaction (H_{SE}) between spins with operators S_1 and S_2 is described by the equation

$$H_{SE} = -2 J_{SE} S_1 S_2 \tag{1.30}$$

where J_{SE} is the SE exchange integral.

All three integrals V_{ET}^2, J_{SE}, and J_{TT} are related to the overlap integral (S_i), which quantitatively characterizes the degree of overlap of orbitals involved in these processes. Thus,

$$V_{ET}^2, J_{SE}, J_{TT} \propto S_i^n \propto \exp(-\beta_i R_i) \tag{1.31}$$

where R_i is the distance between the interacting centers and β_i is a coefficient that characterizes the degree of the integral decay. In the first approximation, $n = 2$ for the ET and SE processes with the overlap of two orbitals and $n = 4$ for the TT process in which four orbitals overlap (ground and triplet states of the donor and ground and triplet states of the acceptor). The spin exchange and TT phenomena may be considered an idealized model of ET without or with only a slight replacement of the nuclear frame (see Figure 1.16). Thus, the experimental dependence of exchange parameters k_{TT} and J_{SE} on the distance between the exchangeable centers and the chemical nature of the bridge connecting the centers may be used for evaluating such dependences for the resonance integral in the ET equation (1.30).

A vast literature is connected with the quantitative investigation of exchange processes [101, 102]. As it seen in Figure 1.17, experimental data on the dependence of k_{TT} and J_{SE} on the distance between the centers (ΔR) lie on two curves, which are approximated by the following equation [91].

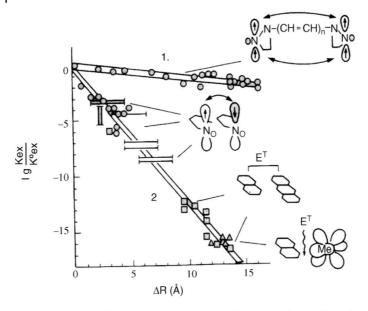

Figure 1.16 Dependence of the logarithm of relative parameters of the exchange interaction on the distance between the interacting centers (ΔR). k_{TT} is the rate constant of triplet–triplet electron transfer and J_{SE} is the spin-exchange integral. Index 0 is related to van der Waals contact. Reproduced with permission from Ref. [91].

$$k_{TT}, J_{SE} \propto \exp(-\beta \Delta R) \tag{1.32}$$

For systems in which the centers are separated by a "nonconductive" medium (molecules or groups with saturated chemical bond), β_{TT} equals 2.6 Å^{-1}. For systems in which the radical centers are linked by "conducting" conjugated bonds, β_{SE} is 0.3 Å^{-1}.

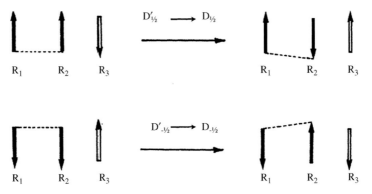

Figure 1.17 Triplet–singlet spin conversion of the radical pair [110, 112].

We can consider the ratios

$$\gamma_{SE} = \frac{J_{SE}^0}{J_{SE}}(\Delta R) \tag{1.33}$$

as parameters of attenuation of the exchange interaction of SE through the given medium. Taking into account Equations 1.31 and 1.32 with value $n = 2$ for SE and ET, we have an expression for the dependence of the attenuation parameters for SE and ET on the distance between remote donor and acceptor centers $D_R D_A$

$$\gamma_{ET} = \gamma_{SE} = \exp(-\beta_i \Delta R) \tag{1.34}$$

with $\beta_{ET}(nc) = 0.5\beta_{TT} = 1.3\,\text{Å}^{-1}$ for a "nonconducting" medium and $\beta_{ET}(c) = 0.3\,\text{Å}^{-1}$ for a "conducting" bridge. The value of β_{ET} (1.3 Å^{-1}) is found to be close to that obtained by analysis of k_{ET} on the distance ΔR in model and biological systems (Figure 2.7).

An examination of the empirical data on the exchange integral values (J_{ET}) for the spin–spin interactions in systems with known structure, that is, biradicals, transition metal complexes with paramagnetic ligands, and monocrystals of nitroxide radicals, allows the value of the attenuation parameter γ_X for the exchange interaction through a given group X to be estimated. By our definition, the γ_X is

$$\gamma_X = \frac{J_{RYZP}}{J_{RYXZP}} \tag{1.35}$$

where R is a nitroxide or organic radical, P is a paramagnetic complex or radical, and X, Y, and Z are chemical groups in the bridge between R and P.

Table 1.1 shows the results of the calculation parameter γ_X from empirical data by Equation 1.35 [9, 19]. The table of values for X, C=O, S=O, P=O and C=C, calculated from independent experimental data, is similar. Data presented in Table 1.1 and

Table 1.1 Values of the attenuation parameter of individual groups (γ_X), van der Waals contact (γ_v), and hydrogen bond (γ_{hb}) for spin exchange in biradicals and paramagnetic complexes of transition metals with nitroxide ligands (see text) (reproduced with permission from Ref. [91]).

Group, X	γ_x	Group, X	γ_x
C_6H_4	6.00 ± 0.03		
C=C	1.7	–NH–CO–	55[a]
C=O	8.4 ± 0.4	γ_v	50
C			
NH	6.5	γ_{hb}	10
O	5	H	12
S=O	2.1	SO_2	2.2
	3.5	RP=O	2.40 ± 0.03[b]

a) Calculated by equation $\gamma_x = \gamma_{CO}\gamma_{NH}$.
b) R ≡ Ph–, CH$_2$=CH–, Ph–CH=CH–, Ph–CCl=CH–.

Equation 1.35 may be used for the analysis of alternative electron transfer pathways in biological systems.

The dynamics of charge transfer from a photoexcited donor to an acceptor coupled through a bridge was investigated by using a correlation function approach in Liouville space that takes into account solvent dynamics with an arbitrary distribution of timescales [108]. A nonadiabatic theory of electron transfer, which improves the standard theory near the inversion point and becomes equivalent to it far from the inversion point, was presented [109]. This theory revealed the existence of an especially interesting marginal case when the linear and nonlinear coefficients of a two electronic states system are appropriately tuned for forming a coherent electron–phonon oscillator. An electron injected in one of the electronic states of a CEPO generates large amplitude charge oscillations associated with coherent phonon oscillations and electronic-level oscillations that may resonate with a third site that captures the electron so ultrafast electron transfer becomes possible (Figure 1.5). Numerical results are shown where two weakly interacting sites, a donor and a catalyst, form a CEPO that catalytically triggers an UFET to an acceptor.

1.2.6
Spin Effects on Charge Separation

Chemical reactions are known to be controlled by two fundamental parameters, energy (both free and activation energy) and angular momentum (spin) of reactants [110, 111]. The latter results in electron and nuclear spin selectivity of reactions: only those spin states of reactants are chemically active whose total spin is identical to that of products. For example, for the triplet radical pair (R_1, R_2) prepared by photolysis, radiolysis, or encounter of freely diffusing radicals to recombine and produce diamagnetic, zero-spin molecule R_1R_2, triplet–singlet spin conversion of the radical pair is required (Figure 1.17).

In a static model of spin catalysis, if the starting spin state of the pair is triplet (it corresponds to D' state of the triad), then the probability to find this pair in the singlet state (it corresponds to D state of the triad) [110, 111]:

$$\varrho_S(t) = (\Delta J/2\Omega)\sin^2 \Omega t \tag{1.36}$$

where

$$\Omega = 2^{-1/2}[(J_{12}-J_{13})^2 + (J_{12}-J_{23})^2 + (J_{13}-J_{23})^2]^{1/2} \tag{1.37}$$

and $\Delta J = J_{13} - J_{23}$; J_{ij} denotes the pair-wise exchange energies for pairs R_i and R_j ($i \neq j$), ($i, j = 1, 2, 3$). Both conjugated processes, triplet–singlet conversion of the pair and doublet–doublet evolution of the triad, oscillate in time with a period $\tau = (2\Omega)^{-1}$.

Photogenerated radical pairs are capable of exhibiting coherent spin motion over microsecond timescales, which is considerably longer than coherent phenomena involving photogenerated excited states. The rate of radical pair intersystem crossing between photogenerated singlet and triplet radical pairs has been shown to increase in the presence of stable free radicals and triplet state molecules. Spin catalysis was proved to operate in radical recombination biradical decay, *cis–trans* isomerization of molecules, primary light-harvesting reactions in photosynthetic centers, charge separation and water oxidation by photosystem II, in particular, paramagnetic quenching of excited molecules, and so on [110]. Another way of controlling the lifetime of the photoseparated charges by the spin chemistry mechanism appears to be the introduction to donor or acceptor molecules an isotope bearing a nuclear spin [111].

1.2.7
Electron–Proton Transfer Coupling

Proton-coupled electron transfer (PCET) reactions play a vital role in a wide range of chemical and biological processes such as the conversion of energy in photosynthesis and respiration, in electrochemical processes, and in solid state materials. Recently, a number of experiments on model PCET systems have been performed [112–123]. A theory of PCET was developed by Cukier and coworkers [113–115]. The authors took in consideration that in PCET the electron and proton may transfer consecutively, electron transfer followed by proton transfer (PT), designated as ET/PT, or they may transfer concertedly, in one tunnel event, designated as ETPT. It was suggested that the proton charge is coupled to the solvent dipoles in a fashion similar to the electron–solvent coupling and the analysis of effect of solvation on the shape of the proton potential energy surface allowed to evaluate the PCET rate constant. The dielectric continuum theory was used to obtain the proton-solvated surfaces. According to this model, the proton can affect the PCET rate via Franck–Condon factors between the proton surfaces for the initial and final electron states and also influence the activation energy via the proton energy levels. The rates corresponding to the ETPT and ET/PT channels were evaluated for several model reaction complexes that mimic electron donor–hydrogen-bonded interface–electron acceptor system parameters. Figure 1.18 displays the charge site geometry and the ellipsoid and spheres with the definitions of the relevant dimensions.

According to the theory, the rate constant is

$$k_{\text{ETPT}} = \frac{V_{\text{el}}^2}{\hbar^2} \sqrt{\pi \hbar^2 / \lambda_{\text{s}}^{\text{ETPT}} k_{\text{B}} T} \times \sum_{n'} \varrho_{\text{in}'} \sum_{n \in b} |\langle \chi_{fn} | \chi_{\text{in}'} \rangle|^2$$

$$\times e^{-(\lambda_{\text{s}}^{\text{ETPT}} + \Delta^{\text{el}} + \varepsilon_{fn} - \varepsilon_{\text{in}'})^2 / 4\lambda_{\text{s}}^{\text{ETPT}} k_{\text{B}} T} \tag{1.38}$$

where χ_i is the protonic wave function, which is dependent on the electron state; λ_{s} is the reorganization energy arising from the solvent–charge coupling and the reaction free energy. The reaction free energy is the sum of the electronic structure ΔE^{el} and the equilibrium solvation ΔG^{sol} contributions, ε_i' and ε_f' are the proton energy in initial and final states, respectively.

In the theory for PCET developed by Hammes-Schiffer and coworkers [116–123], a PCET reaction involving the transfer of one electron and one proton (depicted in Figure 1.19) was described in terms of four diabatic states: the proton and electron on

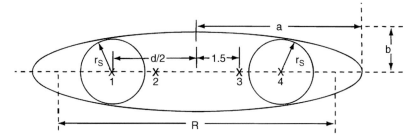

Figure 1.18 Model for the evaluation of the reorganization and solvation free energy of the ET, PT, and ETPT reactions. The donor and acceptor sites are spheres of radius r_s embedded in an ellipsoid with major (minor) axis a (b) and interfocal distance R. The locations 1 and 4 (2 and 3) denote charge sites associated with the electron (proton) states. The proton sites are at a fixed distance of 3 Å and the electron sites are separated by a distance d of 15 Å. The ellipsoid expands to contain the donor and acceptor spheres as the sphere radii increase [113].

their donors, the proton and electron on their acceptors, the proton on its donor and the electron on its acceptor, and the proton on its acceptor and the electron on its donor. The transferring hydrogen nucleus was treated quantum mechanically to include effects such as zero point energy and hydrogen tunneling.

Within this four-state model, the mixed electron/proton vibrational free energy surfaces were obtained as functions of two collective solvent coordinates corresponding to ET and PT and the free energy surfaces for PCET reactions were approximated as two-dimensional paraboloids (Figure 1.20).

In this case, the PCET reaction is viewed as a transition from the reactant set of paraboloids to the product set of paraboloids. Thus, this theory is a multidimensional analogue of standard Marcus theory for single ET and the PCET reaction requires a reorganization of the solvent and involving intramolecular solute modes.

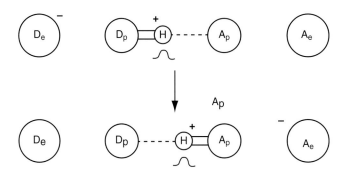

Figure 1.19 Schematic illustration of a PCET reaction, where the electron donor and acceptor are denoted D_e and A_e, respectively, and the proton donor and acceptor are denoted D_p and A_p, respectively. The transferring proton is represented as both a sphere and a quantum mechanical wave function [120].

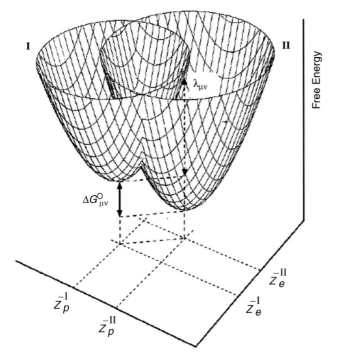

Figure 1.20 Schematic illustration of a pair of paraboloids Iμ and IIν as functions of the solvent coordinates z_p and z_e. The reorganization energy $\lambda_{\mu\nu}$ and the equilibrium free energy difference $\Delta G^0_{\mu\nu}$ are indicated [120].

The theoretical description of the most basic PCET reaction involved the transfer of one electron and one proton requires four diabatic states:

(1a) $D_e^\ominus - {}^\oplus D_p H \cdots A_p - A_e$

(1b) $D_e^\ominus - D_p \cdots HA_p^\oplus - A_e$

(2a) $D_e - {}^\oplus D_p H \cdots A_p - A_e^{\text{Å}}$

(2b) $D_e - D_p \cdots HA_p^\oplus - A_e^{\text{Å}}$

According to Ludlow et al. [122], there are three distinct regimes of PCET:

1) Electronically adiabatic PT and ET, where the coupling between all pairs of the four diabatic states is strong.
2) Electronically nonadiabatic PT and ET, where the coupling between all pairs of the four diabatic states is weak.
3) Electronically adiabatic PT and electronically nonadiabatic ET, where the coupling between PT diabatic states is strong and the coupling between ET diabatic states is weak.

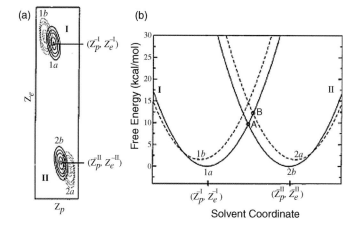

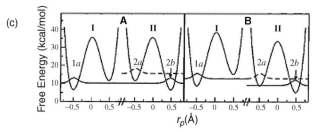

Figure 1.21 (a) Schematic illustration of two-dimensional ET diabatic mixed electronic/proton vibrational free energy surfaces as functions of the solvent coordinates z_p and z_e. The reactant and product ET diabatic surfaces are labeled I and II, respectively. Only two surfaces are shown for each ET diabatic state, and the lower and higher energy surfaces are shown with solid and dashed contour lines, respectively. Each free energy surface is labeled according to the dominant diabatic state, and the minima of the lowest surfaces are labeled $(\bar{z}_p^I, \bar{z}_e^I)$ and $(\bar{z}_p^{II}, \bar{z}_e^{II})$. (b) Slices of the free energy surfaces along the straight-line reaction path connecting solvent coordinates $(\bar{z}_p^I, \bar{z}_e^I)$ and $(\bar{z}_p^{II}, \bar{z}_e^{II})$ indicated in (a). Only the lowest surface is shown for the reactant (I), and the lowest two surfaces are shown for the product (II). (c) The reactant (I) and product (II) proton potential energy curves are functions of r_p at the solvent configurations corresponding to the intersection points A and B indicated in (b) [120].

The regime of electronically adiabatic PT and electronically nonadiabatic ET is suggested to be most relevant for PCET reactions with a well-separated electron donor and acceptor connected by a hydrogen-bonded interface. Schematic illustration of two-dimensional ET diabatic mixed electronic/proton vibrational free energy surfaces as functions of the solvent coordinates z_p and z_e is displayed in Figure 1.21.

A rate expression in the limit of electronically adiabatic PT and electronically nonadiabatic was derived. Application of the golden rule to the two sets of free energy surfaces illustrated in Figure 1.5a leads to the following rate expression:

$$k = \frac{2\pi}{\hbar} \sum_\mu P_{I\mu} \sum_\nu V_{\mu\nu}^2 (4\pi \lambda_{\mu\nu} k_B T)^{-1/2} \exp\left\{\frac{-(\Delta G_{\mu\nu}^0 + \lambda_{\mu\nu})^2}{4\lambda_{\mu\nu} k_B T}\right\} \quad (1.39)$$

where $\sum_\mu$ and $\sum_\nu$ indicate a sum over vibrational states associated with ET states 1 and 2, respectively, and $P_{I\mu}$ is the Boltzmann factor for state I_μ. In this expression, the reorganization energy is defined as

$$\lambda_{\mu\nu} = \varepsilon^I_\mu\left(\bar{z}^{II_\nu}_p, \bar{z}^{II_\nu}_e\right) - \varepsilon^I_\mu\left(\bar{z}^{I_\mu}_p, \bar{z}^{I_\mu}_e\right) = \varepsilon^{II}_\nu\left(\bar{z}^{I_\mu}_p, \bar{z}^{I_\mu}_e\right) - \varepsilon^{II}_\nu\left(\bar{z}^{II_\nu}_p, \bar{z}^{II_\nu}_e\right) \quad (1.40)$$

and the free energy difference is defined as

$$\Delta G^0_{\mu\nu} = \varepsilon^{II}_\nu(\bar{z}^{II_\nu}_p, \bar{z}^{II_\nu}_e) - \varepsilon^I_\mu(\bar{z}^{I_\mu}_p, \bar{z}^{I_\mu}_e) \quad (1.41)$$

where $(\bar{z}^{I_\mu}_p, \bar{z}^{I_\mu}_e)$ and $(\bar{z}^{II_\nu}_p, \bar{z}^{II_\nu}_e)$ are the solvent coordinates for the minima of $\varepsilon^I_\mu(z_p, z_e)$ and $\varepsilon^{II}_\nu(z_p, z_e)$, respectively. As a result of the averaging over, the proton applicability of the Born–Oppenheimer approximation (BOA) for the calculation of the transition probability for a nonadiabatic process of charge transfer in a polar environment with allowance made for temperature effects was theoretically investigated.

The transfer of a quantum particle (proton) that interacts with a local vibration mode in a model of bound harmonic oscillators was considered [124]. The model admitted an exact solution for wave functions of the initial and final states. A calculation showed that the model is applicable even for very large distances of the proton transfer. It was shown that the non-Condon effects are in general temperature dependent and may substantially influence the calculated values of the transition probability.

1.2.8
Specificity of Electrochemical Electron Transfer

The current flowing in either the reductive or oxidative steps of a single electron transfer reaction between two species (O) and (R)

$$O(s) + e^-(m) \underset{k_{ox}}{\overset{k_{red}}{\rightleftharpoons}} R(s) \quad (1.42)$$

can be described using the following expressions:

$$i_a = -FAk_{ox}[R]_0 \quad (1.43)$$

$$i_c = -FAk_{red}[O]_0 \quad (1.44)$$

where i_c and i_a are the reduction and oxidation reaction currents, respectively; A is the electrode area; k_{Red} or k_{Ox} are the rate constant for the electron transfer; and F is the Faraday's constant.

At consideration processes occurring with participation of electrodes, the conception of the Fermi energy, Fermi level, and the Fermi–Dirac (F–D) distribution are widely used (http://en.wikipedia.org/wiki/Fermi_energy) [125–128]. The Fermi energy is the energy of the highest occupied quantum state in a system of fermions at absolute zero temperature (http://en.wikipedia.org/wiki/Fermi_energy). By definition, fermions are particles that obey Fermi–Dirac statistics: when one swaps two fermions, the wave function of the system changes sign. Fermions can be either elementary, like the electron, or composite, like the proton. F–D statistics describes the energies of single particles in a system comprising many identical particles that obey the Pauli exclusion principle. For a system of identical fermions, the average number of fermions in a single-particle state i, is given by the F–D distribution,

$$\bar{n}_i \frac{1}{e^{(\varepsilon_i - \mu)/kT} + 1} \tag{1.45}$$

where k is Boltzmann constant, T is the absolute temperature, ε_i is the energy of the single-particle state i, and μ is the chemical potential. At $T = 0$, the chemical potential is equal to the Fermi energy. For the case of electrons in a semiconductor, μ is also called the Fermi level.

The one-dimensional infinite square well of length L is a model for a one-dimensional box. In the framework of a standard model-system in quantum mechanics for a single quantum number n, the energies are given by

$$E_n = \frac{\hbar^2 \pi^2}{2mL^2} n^2 \tag{1.46}$$

(http://hyperphysics.phy-astr.gsu.edu/HBASE/quantum/disfd.html#c2)

A number of theoretical research studies were devoted to electron transfer reaction from a superconducting electrode [129–142]. The theory of electron transfer reaction from a superconducting electrode that was described in the framework of the resonance valence bond model to a reagent in solution was developed [130]. It was shown that current–overpotential dependence at the boundary between an electrode and solution should be asymmetrical and the current between a superconducting electrode and solution should be substantially suppressed in comparison with that calculated for the same electrode in the normal state.

The quantum theory of electron transfer reactions at metal electrodes was developed [131]. The obtained potential dependence of the electron transfer rate in the weak coupling case was shown to resemble the Butler–Volmer equation of classical electrochemistry. The volcano-shaped dependence of the hydrogen exchange current on the adsorption energy of hydrogen on various metals and the mechanism of hydrogen evolution were explained microscopically. The exchange current density for hydrogen evolution at Pt electrode calculated quantitatively agreed well with the experimental value. Free energy profiles governing electron transfer from a reactant to an electrode surface in water were investigated, based on the reference interaction site model (Figure 1.22) [132]. Three models of a redox pair for charge separation reactions were examined: a pair of atomic solutes and systems consisting of an atom and a surface with a localized or a delocalized electron. It was found that the profile becomes highly asymmetrical when an electron in the electrode is delocalized.

Theoretical analysis of asymmetric Tafel plots and transfer coefficients for electrochemical proton-coupled electron transfer (EPCET) was performed [122]. The input quantities to the heterogeneous rate constant expressions for EPCET were

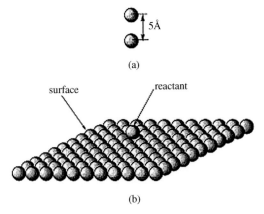

Figure 1.22 Models of CS electron transfer. (a) A one-site electron donor and a one-site electron acceptor. (b) A two-dimensional array consisting of 121 identical atoms whose structure is 1 and a one-site electron donor that is 5 Anstrem apart from the central atom of the surface [132].

calculated with density functional theory in conjunction with dielectric continuum models. The theoretical calculation indicated that the asymmetry of the Tafel plot (the dependence of electric current on electrode potential) and the deviation of the transfer coefficient at zero overpotential from the standard value of one-half arise from the change in the equivalent proton donor–acceptor distance upon electron transfer. The magnitude of these effects was obtained by the magnitude of this distance change, as well as the reorganization energy and the distance dependence of the overlap between the initial and the final proton vibrational wave functions. This theory provided experimentally testable predictions for the impact of specific system properties on the qualitatively behavior of the Tafel plots. A theoretical analysis of EPCET was based on the following approximate expressions for the heterogeneous PCET anodic and cathodic nonadiabatic rate constants:

$$k_a(\eta) = \frac{(V^{el}S)^2}{\hbar}\sqrt{\frac{\pi}{k_B T \Lambda}} \exp\left[2\alpha^2 k_B T / F_R\right] \varrho_M$$
$$\times \int d\varepsilon \, [1-f(\varepsilon)] \exp\left[-\frac{(\Delta\tilde{U}+\varepsilon-e\eta+\Lambda+2\alpha\delta R k_B T)^2}{4\Lambda k_B T}\right]$$

$$k_c(\eta) = \frac{(V^{el}S)^2}{\hbar}\sqrt{\frac{\pi}{k_B T \Lambda}} \exp\left[-2\alpha\delta R+2\alpha^2 k_B T / F_R\right] \varrho_M$$
$$\times \int d\varepsilon f(\varepsilon) \exp\left[\frac{(-\Delta\tilde{U}-\varepsilon+e\eta+\Lambda-2\alpha\delta R k_B T)^2}{4\Lambda k_B T}\right]$$

(1.47)

Here, $f(\varepsilon)$ is the Fermi distribution function for the electronic states in the electrode, ϱ_M is the density of states at the Fermi level, η is the overpotential defined as the difference between the applied and the formal electrode potentials, V_{el} is the

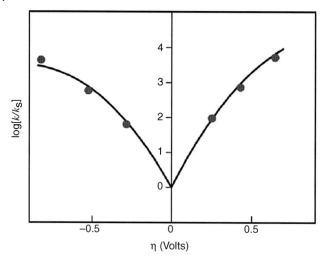

Figure 1.23 Tafel plot of $\log[k_c/k_s]$ for $\eta < 0$ and $\log[k_a/k_s]$ for $\eta > 0$ calculated using Equation 1.1. The experimental data on the of PCET(0) in Osmium Aquo Complex generated at pH 6.0 are shown as circles [122].

electronic coupling, δR is the difference between the equilibrium proton donor–acceptor distances for the oxidized and reduced complexes, F_R is the force constant associated with the proton donor–acceptor mode, S is the overlap integral between the ground reactant and the product proton vibrational wave functions, $\alpha = -\partial \ln S/\partial R$, and $\Lambda = \lambda_s + \lambda_R$ is the total reorganization energy, where λ_s is the solvent reorganization energy and $\lambda_R = F_R \delta R^2/2$ is the reorganization energy of the proton donor–acceptor mode.

According the theory, the associated cathodic transfer coefficient at small overpotential η is

$$\alpha_{PCET}(\eta) = \frac{1}{2} - \frac{\alpha\delta R k_B T}{\Lambda} + \frac{e\eta}{2\Lambda} \tag{1.48}$$

The $\pm 2\alpha\delta R k_B T$ terms in the exponentials of Equation 1.45 lead to asymmetry of the Tafel plots, and the $\alpha\delta R k_B T/\Lambda$ term leads to deviation of $\alpha_{PCET(0)}$ from one-half predicted from classic linear Tafel plot (Figure 1.23).

A study [136] treated the role of the density of electronic states ϱ_F at the Fermi level of a metal in affecting the rate of nonadiabatic electron transfer. The rate constant k_{ET} was calculated for the electron transfer across an alkanethiol monolayer on platinum and on gold. It was shown that the metal bands that are weakly coupled contribute much less to the rate constant than was suggested by their density of states ϱ_F. The authors concluded that k_{ET} is approximately independent of ϱ_F in two cases: (1) adiabatic electron transfer and (2) nonadiabatic electron transfer when the extra ϱ_F is due to the d electrons. The temperature dependence of the electronic contribution to the nonadiabatic electron transfer rate constant (k_{ET}) at metal electrodes was discussed [137]. It was found that this contribution is proportional to the absolute temperature T. The nonadiabatic rate constant for electron transfer at a semiconductor electrode was also considered. Under conditions for the maximum rate constant, the electronic contribution was estimated to be proportional to T, but for different reasons from those the case of metals, that is, Boltzmann statistics and transfer at the conduction band edge for the semiconductor versus Fermi–Dirac statistics and transfer at the Fermi level, which is far from the band edge, of the metal.

Molecular dynamics simulations of electron and ion transfer reactions near a smooth surface were presented [136]. The effect of the geometrical constraint of the surface and the interfacial electric field on the relevant solvation properties of both a monovalent negative ion and a neutral atom was analyzed. The quantum mechanical electron transfer between the metal surface and the ion/atom in solution was done by the MD simulation using a model Hamiltonian. The authors calculated two-dimensional free energy surfaces for ion adsorption allowing for partial charge transfer.

A generalized quantum master equation theory that governs the quantum dissipation and quantum transport was formulated in terms of hierarchically coupled equations of motion for an arbitrary electronic system in contact with electrodes under either a stationary or a nonstationary electrochemical potential bias [137]. The multiple frequency dispersion and the non-Markovian reservoir parameterization schemes were considered. The resulting hierarchical equations of motion formalism was applied to arbitrary electronic systems, including Coulomb interactions, under the influence of arbitrary time-dependent applied bias voltage and external fields. The authors claimed that the present theory provides an exact and numerically tractable tool to evaluate various transient and stationary quantum transport properties of many-electron systems, together with the involving nonperturbative dissipative dynamics.

Effects of electron correlations in a surface molecule model for the adiabatic electrochemical electron transfer reactions with allowance for the electrostatic repulsion of electrons on an effective orbital of metal was considered [138]. It was shown that taking into account the electrostatic repulsion on the effective orbital of the metal and the correlation effects leads to qualitative different forms of adiabatic free energy surfaces in some regions of values of the model's parameters. Approximate method for calculation of electron transition probability for simple outer-sphere electrochemical reactions was developed [139]. The probability of an elementary act in an outer-sphere electrochemical electron transfer reaction was calculated with arbitrary values of the parameter of reactant–electrode electron interaction for diabatic free energy surfaces of the parabolic form. The dependence of effective transmission coefficient on the Landau–Zener parameter was found. Interpolation formulas obtained allowed authors to calculate the electron transition probability using the results of quantum chemical calculation of the electronic matrix element as a function of distance.

Direct electrochemistry of redox enzymes as a tool for mechanistic studies was proposed [140]. The following analytical expression for the rate of redox processes was

$$k_{\text{red/ox}} = \frac{k_{\max}}{\sqrt{4\pi\lambda/RT}} \int_{-\infty}^{\infty} \frac{\exp\left(-(1/4\lambda RT)[\lambda \pm F(E-E^0)-RTx]^2\right)}{1+\exp(x)} dx$$

(1.49)

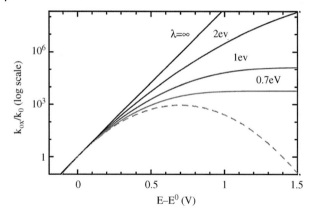

Figure 1.24 Dependence of the rate of interfacial, oxidative ET on the electrode potential according to the BV (black) or Marcus theory of interfacial ET for different values of the reorganization energy λ (solid gray lines). The dashed line is the traditional Marcus paraboloid with $\lambda = 0.7$ eV, showing the inverted region at high driving force. For interfacial ET, the rate levels off at a high driving force instead of decreasing [141].

where k_{max} is the asymptotic value of the rate constant at large overpotential. Equation 1.49 is referred to as Marcus theory applied to interfacial ET kinetics. Dependence of the rate of interfacial, oxidative ET on the electrode potential is displayed in Figure 1.24.

1.3
Concerted and Multielectron Processes

In order to explain the high efficiency of many chemical and enzymatic processes, the concept of energetically favorable, concerted mechanisms is widely used.

In a concerted reaction, a substrate is simultaneously attracted by different active reagents with acid and basic groups, nucleophile and electrophile, or reducing and oxidizing agents. It may, however, be presumed that certain kinetic statistic limitation exists on realization of reactions that are accompanied by a change in the configuration of a large number of nuclei [9, 142–148]. A concerted reaction occurs as a result of the simultaneous elementary transition (taking approximately 10^{-13} s) of a system of independent oscillators, with the mean displacement of nuclei φ_0, from the ground state to the activated state in which this displacement exceeds for each nucleus a certain critical value (φ_{cr}). If $\varphi_{cr} > \varphi_0$ and the activation energy of the concerted process $E_{syn} > nRT$, the theory gives the following expression for the synchronization factor α_{syn}, which is the ratio of preexponential factors synchronous and regular processes:

$$\alpha_{syn} = \frac{n}{2^{n-1}} \left(\frac{nRT}{\pi E_{syn}} \right)^{n-1/2} \tag{1.50}$$

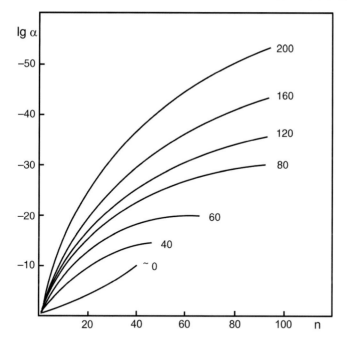

Figure 1.25 Theoretical dependence of the synchronization factor (f_{syn}) on the number of degree of freedom (n) of nuclei involved in a concerted reaction. The curves have been constructed in accordance with Equation 1.50 [9]. Energy activation of the synchronous reaction is given in kJ/mol [156].

Analysis of Equation 1.50 provides a clear idea of the scale of the synchronization factor and dependence of this factor on the number of n and the energy activation (Figure 1.25) [9, 150]. For example, at moderate energy activation 20–30 kJ/mol, the incorporation of each new nucleus into the transition state can lead to a 10-fold decrease in the rate of the process.

The model of concerted processes discussed above is only a crude approximation of the motion of a complex system of nuclei along the reaction coordinate. However, such an approximation apparently permits one to choose from among the possible reaction mechanisms.

In the case of effective concerted mechanism, the decrease in the synchronization probability (α_{syn}) with increasing n must be compensated for by an appreciable decrease in the activation energy. This consideration has led to the formulation of the principle of *optimum motion* [9, 144–147]. According to this principle, the number of nuclei whose configuration is charged in the elementary act of a chemical reaction (including electron transfer) must be sufficiently large to provide favorable energetic of the step and, at the same time, sufficiently small for the maintenance of a high value of the synchronization probability during motion along the reaction pathway to the reaction products.

There are a considerable number of reactions in which the products contain two electrons more than the starting compounds, and the consecutive several-step one-electron transfer process proves to be energetically unfavorable. In such cases, it is presumed that two-electron process occurs in one elementary two-electron step.

Two-electron mechanism may involve the direct transport of two electrons from a mononuclear transition complex to a substrate. Such a transport may alter sharply the electrostatic states of the systems and obviously require a substantial rearrangement of the nuclear configuration of ligands and polar solvent molecules. For instance, the estimation of the synchronization factor (α) for an octahedral complex with low molecular saturated ligands shows a very low value of $\alpha = 10^{-7}$–10^{-8} and, therefore, a very low rate of reaction. The probability of two-electron processes, however, increases sharply if compensation shift of electronic cloud from bulky nonsaturated ligands to central metal atom takes place. Involvement of bi- and, especially, polynuclear transition metal complexes and metal clusters, as well as synchronous proton transfer, may essentially decrease the environment reorganization and, therefore, provide a high rate for the two- and four-electron mechanisms.

The concept of four-electron mechanism was first suggested in Ref. [151] and then developed and applied to such "heavy" enzymatic and chemical reactions as the reduction of molecular nitrogen and the water splitting under mild conditions [144, 147, 149, 151–155].

The multielectron nature of the energetically favorable processes in clusters does not evidently impose any new, additional restrictions on its rate. Within the clusters' coordination sphere, the multiorbital overlap is effective and, therefore, the resonance integral V is high. The electron transfer from (or to) the orbitals of the metal to substrate orbital is accompanied by the simultaneous shift of electron clouds to the reverse direction. Such a transport may prevent significant changing of local charges and does not violate markedly the reaction complex nuclear frame. The strong delocalization electron in clusters and polynuclear complexes reduces to minimum the reorganization of the nuclear system during electron transitions and, therefore, provides low energy activation and a relatively high value for the synchronization factor.

The data presented in this chapter show that a particularly significant contribution has been recently made to the theory of electron transfer, which has formed the theoretical basis for a deeper understanding and effective planning of redox processes in dye-sensitized solar cells and other photochemical systems.

References

1 Jortner, J. and Bixson, M. (eds.) (1999) Electron transfer: from isolated molecules to biomolecules, in *Advances in Chemical Physics*, vol. **107**, Parts 1 and 2, John Wiley & Sons, Inc., NY.

2 Marcus, R.A. (1968) *J. Phys. Chem.*, **72** (8), 891–899.

3 Marcus, R.A. (1999) Electron transfer past and future, in *Advances in Chemical Physics*, vol. **107**, Part 1 (eds. J. Jortner and M. Bixon), John Wiley & Sons, Inc., NY, pp. 1–6.

4 Hush, N.S. and Reimers, J.R. (1998) *Coordin. Chem. Rev.*, **177** (1), 37–60.

5 Marcus, R.A. (2000) *J. Electroanal. Chem.*, **483** (1), 2), 2–6.

6 Sutin, N. (1999) Electron transfer reaction in solution: a historical

perspective, in *Advances in Chemical Physics*, vol. **107**, Part 1 (eds. J. Jortner and M. Bixon), John Wiley & Sons, Inc., NY, pp. 7–33.
7 Gust, D., Moore, T.A., and Moore, A.L. (2009) *Acc. Chem. Res.*, **42** (12), 1890–1898.
8 Rutherford, A.W. and Moore, T.A. (2008) *Nature (London)*, **453** (7194), 449–453.
9 Likhtenshtein, G.I. (2003) *New Trends In Enzyme Catalysis and Mimicking Chemical Reactions*, Kluwer Academic/ Plenum Publishers, Dordrecht.
10 Blankenship, R.E. (2002) *Molecular Mechanisms of Photosynthesis*, Blackwell Science, Oxford.
11 Feskov, S.V., Ivanov, A.I., and Burshtein, A.I. (2005) *J. Chem. Phys.*, **122** (12), 124509/1–124509/11.
12 Hall, D.O. and Krishna, R. (1999) *Photosynthesis*, 6th edn, Cambridge University Press, Cambridge.
13 Collings, A.F. and Critchley, C. (2005) *Artificial Photosynthesis: From Basic Biology to Industrial Application*, Wiley-VCH Verlag GmbH, Weinheim.
14 Lawlor, D. (2000) *Photosynthesis*, Kluwer Academic, Dordrecht.
15 Kenneth, K.C. (1992) *Energy Conversion (West - Engineering Series)*, PWS Pub. Co., electronic edition, revised from the 1992 edition.
16 Wurfel, P. (2009) *Physics of Solar Cells: From Basic Principles to Advanced Concepts*, 2nd edn, Wiley-VCH Verlag GmbH, Weinheim.
17 Marcus, R.A. and Sutin, N. (1985) *Biochim. Biophys. Acta*, **811** (2), 265–322.
18 Mikhailova, V.A. and Ivanov, A.I. (2007) *J. Phys. Chem. C*, **111** (11), 4445–4451.
19 Bixon, M. and Jortner, J. (1999) Electron transfer: from isolated molecules to biomolecules, in *Advances in Chemical Physics*, vol. **107**, Part 1 (eds. J. Jortner and M. Bixon), John Wiley & Sons, Inc., NY, pp. 35–202.
20 Gray, H.B. and Winkler, JR. (1996) *Annu. Rev. Biochem.*, **65**, 537–561.
21 McLendon, G., Komar-Panicucci, S., and Hatch, S. (1999) *Adv. Chem. Phys.*, **107**, 591–600.
22 Khokhlova, S.S., Mikhailova, V.A., and Ivanov, A.I. (2008) *Russ. J. Phys. Chem. A*, **82** (6), 1024–1030.
23 Lluch, J.M. (2000) *Theor. Chem. Acc.*, **103** (3–4), 231–233.
24 Wurfel, P. (2009) *Physics of Solar Cells: From Basic Principles to Advanced Concepts*, 2nd edn, Wiley-VCH Verlag GmbH, Weinheim.
25 Wenham, S.R., Green, M.A., Watt, M.E., and Corkish, R. (2007) *Applied Photovoltaics*, 2nd edn, Earthscan Publications Ltd., London.
26 Green, M.T. (2005) *Generation Photovoltaics: Advanced Solar Energy Conversion, Springer Series in Photonics*, Springer, Heidelberg.
27 Marcus, R.A. (1997) *Pure Appl. Chem.*, **69** (1), 13–29.
28 Chernick, E.T., Mi, Q., Vega, A.M., Lockar, J.V., Ratner, M., and Wasielewski, M.R. (2007) *J. Phys. Chem. B*, **111** (24), 6728–6737.
29 Kurnikov, I.V., Zusman, L.D., Kurnikova, M.G., Farid, R.S., and Beratan, D.N. (1997) *J. Am. Chem. Soc.*, **119** (24), 5690–5700.
30 Egorova, D., Thoss, M., Domcke, W., and Wang, H. (2003) *J. Chem. Phys.*, **119** (5), 2761–2773.
31 Khudyakov, I., Zharikov, A.A., and Burshtein, A.I. (2010) *J. Chem. Phys.*, **132** (1), 014104/1–014104/6.
32 Leontyev, I.V., Vener, M.V., Rostov, I.V., Basilevsky, M.V., and Newton, M.D. (2003) *J. Chem. Phys.*, **119** (15), 8024–8037.
33 Likhtenshtein, G.I., Nakatsuji, S., and Ishii, K. (2007) *Photochem. Photobiol.*, **83** (4), 871–881.
34 Likhtenshtein, G.I., Pines, D., Pines, E., and Khutorsky, V. (2009) *Appl. Magn. Reson.*, **35** (3), 459–472.
35 Gamov, C.A. (1926) *Quanum Theo. Atoms Phys.*, **51** (2), 204–212.
36 Marcus, R.A. (1956) *J. Chem. Phys.*, **24**, 966–978.
37 Hush, N.S. (1958) *J. Chem. Phys.*, **28** (5), 962–972.
38 Fermi, E. (1950) *Nuclear Physics*, University of Chicago Press, Chicago.
39 Landau, L. (1932) *Phyz. Soviet Union*, **2** (1), 46–51.

40 Zener, C. (1933) *Proc. Royal. Soc. London,* **A140**, 660–668.

41 Levich, V.G. and Dogonadze, R. (1960) *Dokl. Acad. Nauk.,* **133** (1), 159–161.

42 Levich, V.G., Dogonadze, R., German, E., Kuznetsov, A.M., and Kharkats, Y.I. (1970) *Electrochem. Acta,* **15** (2), 353–368.

43 German, E.D. and Kuznetsov, A.M. (1994) *J. Phys. Chem.,* **98** (24), 6120–6127.

44 Efrima, S. and Bixon, M. (1974) *Chem. Phys. Lett.,* **25** (1), 34–37.

45 Nitzan, A., Jortner, J., and Rentzepis, P.M. (1972) *Proc. Royal Soc. London,* **A327** (2), 367–391.

46 Neil, R., Kestner, N.R., Logan, J., and Jortner, J. (1974) *J. Phys. Chem.,* **78** (21), 2148–2166.

47 Hopfield, J.J. (1974) *Proc. Natl. Acad. Sci. USA,* **71** (9), 3640–3644.

48 Grigorov, L.N. and Chernavsky, D.S. (1972) *Biofizika,* **17** (2), 195–202.

49 Costentin, C. (2008) *Chem. Rev.,* **108** (7), 2145–2179.

50 Burshtein, A.I. (2000) United theory of photochemical charge separation, in *Advances in Chemical Physics,* vol. **114** (eds I. Prigogine and A. Rice), John Wiley & Sons, Inc., pp. 419–587.

51 Benjamin, I. and Pollak, E. (1996) *J. Chem. Phys.,* **105** (20), 9093–9910.

52 Kurnikov, I.V., Zusman, L.D., Kurnikova, M.G., Farid, R.S., and Beratan, D.N. (1997) *J. Am. Chem. Soc.,* **119** (24), 5690–5700.

53 Freed, K.F. (2003) *J. Phys. Chem. B,* **107** (38), 10341–10343; Egorova, D., Thoss, M., Domcke, W., and Wang, H. (2003) *J. Chem. Phys.,* **119** (5), 2761–2773.

54 Kuznetsov, A.M., Sokolov, V.V., and Ulstrup, J. (2001) *J. Electroanal. Chem.,* **502** (1–2), 36–46.

55 Zhou, H.-X. and Szabo, A. (1995) *J. Chem. Phys.,* **103** (9), 3481–3494; Zhao, Y. and Liang, W.Z., *Phys. Rev. A Atom. Mol. Opt. Phys.,* **74** (3 Pt A), 032706/1–032706/5.

56 Wu, Q. and Van Voorhis, T. (2006) *J. Chem. Phys.,* **125** (16), 164105/1–164105/9.

57 Migliore, A., Corni, S.I., Felice, R., and Molinari, E. (2006) *J. Chem. Phys.,* **124** (6), 064501/1–064501/16.

58 Zhu, C. and Nakamura, H. (2001) *Adv. Chem. Phys.,* **117**, 127–166.

59 Zhao, Y. and Liang, W. (2006) *Phys. Rev. A Atom. Mol. Opt. Phys.,* **74** (3 Pt A), 032706/1–032706/5.

60 Zhao, Y. and Nakamura, H. (2006) *J. Theor. Comput. Chem.,* **5** (Special Issue), 299–306.

61 Aubry, S. (2007) *J. Phys. Condens. Matter,* **19** (25), 255204/1–255204/30.

62 Dougherty, R.C. (1997) *J. Chem. Phys.,* **106** (7), 2621–2626.

63 Newton, M.D. (2004) in *Comprehensive Coordination Chemistry II,* vol. **2**, pp. 573–587.

64 Finklea, H.O. (2007) *Encycloped. Electrochem.,* **11**, 623–650.

65 Newton, M.D. (1999) Control of electron transfer kinetics: models for medium reorganization and donor–acceptor coupling, in *Advances in Chemical Physics,* vol. **107**, Part 1 (eds. J. Jortner and M. Bixon), John Wiley & Sons, Inc., NY, pp. 303–376.

66 VandeVondele, J., Ayala, R., Sulpizi, M., and Sprik, M. (2007) *J. Electroanal. Chem.,* **607** (1–2), 113–120.

67 Wu, Q. and Van Voorhis, T. (2006) *J. Phys. Chem. A,* **110** (29), 9212–9218.

68 Berlin, Y.A., Grozema, F.C., Siebbeles, L.D.A. and Ratner, M.A. (2008) *J. Phys. Chem. C,* **112** (29), 10988–11000.

69 Jang, S. and Newton, M.D. (2005) *J. Chem. Phys.,* **122** (2), 024501/1–024501/15.

70 Basilevsky, M.V., Rostov, I.V., and Newton, M.D. (1998) *J. Electroanal. Chem.,* **450** (1), 69–82.

71 Fletcher, S. (2007) *J. Solid State Electrochem.,* **11** (5), 965–969.

72 Fletcher, S. (2008) *J. Solid State Electrochem.,* **12** (4), 765–770.

73 Liang, K.K., Mebel, A.M., Lin, S.H., Hayashi, M., Selzle, H.L., Schlag, E.W., and Tachiya, M. (2003) *Phys. Chem. Chem. Phys.,* **5** (20), 4656–4665.

74 Rasaiah, J.C. and Zhu, J. (2008) *J. Chem. Phys.,* **129** (21), 214503/1–214503/5.

75 Chen, Y., Xu, R.-X., Ke, H.-W., and Yan, Y.-J. (2007) *Chin. J. Chem. Phys.,* **20** (4), 438–444.

76 de la Lande, A. and Salahub, D.R. (2010) *J. Mol. Struct. Theochem*, **943** (1–3), 115–120.
77 Kharkatz, Y.I. (1976) *Electrokhimia*, **12**, 592–595.
78 Miyashita, O. and Go, N. (2000) *J. Phys. Chem.*, **104**, 7516–7521.
79 Georgievskii, Y., Hsu, C., and Marcus, R.A. (1999) *J. Chem. Phys.*, **110** (11), 5307–5317.
80 Zusman, L.D. (1980) *Chem. Phys.*, **49** (2), 295–304.
81 Zusman, L.D. (1983) *Chem. Phys.*, **80** (1–2), 29–43.
82 Zusman, L.D. (1991) *Electrochim. Acta*, **36** (3–4), 395–399.
83 Zhang, M.-L., Zhang, S., and Pollak, E. (2003) *J. Chem. Phys.*, **119** (22), 11864–11877.
84 Shi, Q., Chen, L., Nan, G., Xu, R., and Yan, Y. (2009) *J. Chem. Phys.*, **130** (16), 164518/1–164518/7.
85 Tanaka, M. and Tanimura, Y. (2009) *J. Phys. Soc. Jpn.*, **78** (7), 073802/1–073802/4.
86 Zhang, M.-L., Zhang, S., and Pollak, E. (2003) *J. Chem. Phys.*, **119** (22), 11864–11877.
87 Schröder, M., Schreiber, M., and Kleinekathöfer, U. (2007) *J. Chem. Phys.*, **126**, 114102.
88 Casado-Pascual, J., Goychuk, I., Morillo, M., and Hanggi, P. (2002) *Chem. Phys. Lett.*, **3603** (4), 333–339.
89 Casado-Pascual, J., Morillo, M., Goychuk, I., and Hanggi, P. (2003) *J. Chem. Phys.*, **118** (1), 291–303.
90 Kuznetsov, A.M. and Ulstrup, J. (2001) *Electrochim. Acta*, **46** (20–21), 3325–3333.
91 Likhtenshtein, G.I. (1996) *J. Phochem. Photobiol. A Chem.*, **96** (1), 79.
92 Zhou, H.-X. and Szabo, A. (1995) *J. Chem. Phys.*, **103** (9), 3481–3494.
93 Banerjee, S. and Gangopadhyay, G. (2007) *J. Chem. Phys.*, **126** (3), 034102/1–034102/14.
94 Beratan, D.N., Onuchic, J.N., Betts, J.N., Bowler, B.E., and Gray, G.H. (1990) *J. Am. Chem. Soc.*, **112** (34), 7915–7921.
95 Cook, W.R., Coalson, R.D., and Evans, D.G. (2009) *J. Phys. Chem. B*, **113** (33), 11437–11447; Marcus, R.A. (1997) *Pure Appl. Chem.*, **69** (1), 13–29.
96 Stuchebrukov, A.A. and Marcus, R.A. (1995) *J. Phys. Chem.*, **99** (19), 7581–7590.
97 Scourotis, S.S. and Beratan, D.N. (1999) Theories of structure–function relationships for bridge-mediated electron transfer reaction, in *Advances in Chemical Physics*, vol. **107**, Part 1 (eds J. Jortner and M. Bixon), John Wiley & Sons, Inc., NY, pp. 377–452.
98 Balabin, I.A. and Onuchic, J.N. (2000) *Science*, **290**, 114–117.
99 Kosloff, R. and Ratner, M.A. (1990) *Isr. J. Chem.*, **30** (1–2), 45–58.
100 Likhtenshtein, G.I. (2008) *Pure and Appl. Chem.*, **80** (10), 2125–2139.
101 Zamaraev, K.I., Molin, Y.N., and Salikhov, K.M. (1981) Spin exchange, in *Theory and Physicochemical Application*, Springer, Heidelberg.
102 Ermolaev, V.L., Bodunov, E.N., Sveshnikova, E.B., and Shakhverdov, T.A. (1977) *Radiationless Transfer of Electronic Excitation Energy*, Nauka, Leningrad.
103 Chong, S.-H. and Hirata, F. (1995) *J. Mol. Liq.*, **65–66**, 345–348.
104 Siddarth, P. and Marcus, R.A. (1993) *J. Phys. Chem.*, **97** (11), 2400–2405.
105 Siddarth, P. and Marcus, R.A. (1993) *J. Phys. Chem.*, **97** (23), 6111–6114.
106 Siddarth, P. and Marcus, R.A. (1993) *J. Phys. Chem.*, **97** (50), 13078–13082.
107 Tanimura, Y., Leite, V.B.P., and Onuchic, J.N. (2002) *J. Chem. Phys.*, **117** (5), 2172–2179.
108 Okada, A., Chernyak, V., and Mukamel, S.J. (1998) *Phys. Chem. A*, **102** (8), 1241–1251.
109 Aubry, S. and Kopidakis, G. (2005) *J. Biol. Phys.*, **31** (3–4), 375–402.
110 Buchachenko, A.L. and Berdinsky, V.L. (2002) *Chem. Rev.*, **102** (3), 603–612.
111 Buchachenko, A.L. and Kuznetsov, D.A. (2006) *Mol. Biol.*, **40** (1), 9–15.
112 Okamura, M.Y. and Feher, G. (1986) *Proc. Natl. Acad. Sci. USA*, **83** (21), 8152–8156.
113 Cukier, R.I. (2002) *J. Phys. Chem. B*, **106** (7), 1746–1757.
114 Cukier, R.I. (1996) *J. Phys. Chem.*, **100** (38), 15428–15443.
115 Cukier, R.I. and Nocera, D.G. (1998) *Annu. Rev. Phys. Chem.*, **49**, 337–369.

116 Chakraborty, A. and Hammes-Schiffer, S. (2008) *J. Chem. Phys.*, **129** (20), 204101/1–204101/16.
117 Hammes-Schiffer, S., and Soudackov, A.V. (2008) *J. Phys. Chem. B*, **112** (45), 14108–14123.
118 Venkataraman, C., Soudackov, A.V., and Hammes-Schiffer, S. (2008) *J. Phys. Chem. C*, **112** (32), 12386–12397.
119 Hatcher, E., Soudackov, A., and Hammes-Schiffer, S. (2005) *J. Phys. Chem. B*, **109** (39), 18565–18574.
120 Hammes-Schiffer, S. (2001) *Acc. Chem. Res.*, **34** (4), 273–281.
121 Moore, D.B. and Martinez, T.J. (2000) *J. Phys. Chem. A*, **104** (11), 2525–2531.
122 Ludlow, M.K., Soudackov, A.V., and Hammes-Schiffer, S. (2010) *J. Am. Chem. Soc.*, **132** (4), 1234–1235.
123 Venkataraman, C., Soudackov, A.V., and Hammes-Schiffer, S. (2010) *J. Phys. Chem. C*, **114** (1), 487–496.
124 Kuznetsov, A.M. and Medvedev, I.G. (2008) *Russ. J. Electrochem.*, **44** (2), 167–186.
125 Fermi, E. (1926) *Rend. Lincei*, **3**, 145–149.
126 Dirac, P.A.M. (1926) *Proc. R. Soc. Series A*, **112**, 661–677.
127 Dirac, P.A.M. (1967) *Principles of Quantum Mechanics* (revised 4th ed.), Oxford University Press, London, pp. 210–211.
128 Sommerfeld, A. (1927) *Naturwissenschaften*, **15** (41), 824–832.
129 Kittel, C. and Kroemer, H. (1980) *Thermal Physics*, 2nd edn, W.H. Freeman, San Francisco, p. 357.
130 Zusman, L.D. (1992) *Chem. Phys. Lett.*, **200** (4), 379–381.
131 Bramwell, S.T., Giblin, S.R., Calder, S., Aldus, R., Prabhakaran, D., and Fennell, T. (2009) *Nature*, **461** (7266), 956–960.
132 Ryo, A., Masahiro, K., and Fumio, H. (1999) *Chem. Phys. Lett.*, **305** (3), 251–257.
133 Sakata, T. (1996) *Bull. Soc. Chem. Jpn.*, **69** (9), 2435–2446.
134 Gosavi, S. and Marcus, R.A. (2000) *J. Phys. Chem. B*, **104** (9), 2067–2072.
135 Gosavi, S., Gao, Y.Q., and Marcus, R.A. (2001) *J. Electroanal. Chem.*, **500** (1–2), 71–77.
136 Hartnig, C. and Koper, M.T.M. (2003) *J. Am. Chem. Soc.*, **125** (32), 9840–9845.
137 Jin, J., Zheng, X., and Yan, Y. (2008) *J. Chem. Phys.*, **128** (23), 234703/1–234703/15.
138 Kokkanen, A.A., Kuznetsov, A.M., and Medvedev, I.G. (2007) *Russ. J. Electrochem.*, **43** (9), 1033–1046.
139 Kokkanen, A.A., Kuznetsov, A.M., and Medvedev, I.G. (2008) *Russ. J. Electrochem.*, **44** (4), 397–407.
140 Leger, C. and Bertrand, P. (2008) *Chem. Rev.*, **108** (7), 2379–2483.
141 Chidsey, C.E.D. (1991) *Science*, **251** (4996), 919–922.
142 Bordwell, F.G. (1970) *Acc. Chem. Res.*, **3**, 281–290.
143 Likhtenshtein, G.I. (1974) *Spin Labeling Method in Molecular Biology*, Nauka, Moscow.
144 Likhtenshtein, G.I. (1976) *Spin Labeling Method in Molecular Biology*, Wiley Interscience, NY.
145 Likhtenshtein, G.I. (1977) *Kinet. Catal.*, **28**, 878–882.
146 Likhtenshtein, G.I. (1977) *Kinet. Catal.*, **28**, 1255–1260.
147 Likhtenshtein, G.I. (1988) *Chemical Physics of Redox Metalloenzymes*, Springer, Heidelberg.
148 Bernasconi, C.F. (1992) *Acc. Chem. Res.*, **25**, 9–16.
149 Denisov, E.T., Sarkisov, O.M., and Likhtenshtein, G.I. (2003) *Chemical Kinetics: Fundamentals and Recent Developments*, Elsevier Science.
150 Alexandrov, I.V. (1976) *Theor. Exper. Khim*, **12**, 299–306.
151 Likhtenshtein, G.I. and Shilov, A.E. (1970) *Zhurnal Fiz. Khem.*, **44**, 849–856; Semenov, N.N., Shilov, A.E., and Likhtenshtein (1975) *Dokl. Acad. Nauk SSSR*, **221**, 1374–1377.
152 Likhtenshtein, G.I., Kotel'nikov, A.I., Kulikov, A.V., Syrtsova, L.A., Bogatyrenko, V.R., Mel'nikov, A.I., Frolov, E.N., and Berg, A.J. (1979) *Int. J. Quantum Chem.*, **16** (3), 419–435.
153 Likhtenshtein, G.I. (1988) *J. Mol. Catal.*, **48**, 129–138.
154 Likhtenshtein, G.I. (1990) *Pure Appl. Chem.*, **62**, 281–288.
155 Likhtenshtein, G.I. (1985) in (ed. Y.A. Ovchinnikov), Proceedings of the 16th FEBS Congress (1985), A, p. 9.

2
Principal Stages of Photosynthetic Light Energy Conversion

2.1
Introduction

Nature's solar energy storage system is found in photosynthetic organisms, including plants, algae, and a variety of bacteria. All these organisms utilize sunlight to power cellular processes and ultimately derive biomass through chemical reactions driven by light. The main outcome of plant photosynthesis is the oxidation of water and the synthesis of glucose from carbon dioxide at the expense of sunlight energy. On the scale of the Earth, annually about 50 billion tons of carbon from carbon dioxide is bound into forms that provide energy and structural material for all living organisms.

Plants, algae, and some bacteria use two photosystems, PS I with chlorophyll P_{700} and PS II with chlorophyll P_{680}. After the initial electron transfer event, a series of electron transfer reactions take place that eventually stabilize the stored energy in forms that can be used by cells. Using light energy, PS II acts first to channel an electron through a series of acceptors that drive a proton pump to generate ATP, before passing the electron on to PS I. Once the electron reaches PS I, it has used most of its energy in producing ATP, but a second photon of light captured by P_{700} provides the required energy to channel the electron to ferredoxin, generating reducing power in the form of NADPH. The ATP and NADPH produced by PS II and PS I, respectively, are used in the light-independent reactions for the formation of organic compounds. In these organisms, the eventual electron donor is water, liberating molecular oxygen, and the ultimate electron acceptor is carbon dioxide, which is reduced to sugars. Other types of photosynthetic organisms contain only a single photosystem, which in some cases is more similar to photosystem II and in other cases to photosystem I of the oxygen-evolving organisms [1–14].

The problems of photosynthesis embrace practically all aspects of modern biochemistry, biophysics, and molecular biology. Here, we shall briefly consider three aspects of fundamental importance not only for biology but also for modern chemistry and artificial photosynthesis, in particular, (1) light-harvesting (antenna) complex; (2) the structure and action mechanism of the system of conversion of light energy into chemical energy in the primary charge photoseparation in bacterial

Solar Energy Conversion. Chemical Aspects, First Edition. Gertz Likhtenshtein.
© 2012 Wiley-VCH Verlag GmbH & Co. KGaA. Published 2012 by Wiley-VCH Verlag GmbH & Co. KGaA.

and plant photosynthesis; and (3) structure and the possible mechanisms of the participation of polynuclear manganese systems in the photooxidation of water. The first system is a very effective device for collection and goal direction of light; the second is a remarkable example of the appearance of qualitatively new properties upon combination of active groups into an ordered structure; and the third system accomplishes one of the most surprising reactions that occur in nature, the production of a strong reducing agent from water on account of the quanta of low energy.

2.2
Light-Harvesting Antennas

2.2.1
General

Photosynthesis begins when light is absorbed by an antenna pigment. This pigment can be a (bacterio)chlorophyll, carotenoid, or bilin (open-chain tetrapyrrole) depending on the type of organism. A wide variety of antenna complexes are found in different photosynthetic systems [1, 6, 7].

Antennas permit an organism to increase greatly the absorption cross section for light energy transfer processes that may involve transfers to many intermediate pigments eventually resulting in the electronic excitation of a closely coupled pair of (bacterio)chlorophyll molecules in the photochemical reaction center. Each antenna complex has between 250 and 400 pigment molecules. In many photosynthetic systems, the size of the antenna can be adjusted to suit the intensity of light. Their absorption spectra are nonoverlapping in order to broaden the range of light that can be absorbed in photosynthesis. Transfer of excitation from the antenna to the reaction center (RC) occurs with high efficiency. For an efficient energy transfer, the system is usually organized with a thermodynamic gradient from antenna to reaction center. The peripheral antenna absorbs the light of higher energy and transfers the excitation to pigments with lower energy absorbance bands with the photochemical trap (the reaction center) at the lowest energy.

The light-harvesting complex (LHC) of plants is an array of protein and chlorophyll molecules embedded in the thylakoid membrane that transfer light energy to one chlorophyll a molecule at the reaction center of a photosystem. The photosystems I and II light-harvesting system carry out two essential functions, the efficient collection of light energy for photosynthesis, and the regulated dissipation of excitation energy in excess of that which can be used [1, 15–18].

According to the Förster theory [19], the efficiency (E) of Förster resonance energy transfer (FRET) between an excited fluorophore (donor, D) and a quenching chromophore (acceptor, A)

$$E = \frac{k_{ET}}{k_f + k_{ET} + \Sigma k_i} \tag{2.1}$$

where k_{ET} is the rate of energy transfer, k_f the radiative decay rate, and the k_i are the rate constants of any other deexcitation pathway and

$$E = \frac{1}{1+(r/R_0)^6} \quad (2.2)$$

where r is the distance between the centers and R_0 is the Förster distance (critical radius) of this pair of donor and acceptor, that is, the distance at which the energy transfer efficiency is 50%.

$$R_0^6 = \frac{9 Q_0 (\ln 10) k^2 J}{128 \pi^5 n^4 N_A} \quad (2.3)$$

where Q_0 is the fluorescence quantum yield of the donor in the absence of the acceptor, κ^2 is the dipole orientation factor, n is the refractive index of the medium, N_A is Avogadro's number, and J is the overlap integral, which expresses the degree of spectral overlap between the donor emission and the acceptor absorption:

$$J = \int f_D(\lambda) \varepsilon_A(\lambda) \lambda^4 d\lambda \quad (2.4)$$

Therefore, the efficient energy transfer in the light-harvesting antenna should fulfill a number of requirements:

1) The light must have sufficient energy for useful chemistry, but not be destructive (the former limits the range to <1000 nm, the near IR, and the latter to >340 nm, the near UV).
2) For energy transfer, the molecules must (a) be the optimal distance apart, (b) be appropriately aligned, and (c) have overlapping spectra for fluorescence and absorption.
3) The system should be organized with a thermodynamic gradient from antenna to reaction center (an energy "funnel").

2.2.2
Bacterial Antenna Complex Proteins

2.2.2.1 The Structure of the Light-Harvesting Complex

The structure of the LHC from a photosynthetic bacterium has been solved by X-ray crystallography and other methods in a number of works [20–31]. The bacterial antenna complexes are generally composed of two types of polypeptides (alpha and beta chains) that are arranged in a ring-like fashion creating a cylinder that spans the membrane; the proteins bind two or three types of bacteriochlorophyll (BChl) molecules and different types of carotenoids depending on the species. Both chains are small proteins of 42–68 residues that share a three-domain organization.

Figure 2.1 shows the refinement of the X-ray structure of the RC–LH1 core complex from *Rhodopseudomonas palustris* obtained by single-molecule spectroscopy [30].

Figure 2.1 X-ray structure of the RC–LH1 complex from *R. palustris*. This structure was determined to a resolution of 4.8 Å. In this complex the RC is enclosed by the LH1, which are arranged in an overall elliptical shape and accommodate two BChl *a* molecules each. The positions and orientations of the BChl *a* molecules were only placed in the model as a guide for the eye. These positions were not absolutely determined at this resolution. The LH1 ring features a gap where one dimer is replaced by another small protein that has been termed W [30]. Reproduced with permission. Copyright (2007) National Academy of Sciences, USA.

2.2.2.2 Dynamic Processes in LHC

Energy transfer processes in photosynthetic light-harvesting 2 (LH2) complexes isolated from purple bacterium *R. palustris* grown at different light intensities were studied by ground-state and transient absorption spectroscopy [32]. A spectral analysis of picosecond exciton relaxation revealed strong inhomogeneity of the B850 excitons in the low light adaptation (LL) samples. Energy transfer from B800 to B850 for high light adaptation (HL) samples occurred in a monoexponential process, while in LL samples, spectral relaxation of the B850 exciton follows strongly nonexponential kinetics. The authors explained obtained spectral changes by picosecond exciton relaxation as being caused by a small coupling parameter of the excitonic splitting of the BChl a molecules to the surrounding bath. Photophysical model of energy transfer paths and associated rate constants between BChl a molecules in the B800 and the B850 ring was presented. Figure 2.2 shows the differential absorption spectra of LH2 complexes at individual pump–probe delays. Data on rate constants (k_i) from global fitting of transient absorption spectra indicated heterogeneous kinetics. For example, the values of $1/k_i$ for LL1 were found to be 1.10, 2, and 280 ps.

The dynamics of excitation energy transfer within the B850 ring of light-harvesting complex from *Rhodobacter sphaeroides* and between neighboring B850 rings was

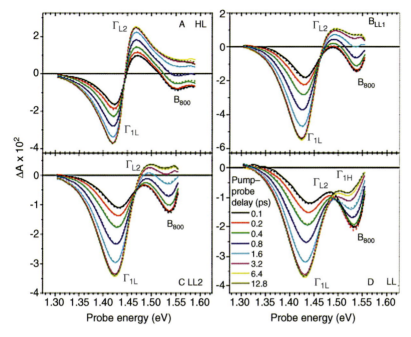

Figure 2.2 Differential absorption spectra of LH2 complexes at individual pump–probe delays. HL complex (A) shows G1L transition, there is no change after 3.2 ps; LL complex (D) presents an additional transition G1H changing till ∼10 ps panels of intermediate samples; LL1 (B) and LL2 (C), respectively, represent increasing contribution of G1H band to the spectra. All panels: solid lines, measured values; dashed lines, global values [32]. Reproduced with permission from Elsevier.

calculated by dissipative quantum mechanics [33]. The assumption of Boltzmann populated donor states for the calculation of intercomplex excitation transfer rates by generalized Förster theory showed that intracomplex exciton relaxation to near-Boltzmann population exciton states occurs within a few picoseconds. Data on ultrafast time-resolved carotenoid-to-bacteriochlorophyll energy transfer in LH2 complexes from photosynthetic bacteria was reported [34]. Steady-state absorption and fluorescence excitation experiments have revealed that the total efficiency of energy transfer from the carotenoids to bacteriochlorophyll is independent of temperature and close to 90% for the LH2 complexes. Ultrafast transient absorption spectra in the near-infrared region of the purified carotenoids in solution showed the energies of the S_1 ($2^1A_g^-$) → S_2 ($1^1B_u^+$) excited state.

2.2.3
Photosystems I and II Harvesting Antennas

The light-harvesting complex of plants is an array of protein and chlorophyll molecules embedded in the thylakoid membrane that transfer light energy to one chlorophyll a molecule at the reaction center of a photosystem [1, 17, 18, 26, 31, 35–42]. Under normal light conditions, plants convert the absorbed solar energy into biologically useful chemical energy by photosynthesis. However, under high-light conditions, the excess excitation energy is dissipated in the form of heat by nonphotochemical quenching (NPQ), without the risk of causing damage to the photosynthetic apparatus.

Photosystem I (PS I) is a large, membrane protein complex, consisting of 12–15 proteins and more than 100 bound cofactors, which catalyzes light-driven electron transfer across the photosynthetic membrane in cyanobacteria, green algae, and plants [1, 43]. Photosystem I can therefore be regarded as a joint reaction center–core antenna system. The structural and functional organization of the light-harvesting system in the peripheral antenna of photosystem I (LHC I) and its energy coupling to the photosystem I (PS I) core antenna network in the eukaryotic photosystem I–LHC I complex, eukaryotic LHC II complexes, and the cyanobacterial Photosystem I core have been considered [25]. In the PS I core antenna, the Chl clusters with a different magnitude of low-energy shift contribute to better spectral overlap of Chls in the reaction center and participate in downhill enhancement of energy transfer from LHC II to the PS I core. Chlorophyll clusters forming terminal emitters in LHC I were suggested to be involved in photoprotection against excess energy. A model of three-dimensional structure of the Lhca4 polypeptide of LHC I and LHC II is presented in Figure 2.3.

The light-harvesting antenna of photosystem II consists of three main groups of proteins [44, 45]. The isolated photosystem II reaction center contains two main subunits, CP43 and CP47, which bind chlorophyll a as a "core" antenna. LHC II consists of CP29, 26, and 24 (CP – chlorophyll protein, where the number indicates the M_r on gels) [44]. These polypeptides bind to chlorophyll a and beta-carotene and pass the excitation energy on to the reaction center. Architecture of a charge transfer state regulating light harvesting in a plant antenna protein was reported [31]. Evidence for charge transfer quenching in all three of the individual minor antenna complexes of PS II (CP29, CP26, and CP24) was found. The authors concluded that charge transfer quenching in CP29 involves a delocalized state of an excitonically coupled chlorophyll dimer. The 2.5 Å structure of pea light-harvesting complex of photosystem II (LHC II) was determined by X-ray crystallography of stacked two-dimensional crystals (Figure 2.4) [28]. Figure 2.5 shows how membranes interact to form chloroplast grana and reveals the mutual arrangement of 42 chlorophylls a and b, 12 carotenoids, and 6 lipids in the LHC II trimer. The closest π–π distance of an intertrimer Chl pair (Chl 2:Chl 2′) is 14.8 Å yielding a dipole coupling constant of 11 cm^{-1}.

In the rhombohedral crystal form of spinach major light-harvesting complex at 2.72 Å resolution, the shortest π–π distance between Chl 8 and Chl 14′ was found to be 5.4 Å, well within the range of Chl distances in the LHC II monomer [38]. The center-to-center distance was found to be 12.3 Å, so that the dipole coupling constant is 60 cm^{-1}. From these distances, the authors estimated the Förster transfer rates between LHC II trimers of 1065 ns^{-1} for rhombohedral crystals and 80 ns^{-1} for type-I crystals. In the thylakoid membrane, the shortest center-to-center distance between Chls in adjacent trimers was ~13.5 Å yielding a transfer rate of ~100 ns^{-1}.

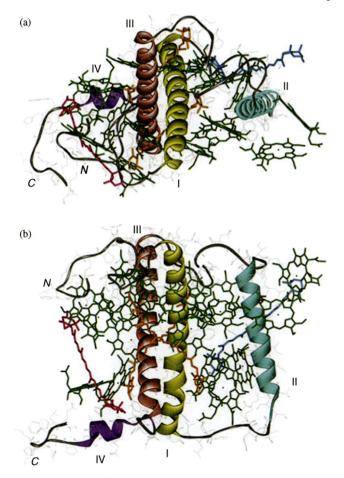

Figure 2.3 Model of three-dimensional structure of the Lhca4 polypeptide based on secondary structure prediction, and structural homologies of LHC I and LHC II. (a) Top view. (b) Side view. N and C label the N- and C-termini, respectively. I–IV, helices [25].

A number of investigations on dynamics of effects on the LHC of photosystems I and II were performed including the time-resolved fluorescence analysis of the recombinant photosystem II antenna complex CP29 [46]: fluorescence lifetime imaging (FLI) in real time [47] and a major light-harvesting polypeptide of photosystem II functions in thermal dissipation [48]. In photosystem II, the energy transfer processes between carotenoids and Chls have been studied by femtosecond transient absorption in the CP29-WT, which contains only two carotenoids per polypeptide located on the L1 and L2 sites, and in the CP29-E166V mutant in which only the L1 site is occupied [49]. It was shown that the main transfer occurs from the S_2 state of the carotenoid. The transfers take place with lifetimes of 80–130 fs. Two additional transfers were observed with 700 fs and 8–20 ps lifetimes. For the authors, the faster

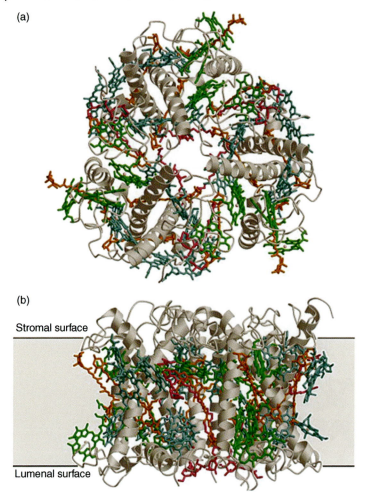

Figure 2.4 The LHC II trimer: (a) top view from stromal side; (b) side view. LHC II protrudes from a 35 Å lipid bilayer by 13 Å on the stromal side and by 8 Å on the lumenal side [28]. Reproduced with permission from American Chemical Society.

lifetime is due to energy transfer from a vibrationally unrelaxed S_1 state, whereas the 8–20 ps component is due to a transfer from the $S_{1,0}$ state of violaxanthin.

Little light reaches algae residing at a depth of 1 m or more in seawater. A phycobilisome is a light-harvesting protein complex present in cyanobacteria, glaucocystophyta, and red algae functioning as an antenna ([50] and references therein). The pigments, which are present in the phycobilisome, such as phycocyanobilin and phycoerythrobilin, enhance the amount and spectral window of light absorption and result in 95% efficiency of energy transfer. A central core of allophycocyanin sits above a photosynthetic reaction center. There are phycocyanin and phycoerythrin subunits that radiate out from this center like thin tubes. Such

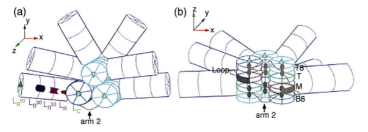

Figure 2.5 Representation of a hemidiscoidal PBS, as seen from side (a) and from the thylakoid membrane upward (b). The architecture exhibits a tricylindrical core, from which radiate six rods composed of three PC hexamers each. The two core-membrane linkers (LCM) are represented, with their different domains: the PB domain (forming part of the M disks), the PB loop, arm 2, and the Reps (filled ellipses). Triangles represent the small core and rod linkers LC and LR 10 [50]. Reproduced with permission from Elsevier.

an arrangement concentrates light energy down into the reaction center. The energy absorbed by pigments in the phycoerythrin subunits at the periphery is transferred to the reaction center in less than 100 ps. Excited state lifetimes of isolated antenna complexes, where the reaction centers have been removed, are typically in the 1–5 ns range.

In cyanobacteria, the harvesting of light energy for photosynthesis is mainly carried out by the phycobilisome, a giant, multisubunit pigment–protein complex. This complex is composed of heterodimeric phycobiliproteins that are assembled with the aid of linker polypeptides such that light absorption and energy transfer to photosystem II are optimized. In the work [50], the phycobilisome structure in mutants lacking either two or all three of the phycocyanin hexamers has been studied using single-particle electron microscopy. The images presented in Figure 2.5 give two-dimensional projection maps of a resolution of 13 Å.

2.3
Reaction Center of Photosynthetic Bacteria

2.3.1
Introduction

The primary photochemical and photophysical processes in the reaction center of photosynthetic bacteria (RCPB) donor–acceptor pair D-A lead to charge photoseparation, that is, an appearance of the pair of two charges ($D^+ A^-$), where (D^+) the cation radical is a strong oxidant and (A^-) the anion radical is a strong reducing agent. Therefore, in the D-A pair, the light absorption energy is converted to chemical energy accumulated in the photoseparated pair. The most important problems are the structure and action mechanisms of biological photosynthesis, which prevent fast recombination of D^+ and A^- centers of high chemical reactivity and provide relatively long lifetime for these centers, involving them in subsequent chemical reactions eventually resulting in the formation of stable compounds such as ATP and NADPH.

The primary photochemical processes of photosynthesis take place within membrane-bound complexes of pigment and protein reaction centers [1, 6, 7, 51–54]. One mole of a reaction center from different bacteria contains four moles of bacteriochlorophyl, two moles of bacteriopheophytin (Bph), two moles of ubiquinone (Q), and a nonheme Fe atom. In RC from *R. sphaeroides*, 11 hydrophobic α-helices create a framework that organizes the cofactor and a hydrophobic band approximately 35 Å wide. RCPB from *Rhodopseudomonus viridus* has three polypeptides possessing pronounced hydrophobic properties. The molecular mass of the polypeptides are 37 571 (L), 35 902 (M), and 28 902 (H). The H subunit does not carry pigments, but it is sufficient for the photochemical activity. The protein components of reaction centers from different bacteria are similar.

A large series of studies [5, 55–81] were conducted on the use of a whole arsenal of biochemical, physicochemical, and physical methods including X-ray diffraction, ESR, electron–nuclear double resonance (ENDOR), electron–nuclear–nuclear triple resonance (TRIPPLE), pulsed electron–electron double resonance (PELDOR), ESSEM, extended X-ray absorption fine structure (EXAFS}, X-ray absorption near-edge structure (XANES), time-resolved resonance infra-red and Raman spectroscopy, Mössbauer spectroscopy, optically detected magnetic resonance (ODMAR), adsorption-detected magnetic resonance (ADMAR), reaction yield detection magnetic resonance (RYDMER), magnetic field effect on reaction yield (MARY), scanning tunneling microscopy, and pico- and femtosecond optical spectroscopy. These methods have established the main features of the structure of RCs and the kinetics of electron transfer during photoseparation of charges.

In the 1970s, the author of this monograph suggested that rapid electron transfer in reaction centers in the forward direction and significantly slower transfer in the reverse direction may account for the cascade structure of RCPB, which provides tunneling (long-distance) mechanism of the photoseparated charges [74, 76, 80]. A key of this concept was an assumption that RCPB bring about positioning of donor and acceptor pigments in the cascade at optimum distances. Later on, the author and his colleagues reported experimental evidences of this idea [74–82]. The concepts of tunneling mechanisms of electron transfer in photosynthetic systems were originally proposed in the classical works of Chance and DeVault (1967) [83] for the electron transfer reaction between oxidized chlorophyll and reduced cytochrome *c* in photosynthetic bacteria. But the new basic idea underlying the above-mentioned mechanism is an assumption that the primary donor (D) and several acceptor (A_i) centers compose a cascade in an ordered structure, in which all these centers are placed at an optimum distance from each other and are separated by a nonconducting protein medium. Such a separation slows down the forward electron transfer between adjacent $D–A_1$ and $A_i–A_{i+1}$ pairs compared to electron transfer in a system with close contacts between the centers. Nevertheless, the direct transfer can be sufficiently fast if the optimum distances do not exceed 6–10 Å (see Section 2.1). What is important is that the recombination of each $D^+–A_i^-$ pair becomes slower and slower as A_i moves away from the donor. In contrast, in the system of tightly packed centers, the recombination rate is expected to be very fast.

From this analysis, the main two conclusions are (1) an effective fast conversion of light energy to energy of a chemical compound of high quantum yield can take place only in biological and model cascade photochemical systems in which photo- and chemically active centers (aromatic photochromes and transition metal clusters) are separated by "nonconducted" zones of 6–10 Å width, consisting of nonsaturated molecules and bonds, and (2) the electron transfer between the donor and the acceptor centers has to occur by a long-range, most probably, nonadiabatic mechanism.

The cascade tunneling hypothesis has been supported in subsequent experiments. By the electron paramagnetic resonance measurements [74–76], the distances between $(Bchl)_2^+$ and Q_A^- in RC from *R. sphaeroidos* was shown to be 32–35 Å. This value was obtained employing the method based on measurement of the effect of Q_A^- on the spin–lattice relaxation time of $(Bchl)_2^+$. On the basis of this result and analysis of quantitative data on exchange interactions (exchange integral J values) between other RC components and the experimental dependence of the spin exchange integral on the distance between paramagnetic centers (Section 1.2.5, Figure 1.16), a scheme of the spatial localization of the electron donors and acceptors in reaction centers has been composed (Figure 2.6) [82].

As one can see from Figure 2.6, the RC model proposed on the basis of a physicochemical investigation shows similar principal features just as a subsequent crystallographic model [55] does, namely, that the pigments in the reaction center from *Rhacophorus viridis* are located at a distance of 7–11 Å and are separated by nonconducting protein media. In both the models, the center–center distance between $(Bchl)_2$ and Q_A in the X-ray structure is about 30 Å.

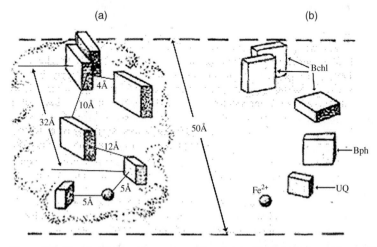

Figure 2.6 Model of the reaction center of the photosynthetic bacterium: (a) according to data obtained from physical methods; (b) from X-ray structure [82]. Reproduced with permission from Elsevier.

Four main tendencies have been underlined in studies done in recent decades of structure and action mechanism of bacterial photosynthetic reaction centers. Resolution and refinement of these structures have been subsequently extended. Investigations of the electronic structure of donor and acceptor centers in the ground and exited states by modern physical methods with a combination of pico- and femtosecond kinetic techniques have become more precise and elaborate. Extensive experimental and theoretical investigations on the role of orbital overlap and protein dynamics in the processes of electron and proton transfer have been done. All the above-mentioned research directions are accompanied by extensive use of methods of site-directed mutagenesis and substitution of native pigments for artificial compounds of different redox potential.

2.3.2
Structure of RCPB

The crystallographic structure of the reaction centers from *R. viridis* and *R. sphaeroids* was initially determined to be of 2.8 and 3 Å resolutions [55]. The recent structure of reaction center from *R. sphaeroides* is shown in Figure 2.7 [84].

A review of photosynthetic reaction centers from anoxygenic bacteria was recently published [85]. On the basis of data quality and quantity, maximal resolution limits, and structural features, the study compared over 50 X-ray crystal structures of reaction centers from *R. (Blastochloris) viridis*, *R. sphaeroides*, and *Thermochromatium tepidum*. Using time-dependent density functional theory (TDDFT), the authors of the work [86] calculated eight low-lying (1.3–1.7 eV energy region) electronic excited states in accordance with the absorption and CD spectroscopic properties of PSRC from *R. sphaeroides*. It was demonstrated that only when the interactions among the prosthetic groups were taken into account could a set of satisfactory assignments for both absorption and CD spectra of PSRC from *R. sphaeroides* be achieved. All electron calculations were performed on the photosynthetic reaction center of *Blastochloris viridis*, using the fragment MO (FMO) method [87]. It was demonstrated that despite the structural symmetry of the system, asymmetric excitation energies were observed especially on the bacteriopheophytin molecules. The authors suggested that the asymmetry was attributed to electrostatic interaction with the surrounding proteins, in which the cytoplasmic side plays a major role.

The influence of the protein environment on the primary electron donor, P, a bacteriochlorophyll a dimer, of reaction centers from *R. sphaeroides*, has been investigated using ESR and electron nuclear double resonance spectroscopy [88]. These techniques were used to probe the effects on P that are due to alteration of three amino acid residues, His L168, Asn L170, and Asn M199. The introduction of Glu at L168, Asp at L170, or Asp at M199 changes the oxidation/reduction midpoint potential of P in a pH-dependent manner. According to authors, these results indicate that the energy of the two halves of P changes by about 100 meV due to the mutations and are consistent with the interpretation that electrostatic interactions involving these amino acid residues contribute to the switch in pathway of electron transfer.

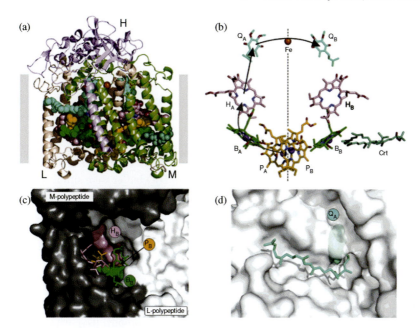

Figure 2.7 Structure and mechanism of the *R. sphaeroides* RC. (a) The L, M, and H polypeptides (ribbons) enclose 10 cofactors (spheres). Boxes indicate the approximate position of the membrane. (b) The cofactors are arranged around an axis of twofold symmetry running perpendicular to the plane of the membrane (dotted line) and comprise four BChls (BA, BB, PA, and PB), two BPhes (HA and HB), two quinones (QA and QB), a carotenoid (Crt), and an iron atom (Fe). Arrows show the route of electron transfer. (c) The BChl and BPhe cofactors are located in an extended binding cavity between the L- and M-polypeptides). The HB BPhe, BB BChl, and PB BPhe are partially exposed at the intramembrane surface of the protein; bacteriochlorin macrocycles are shown as solid objects and phytol side chains as sticks. (d) The headgroup of the QA ubiquinone is embedded in the M-polypeptide with the isoprenoid side chain emerging into the membrane interior [84]. Reproduced with permission from Elsevier.

The directionality of the charge separation process in the LDHW quadruple mutant, HL(M182)/GD(M203)/LH(M214)/AW(M260), was investigated under light or dark freezing conditions first directly by 95 GHz (W-band) high-field EPR spectroscopy [89]. The charge-separated radical pairs ($P_{865}{}^+Q^-$) of the primary donor P_{865}, a bacteriochlorophyll dimer, and the terminal acceptor, QB, a ubiquinone-10, were also studied indirectly by 34 GHz (Q-band) EPR, examining the triplet states of the primary donor ($3P_{865}$) that occur as a by-product of the photoreaction. B-branch charge separation and formation of the triplet-state $3P_{865}$ via a radical pair mechanism was induced with a low yield at 10 K by direct excitation of the bacteriopheophytins in the B-branch at 537 nm. It was shown that about 70% of RCs illuminated upon freezing are trapped in the long-lived ($\tau > 10^4$ s) charge-separated state $P^+Q_B{}^-$. The temperature dependence of the EPR signals from P^+Q^- pointed to two factors responsible for the forward electron transfer to the terminal acceptor Q_B and for the charge recombination reaction, a significant protein conformational change to initiate $P^+Q_B^-$ charge separation and protein relaxation, which governs the charge recombination process along the B-branch pathway of the LDHW mutant.

The geometry of the hydrogen bonds to the two carbonyl oxygens of the semiquinone Q_A in the reaction center from the photosynthetic purple bacterium *R. sphaeroides* R-26 was detected by fitting a spin Hamiltonian to the data derived from 1H and 2H ENDOR spectroscopies at 35 GHz and 80 K [90]. The experiments were performed on RCPB in which the native Fe^{2+} (high spin) was replaced by diamagnetic Zn^{2+} to prevent spectral line broadening due to magnetic coupling with the iron. The ENDOR spectra at different magnetic fields on frozen solution of deuterated Q_A in H_2O buffer and protonated Q_A in D_2O buffer were simulated. It was found that the asymmetric hydrogen bonds of Q_A affect the spin density distribution in the quinone radical and its electronic structure.

2.3.3
Kinetics and Mechanism of Electron Transfer in RCPB

The process starts with the accumulation of light quanta by the light-harvesting complex (Section 2.2). The singlet electronic excitation migrates along the LHC and enters the primary acceptor P_{870}, the dimer of bacteriochlorophyl (D_A), which also passes over to the singlet state. This is followed by a chain of events (Figure 2.7) [56–60]. During the time of the order of a picosecond, an electron from the excited P_{870} is transferred to bacteriochlorophyl, DA 1, in picoseconds to Bph, and in about 200 ps to the primary acceptor ubiquinone (Q_A). The next electron transfer from Q_A to the secondary acceptor Q_B occurs at a rate in the millisecond range. During this time, the electron from the secondary donor, that is, type c cytochrome, has the chance to be transformed from reduced cytochrome c to P_{870}^+. As a result, the energy of a solar quantum is transformed into chemical energy of the reduced secondary acceptor, which can be involved in consequence reactions.

The coupling between electron transfer and the conformational dynamics of the cofactor–protein complex in photosynthetic reaction centers from *R. sphaeroides* in water/glycerol solutions or embedded in dehydrated poly(vinyl alcohol) (PVA) films or trehalose glasses (Figure 2.8) is reported [10]. Matrix effects were studied by time-resolved 95 GHz high-field electron paramagnetic resonance (EPR) spectroscopy at room (290 K) and low (150 K) temperature. It was shown that ET from the photoreduced quinone acceptor ($Q_A^{\bullet -}$) to the photooxidized donor ($P_{865}^{\bullet +}$) strongly depends on matrix at room temperature. Despite the matrix dependence of the ET kinetics, continuous wave EPR and electron spin echo (ESE) analyses of the photogenerated $P_{865}^{\bullet +}$ and $Q_A^{\bullet -}$ radical ions and $P_{865}^{\bullet +}Q_A^{\bullet -}$ radical pairs do not reveal significant matrix effects, at either 290 or 150 K, indicating no change in the molecular radical pair configuration of the $P_{865}^{\bullet +}$ and $Q_A^{\bullet -}$ cofactors. The authors inferred that the relative geometry of the primary donor and acceptor, as well as the local dynamics and hydrogen bonding of Q_A in its binding pocket, are not involved in the stabilization of $P_{865}^{\bullet +}Q_A^{\bullet -}$ and the RC relaxation occurs rather by changes throughout the protein/solvent system.

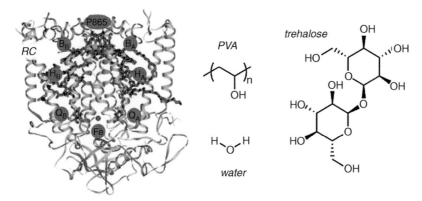

Figure 2.8 X-ray structure of the RC from *R. sphaeroides* R26 (61) and solvent molecules used as matrixes (PVA stands for poly(vinyl alcohol)). Cofactor abbreviations (indices A and B refer to the two electron transfer protein branches in the RC): P_{865}, bacteriochlorophyll a dimer; B, BChl monomer; H, bacteriopheophytin a; Q, ubiquinone-10 [10]; Fe, nonheme iron Fe^{2+} [10]. Reproduced with permission from American Chemical Society.

Detailed analysis of kinetics and the thermodynamics of electron transfer from the excited primary donor P^* to Bph via Bchl has been performed ([91] and references therein). Levels of energy for the primary charge separation in photosynthetic RC are presented in Figure 2.9. Accordingly, the primary ET in RC takes place as a transition in Franck–Condon systems with two quasi-continua. Transfer P^* (Bchl)(Bph) → P^+ (Bchl)(Bph)$^-$ can occur as a two-step process via an intermediate P^+ (Bchl)$^-$ (Bph) or as a one-step process, in which Bchl provides a superexchange bridge for the direct P^* (Bchl)(Bph) → P^+ (Bchl)(Bph)$^-$ transfer. The possibility of superposition of both sequential and superexchange mechanisms has also been advanced. According to Bixon and Jortner [91], P^* (Bchl)(Bph) → P^+ (Bchl)(Bph)$^-$ transfer and electron transfer from (Bph)$^-$ to the primary quinone acceptor Q_A are activationless processes. The mean characteristic vibrational energy of the former process was estimated as $\hbar\omega = 80$–$100\,cm^{-1}$, where ω corresponds to the vibrational mode of the dimer P. Other quantitative characteristics of the primary processes derived from the theoretical analysis are as follows: (1) the spread of the energy of the P^+ (Bchl)(Bph)$^-$ relative to P^* is accounted for in terms of a Gaussian contribution with width parameter $\sigma \approx 0.05$ eV; (2) for the superexchange route, the reorganization energy $\lambda \approx 0.1$ eV; and (3) the energy gap between P^* (Bchl)(Bph) and P^+ (Bchl)(Bph)$^-$ was estimated as $\Delta G_1 \approx 0.06$ eV.

The effect of the redox potential of the electron acceptor Bchl in RC from *R. sphaeroides* on the initial electron transfer rate and on the P^+ (Bchl)$^-$ population was investigated [17]. Analysis of experimental data estimated the free energy as ($\Delta G^0 = 0.104$ eV), the energy of reorganization as ($\lambda = 0.065$ eV), and the coupling factor as ($V = 6.5 \times 10^{-3}$ eV) of this nonadiabatic process. The recombination rate was found to be increased by a factor of 10 for each 0.060 eV increase in the difference of the redox potential of quinines replacing the native ubiquinone-10. The Gibbs

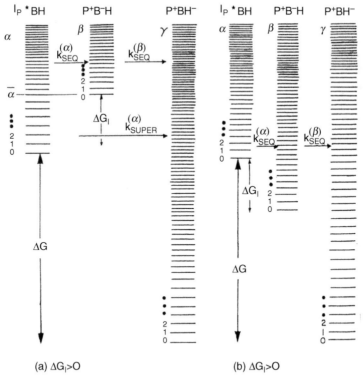

Figure 2.9 Level structure for the primary charge separation in photosynthetic reaction centers. (a) Unistep superexchange dynamics; (b) two-step sequential dynamics. Adapted from Ref. [91].

energy of the trap state at 10 K was estimated as about 0.200 eV higher than the relaxed form at room temperature.

In pheophytin-modified reaction centers of *R. sphaeroides* R-26, the fs oscillations with frequency around 130 cm^{-1} in the P*-stimulated emission and in the BA absorption band at 800 nm were observed [92]. This process was accompanied by reversible formation of the 1020 nm absorption band, which is a characteristic of the radical anion band of bacteriochlorophyll monomer BA$^-$. These results were discussed in terms of a reversible electron transfer between P* and BA induced by a motion of the wave packet near the intersection of potential energy surfaces of P* and P$^+$BA$^-$, when a maximal value of the Franck–Condon factor is created. For the purpose of establishing a relationship between the nuclear subsystem motion and the charge transfer, the fs and ps oscillations in the bands of stimulated emission of P* and in the band of reaction product BA$^-$ at 1020 nm were investigated [93]. The reversible formation of the P$^+$BA$^-$ state was characterized by two vibration modes (130 and 320 cm^{-1}) and connected with an arrival of the wave packet induced by fs excitation to the intersection of potential surfaces P*BA and P$^+$BA$^-$. It was shown that the irreversible formation of the P$^+$BA$^-$ state with the time constant of 3 ps is followed by oscillations with frequencies of 9 and 33 cm^{-1}. The authors suggested

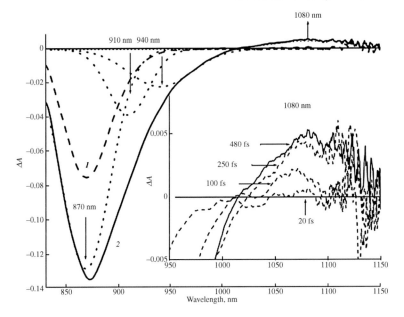

Figure 2.10 Light minus dark difference absorption spectra of R. sphaeroides reaction centers, recorded at a delay time of (1) 20 fs and (2) 480 fs at room temperature. The inset shows the light-induced absorption changes in the longwave region of the spectrum measured at delay times of 20, 100, 250, and 480 fs [94].

that the irreversibility of electron transfer is determined by two factors: (1) by a difference between the energy width of the wave packet and the gap between the named surface and (2) by a difference between the duration of wave packet residence near the intersection of the surfaces and the relaxation time of the P^+BA^- state. Femtosecond absorption and formation at 1080 and 1020 n as an indication of charge-separated S and P^+ in photosynthetic reaction centers of the purple bacterium R. sphaeroides were investigated (Figure 2.10) [94].

Time-resolved electron–nuclear spin dynamics data of reaction centers of depleted quinone have been reported [11]. It is demonstrated that the buildup of nuclear polarization on the primary donor and the bacteriopheophytin acceptor depends on the presence and lifetimes of the molecular triplet states of the donor and carotenoid.

The linear dependence of the logarithm of the rate constant of the electron transfer in RCs of purple bacteria and plant photosystem I ($\log k_{ET}$) on the edge–edge distance between the donor and the acceptor centers (R) was observed (Figure 2.11) [79, 80, 95, 96].

The slope of the dependence corresponds to the slope predicted for long-distance spin superexchange orbital overlap through nonconducting media by the shortest pathway (Equation 1.31 with $\beta_{DA} = 1.3\,\text{Å}^{-1}$) (Section 1.2.5). As one can see from Figure 2.11, the values of k_{ET} for the transfer from $(Bchl)_2^+$ to Bph and from P_{700}^+ to pheophytin acceptor (Ph) markedly deviate from the general $\log k_{ET} - R$ plot. Such deviation is explained by assuming the participation of intermediate acceptors located between $(Bchl)_2^+$ and Bph, and between P_{700}^+ and (Ph) [95, 96]. Another

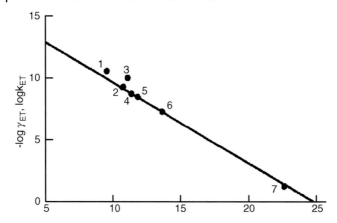

Figure 2.11 Dependence of maximum rate constant of ET on the edge–edge in photosythetic RCs of bacteria and plants: (1) $A_0^- - A_1$; (2) $H^- \, Q_A$; (3) $A_1 - F_x$; (4) $H^- - P^+$; (5) $C_{559} - P^+$; (6) $F_x - F_A$; (7) $Q_A^- - P^+$; (8) $P^- - Bcl$; (9) $P_{700} - A_0$; (10) $Q_A^- - Q_B$. P_{700} is the chlorophyll dimer; A_0 is chlorophyll; A_1 is phylloquinone; and F_x and F_A are 4Fe–4S clusters. The straight line is related to the dependence of the attenuation parameter for spin exchange in "nonconducting bridges." Filled circles correspond to a "regular" dependence and open circles to deviation [95]. Reproduced with permission from Elsevier.

deviation is related to ET from the primary quinone acceptor Q_A to the secondary quinone acceptor Q_B. The process takes place at an edge–edge distance of about 14 Å, but the centers are connected with two hydrogen bonds and two aromatic imidazol groups. On the basis of estimation of the resonance integral of the ET, it was concluded that the process run is controlled by media reorganization [82, 95]. Because the rate constant of electron transfer from Q_A^- to Q_B ($k_{ET} \approx 10^4 \, s^{-1}$) [56] is essentially less than expected for a nonadiabatic activationless ET and the k_{ET} values considerably deviate from the dependence of the superexchange attenuation parameter (γ_{ET}) on the distance between donor and acceptor centers in RCs (Figure 2.11), we can conclude that the ET requires thermal activation. This conclusion was confirmed by Calvo et al. [97]. In the high-resolution ESR (326 GHz) study of the biradical state $Q_A^- \, Q_B^-$ in the R. spheroids RC, the exchange integral in the biradical $J_0 = 10^9 \, s^{-1}$. This value is significantly higher than the rate of electron transfer between Q_A^- and Q_B^-.

Trapping conformational intermediate states in the reaction center protein from photosynthetic bacteria was reported [98]. Two functionally relevant conformational intermediate states of photosynthetic reaction center protein were trapped and characterized at low temperature. RCs frozen in the dark did not allow electron transfer from the reduced primary quinone, Q_A^-, to the secondary quinone, Q_B. In contrast, RCs frozen under illumination in the product $(P^+ Q_A Q_B^-)$ state, with the oxidized electron donor, P^+, and reduced Q_B^-, returned to the ground state at cryogenic temperature in a conformation that allows a high yield of Q_B reduction. When the temperature was raised above 120 K, the protein relaxes to an inactive conformation that is different from the RCs frozen in the dark. The activation energy for this change was found to be 87 meV, and the active and inactive states differ in energy by only 16 meV. The authors concluded that there are several conformational substrates along the reaction coordinate with different transition temperatures. The electron transfer rate from Q_B^- to P^+ was measured at cryogenic temperature and was similar to the rate at room temperature, as expected for an exothermic electron tunneling reaction in RCs.

Another matter of interest is the detailed mechanism of electron transfers with participation of primary Q_A and secondary Q_B acceptors and the role of the coupling proton transfer in these processes. The crystallographic structures of RCs from R. spheroides at cryogenic temperature (90 K) in the dark and under illumination, at resolution 2.2 and 2.6 Å, respectively, have been reported [60]. The main difference in the two structures was the charge-separated state within an area of the primary (Q_A) and secondary (Q_B) acceptor location. In the charge-neutral state PQ_AQ_B, the distance between two ubiquinones is approximately 5 Å. In the "light" structure $PQ_AQ_B^-$, the Q_B has moved about 4.5 Å and undergone a 180°-propeller twist. It was proposed that a hydrogen bond of ubiquinone with HisL190 prompts the electron transfer from Q_A^- to Q_B and Q_BH^-. These results give evidence in favor of the gating model of the protein dynamics, which suggests that electron transfer occurs only in an active conformational state of the medium, promoting electron transfer [99].

ESR and ENDOR were used to study the magnetic properties of the protonated semiquinone, an intermediate proposed to play a role in proton-coupled electron transfer to Q_B [100]. To stabilize the protonated semiquinone state, a ubiquinone derivative, rhodoquinone, was used. To reduce this low-potential quinone, the authors prepared mutant RCs modified to directly reduce the quinone in the Q_B site via B-branch electron transfer. EPR and ENDOR signals were observed upon illumination of mutant RCs in the presence of rhodoquinone. The EPR signals had g-values characteristic of rhodosemiquinone ($g_x = 2.0057$, $g_y = 2.0048$, $g_z = 2.0018$) at pH 9.5 and were changed at pH 4.5. The ENDOR spectrum showed couplings due to solvent exchangeable protons typical of hydrogen bonds similar to, but different from, those found for ubisemiquinone.

Ren et al. [101] calculated eight low-lying (1.3–1.7 eV energy region) electronic-excited states in accordance with the absorption and circular dichroism (CD) of the reaction center from R. shpaeroides by using TDDFT. The calculations demonstrated that only when the interactions among the prosthetic groups were taken into account, a set of satisfactory assignments for both absorption and CD spectra of the RC can be achieved simultaneously. Exploring the energy landscape for Q_A^--to-Q_B electron transfer in bacterial photosynthesis reaction centers, the effect of substrate position and tail length on the conformational gating step was analyzed [102]. The importance of quinone motion was examined by shortening the Q_B tail from 50 to 5 carbons. No change in the rate was found from 100 to 300 K!

On the basis of time-resolved FTIR step-scan measurements in native photosynthetic reaction center from R. sphaeroides RCs, involvement of an intermediary component in the electron transfer step Q_A^- Q_B was suggested [103]. Nevertheless, by a kinetic X-ray absorption experiment at the Fe K-edge in isolated reaction centers with a high content of functional QB, at a time resolution of 30 μs and at room temperature, no evidence for transient oxidation of Fe was obtained [104]. The steady-state FTIR spectra of the photoreduction of Q_A and Q_B in R. sphaeroides reaction centers provided evidence against the presence of a proposed transient electron acceptor X between the two quinines and allowed authors to rule out the possibility that Q_B^- formation precedes Q_A^- reoxidation in native RCs [105].

Several publications were devoted to involving protons in the electron transfer process [106–109]. The electron transfer from the primary donor P to the secondary acceptor Q_B was shown to be coupled to the uptake of two protons followed by exchange of doubly reduced Q_BH_2 for Q_B from the cytoplasm [106]. The high-resolution X-ray diffraction study revealed in the "dark" structure two water channels, P1 and P2, leading from the Q_B pocket to the surface of the protein on the cytoplasmic side of the RC. These channels have been assumed to deliver protons to photoreduced states Q_B and Q_BH^-. The GluH173 in the "light" structure, located in the "dark" structure along the P2 channel, is disordered compared to this.

In the RCs from R. sphaeroides, the reduction of a bound quinone QB is found to be coupled with proton uptake [108]. When Asp-L213 was replaced by Asn, proton transfer is inhibited. Proton transfer was restored by two second-site revertant mutations, Arg-M233 Cys and Arg-H177 His. The structures of the parental and two revertant RCs were detected at resolutions of 2.10, 1.80, and 2.75 Å. From the structures, the authors were able to delineate alternative proton transfer pathways in the revertants (Figure 2.12). The main changes occur near Glu-H173, which allow it to substitute for the missing Asp-L213. It was suggested that the electrostatic changes near Glu-H173 cause it to be a good proton donor and acceptor, and the structural changes create a cavity that accommodates water molecules that connect Glu-H173 to other proton transfer components.

To analyze whether the RCs of different species of B. viridis use the same key residues for proton transfer to Q_B, the conservation of these residues was detected using hidden Markov models [109]. It was found that residues involved in proton transfer, but not located at the protein surface, are conserved, whereas potential proton entry points are not conserved to the same extent. The analysis of the hydrogen bond network of the RCs from R. sphaeroides and B. viridis showed that a large network connects Q_B to the cytoplasmic region in both RCs, and all nonsurface key residues are part of the network. The results of the analysis allowed authors to suggest that the proton transfer to Q_B is not mediated by distinct pathways but by a large hydrogen bond network.

2.3.4
Electron Transfer and Molecular Dynamics in RCPB

The first experimental evidences that electron transfer from Q_A^- to P^+ and from Q_A^- to Q_B in reaction centers is controlled by the protein conformational dynamics was obtained in the late 1970s [79, 83, 95, 110–118]. This conclusion was confirmed in subsequent experimental studies in which molecular dynamics of RC and the

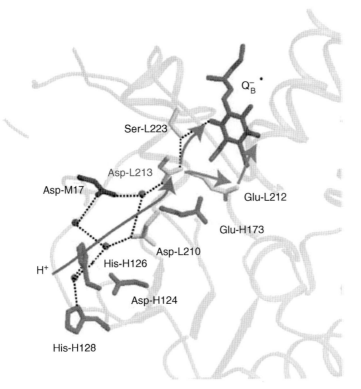

Figure 2.12 The dominant proton transfer pathways to reduced Q_B in the native reaction center of *R. sphaeroides*. The actual proton pathways are indicated by the dotted lines, while the arrows give the general direction of the proton transfer. The arrow indicates the path for both the first and the second proton uptake from solution to Asp-L213. The pathways for the two protons diverge after Asp-L213. The Glu-H173 side chain has a relatively large B factor of 70 Å² compared to the average for the whole protein of 55 Å², indicating disorder [108]. Reproduced with permission from Elsevier.

photosynthetic membrane were determined with a whole set of spin, Mössbauer, fluorescence, and phosphorescence labels. It was shown that the electron transfer from reduced primary acceptor Q_A^- to secondary acceptor Q_B takes place only under conditions in which the labels record the mobility of the protein moiety in the membrane with the correlation frequency $v_c > 10^7\,s^{-1}$ (Figure 2.13). This fact was explained in the framework of two models. The first model is based on the concept of dynamic adaptation of a protein matrix in every step of an enzymatic reaction. Concerning the $Q_A^- \rightarrow Q_B$ transition, fast reversible conformational transitions can provide dipolar relaxation favorable for the media reorganization process. Such reorganization is necessary to release electrons of Q_A^- from the stabilizing elecrostatic frame and to stabilize the Q_B^- anion. The second model [99] suggested conformational gating as a mechanism for providing the shortest, most effective pathway for this transition. According to this model, at temperatures lower than

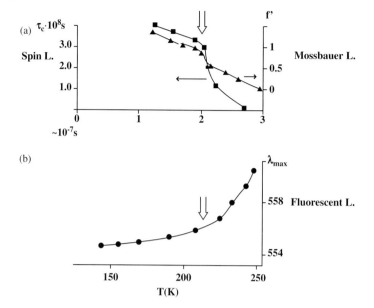

Figure 2.13 Temperature dependence of the parameters of the characteristic time of electron transfer from the reduced primary acceptor to the oxidized primary donor ($\tau_{1/2}$). (a, b) spin and Mössbauer labels and fluorescence label in membranes from *R. rubrum* [95]. Reproduced with permission from Elsevier.

210–220 K, the position of protein groups between Q_A and Q_B is not favorable for electron transfer due to weak superexchange conductivity. Such conductivity is essentially improved under physiological conditions when the intermediate group stands in a position favorable for electron transfer.

The rate of another important process, the recombination of the primary product of the charge separation, that is, the reduced primary acceptor (Q_A^-) and oxidized primary donor, bacteriochlorophyl dimer (P^+), falls from 10^3 to $10^2 \, s^{-1}$ when dynamic processes with $\nu_c = 10^3 \, s^{-1}$ monitored by the triplet labeling method occur. Very fast electron transfer from P^+ to Bchl and from $(Bchl)^-$ to Q_A does not depend on media dynamics and occurs via conformationally nonequilibrium states (Figure 2.14).

Subsequent theoretical studies have confirmed the above-mentioned conclusions that electron transfer between the two quinines Q_A and Q_B in the bacterial photosynthetic centers is coupled to conformational rearrangement and provide added important details [119–122]. For estimation of the quantomechanical coupling factor V_{DA}, it was assumed that the electron transfer involves multiple pathway tubes of different V_{DA}, the population of which is controlled by conformational and nuclear dynamics [122]. The molecular dynamics simulation performed for both the "dark" and the "light" structures indicated that dominant pathway tubes are similar for light and dark RC structures, except the position of Q_B. According to the calculation [60],

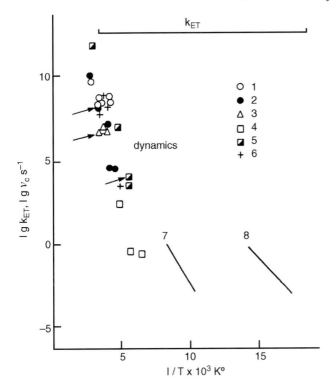

Figure 2.14 Data on the correlation frequency of the mobility of physical labels and their environment in bovine and human serum albumins and in the photosynthetic RC (I) and the rate constants of ET's primary donor (P) to the bacteriaophephytin acceptor (II) in the Arrhenius coordinates [78]. Reproduced with permission from Springer.

the transition from "dark" to "light" states is accompanied by the flipping and moving of Q_B, which shortens the ET pathway by five covalent steps and replaces a through-space jump by a hydrogen bond. As a result of this transition, the ET rate increases by about three orders of magnitude.

A role of dynamic effects on primary photophysical and photochemical events in reaction centers from *R. sphaeroides* and *Chloroflexus aurantiacus* has been a matter of interest to researchers [123]. Primary photochemical events in two site-directed mutants YF(M208) and YL(M208) of RC from *B. viridis*, in which tyrosine at position M208 was replaced by phenylalanine and leucine, respectively, were investigated with the use of 1H-ENDOR as well as optical absorption spectroscopy [123]. It was shown that the residue at M208 is close to the primary electron donor, P, the BChl, and the BPh. Analysis of the experimental data revealed two torsional isomers of the 3-acetyl group of Bph. Enzymes in the $Bph^- Q_A^-$ state accumulate at 100 K and undergo an irreversible change between 100 and 200 K. It was shown [110, 114, 115] that within this temperature range the phosphorescence probes detect animation of millisecond dynamics in the RC from *R. rubrum*.

The initial electron transfer rate and protein dynamics in wild type and five mutant reaction centers from *R. sphaeroides* have been studied as a function of temperature (10–295 K) [124]. In the framework of a model based on reaction diffusion formalism, the time course of electron transfer was detected by the ability of the protein to interconvert between conformations until one is found where the activation energy for electron transfer is near zero. In reaction centers of wild type, the reaction proceeds at least as fast at cryogenic temperature as at room temperature. Nevertheless, the methods of physical labels indicated (Figure 2.14) that in the range of temperature mentioned above, conformational dynamics in photosynthetic reaction centers from *R. rubrum* is frozen.

In a wide range of temperatures, both processes occur significantly faster than the media relaxation (Figure 2.14) and, therefore, the media around the intermediates (Bph)$^-$ and QA$^-$ exist in the conformationally nonequilibrium state [95]. Under such a condition, as was mentioned above, the energy gap ΔG_1 and the reorganization energy λ for primary ET is small. Hence, this activationless process is controlled by the orbital overlap factor but not by the Franck–Condon factor. The linear plot of log k_{ET} and the logarithm of the attenuation parameter for superexchange processes (γ_{ET}) versus the distances between the donor and the acceptor centers (Figure 2.11) support independently this conclusion.

One of the enigmatic problems of photosynthesis is the drastic difference between the rate of photoelectron transfer in the active (A) and the inactive (B) branches of bacterial reaction centers. The quantum mechanical calculation [125] showed that the square of electronic matrix element V_A^2 for the electron transfer from the excited primary donor, P*, to bacteriochlorophyll in the active branch is larger by three orders of magnitude than that in the inactive part V_B^2. Therefore, the electron transfer rate in the RC inactive L-branch should be essentially slower than that in the M-branch. The focus of study [126] was to gain insight into the temperature dependence of rate of initial electron transfer to the parallel cofactor chain associated mainly with the M polypeptide (B branch), which is inactive in the native RC. To this goal, picosecond transient absorption measurements have been carried out on RCs of the YFH mutant of the photosynthetic bacterium *R. capsulatus* at 77 K. It was found that the excitation of P to its lowest singlet excited state (P*) elicits complex kinetic behavior involving two P* cofactor protein populations: one that is capable of charge separation (active) and one that is not (inactive).

2.4
Reaction Centers of Photosystems I and II

In oxygenic photosynthetic organisms, plant and green bacteria, the reaction centers of two systems, photosystem I (PS I) and photosystem II (PS II), convert the absorbed light energy into energy of stable products, that is, ferredoxin in reduced state and dioxygen [1, 106]. PS I from plants and cyanobacteria mediates light-induced electron transfer from plastocyanin to ferredoxin (flavodoxin) at the stromal membrane site, while PS II is a photoenzyme that catalyzes oxidation of the water in a water splitting

Mn-containing system. Subsequent absorption of four light quanta by PS I and PS II results in evaluation of dioxygen from a two-water molecule. The overall process occurs by the following scheme:

$$2H_2O + 4\,[\text{ferredoxin (oxidized)}] \xrightarrow{4h\nu} O_2 + [\text{ferredoxin (reduced)}] + 4H^+$$

The key step of the process is the water splitting under absorption of light quanta of relative low energy. Here, we will focus mainly on the latter process that appears to be one of the most enigmatic reactions in chemistry and photochemistry and will only briefly consider the light energy conversion reaction centers of PS I and PS II.

2.4.1
Reaction Centers of PS I

PS I from cyanobacteria consist of 11 protein subunits and several cofactors. After the photoexcitation of the primary donor, a dimer of chlorophyll a (Chla)$_2$, P$_{700}$, an electron is transferred via a chlorophyll (A$_0$) to a phylloquinone (A$_1$) and then to the iron sulfur clusters, F$_x$, F$_A$, and F$_B$ [106, 127–136].

Data on chrystallographic models PS I from cyanobacteria have been reported [137]. PS I from *Synechococcus elongatus* contains nine protein subunits featuring transmembrane α-helices and three stromal subunits. The organic cofactors are arranged in two branches along the pseudo-C2 axis. The distances between adjacent donor and acceptor centers of the system vary from 8.2 to 14.9 Å for the "right" branch (A) and from 8.6 to 22 Å for the "left" branch (B). Therefore, "nonconducting" zones similar to those in the bacterial RC separate the centers. The crystal structure of the complete photosystem I (PS I) from a higher plant (*Pisum sativum* var. *alaska*) to 4.4 Å resolution was determined [138]. The structure shows 12 core subunits, 4 different light-harvesting membrane proteins (LHC I) assembled in a half-moon shape on one side of the core, 45 transmembrane helices, 167 chlorophylls, 3 Fe–S clusters, and 2 phylloquinones (Figure 2.15). About 20 chlorophylls are positioned in strategic locations in the cleft between the LHC I and the core. The authors underlined that this structure provides a framework for exploration not only of energy and electron transfer but also of the evolutionary forces that shaped the photosynthetic apparatus of terrestrial plants after the divergence of chloroplasts from marine cyanobacteria 1 billion years ago.

Authors of the study [139] reported an improved crystallographic model at 3.3 Å resolution, which allows analysis of the structure in more detail. The obtained plant photosystem I structure revealed the locations of and interactions among 17 protein subunits and 193 noncovalently bound components. The molecular structure of the iron stress-induced A protein IsiA–photosystem I supercomplex, inferred from high-resolution crystal structures of PS I and the CP43 protein, was presented [140]. The optimal structure of the IsiA–PS I supercomplex was derived by systematically rearranging the IsiA monomers and PS I trimer in relation to each other. For each of the 6 969 600 structural configurations considered, the authors counted the

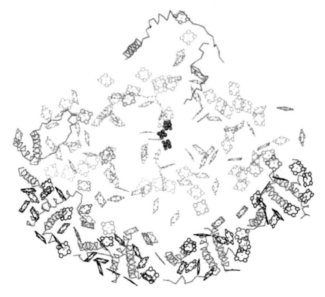

Figure 2.15 The arrangement of 167 chlorophyll molecules of plant PS I as seen from the stromal side (see details in Ref. [138]). Reproduced with permission from *Nature*.

number of optimal Chl–Chl connections (i.e., cases where Chl-bound Mg atoms are $\leq 25\,\text{Å}$ apart). Fifty of these configurations were found to have optimal energy transfer potential.

ENDOR and transient absorption studies of $P_{680}{}^+$ and other cation radicals in PS II reaction centers before and after inactivation of secondary electron donors were performed [141]. It was shown that using prebleached PS II RC offers the advantage that cation radicals from secondary donors, $Chlz^+$ and Car^+, which accumulate during prolonged illumination of control RC, are effectively suppressed.

The three-dimensional structure of higher plant PS I as obtained by electron microscopy of two-dimensional crystals formed at the grana margins of thylakoid membranes was described [142]. It was found that the negatively stained crystal areas displayed unit cell dimensions, $a = 26.6\,\text{nm}$, $b = 27.7\,\text{nm}$, and $\gamma = 90°$, and p22121 plane group symmetry consisting of two monomers facing upward and two monomers facing downward with respect to the membrane plane. Higher plant PS I shows several structural similarities to the cyanobacterial PS I complex, with a prominent ridge on the stromal side of the complex.

Properties of PscB and the F_A, F_B, and F_X iron–sulfur clusters in green sulfur bacteria unifying principles in homodimeric type I photosynthetic reaction centers were reported [143]. The photosynthetic reaction center from the green sulfur bacterium *Chlorobium tepidum* (CbRC) was solubilized from membrane. The CbRC cores treated with NaCl solution did not exhibit the low-temperature EPR resonances from $F_A{}^-$ and $F_B{}^-$ and were unable to reduce $NADP^+$. SDS-PAGE and mass spectrometric analysis showed that the PscB subunit, which harbors the F_A and F_B clusters, had become dissociated. Mössbauer spectroscopy showed that recombinant PscB contains a heterogeneous mixture of $[4Fe-4S]^{2+,1+}$ and other types of Fe/S clusters tentatively identified as $[2Fe-2S]^{2+,1+}$ clusters and rubredoxin-like $Fe^{3+,2+}$ centers, and that the $[4Fe-4S]^{2+,1+}$ clusters present were degraded at high ionic strength. A heme-staining assay indicated that cytochrome c_{551} was firmly attached to the CbRC cores. Low-temperature EPR spectroscopy of photoaccumulated CbRC complexes and CbRC cores showed resonances between $g = 5.4$ and 4.4 assigned to a $S = 3/2$ ground spin state $[4Fe-4S]^{1+}$ cluster and at $g = 1.77$ assigned to a $S = 1/2$ ground spin state $[4Fe-4S]^{1+}$ cluster, both from F_x^-. The authors underlined that these results unify the properties of the acceptor side of the type I homodimeric reaction centers found in green sulfur bacteria and heliobacteria: in both, the F_A and F_B iron–sulfur clusters are present on a salt-dissociable subunit, and F_x is present as an interpolypeptide $[4Fe-4S]^{2+,1+}$ cluster with a significant population in a $S = 3/2$ ground spin state.

The reduction of ferredoxin and flavodoxin by photosystem I in cascade system is shown to be similar to those of the electron jump in bacterial RCs [59, 144–147]. The primary transfer from the excited chlorophyll dimer, primary donor P^*, to A_0 takes place with a time constant of about 25 ps. The next step from A_0 to a secondary acceptor occurs in 200–600 ps. The recombination time constants of P^+ with reduced intermediate acceptors increase as the electron moves along the chain and range from nanoseconds for transition $A_0^- \rightarrow P^+$ to milliseconds for transition FX to P^+. According to kinetic investigation [145], electron transfer in PS I involves both branches with different rate constants of $35 \times 10^6 \, s^{-1}$ (branch B) and $4.4 \times 10^6 \, s^{-1}$ (branch A) for the ET from each phylloquinone to the iron–sulfur cluster F_X.

The FS I primary donor has a very high positive redox potential of about 1.17 V in contrast to 0.4–0.6 V for other oxidized primary donors in photosynthesis and in solution [148]. Values of enthalpy ($\Delta H^0 \approx -0.33$ eV) and entropy ($\Delta S^0 \approx +0.4$ e.u.) for the formation of ion pair $P_{700}^+ F_{AB}^-$ in the intact cells of Synechocystis PCC 6803 and in vitro were determined using pulsed time-resolved photoacoustics [148]. To investigate the effect on thermodynamics, the enthalpy and volume changes of charge separation in PS I in the menA and menB mutants were measured using pulsed time-resolved photoacoustics on the nanosecond and microsecond timescales [149]. The observed thermodynamic data were found to be the same for the menA and menB mutants. The observed reaction was assigned to the formation of $P_{700}^+ A_P^-$ from $P_{700}^* A_P$. An enthalpy change (ΔH) of -0.69 ± 0.07 eV was obtained for this reaction. In contrast, a larger enthalpy change -0.8 eV for the formation of $P_{700}^+ A_1^-$ from P_{700}^* and an apparent entropy change ($T\Delta S$, $T = 25\,°C$) of -0.2 eV were obtained in wild-type PS I. This implies that the reaction of $P_{700}^+ A_P^- F_{A/B} \rightarrow P_{700}^+ A_P F_{A/B}^-$ in the mutants is almost completely entropy driven ($\Delta G = -0.07$ eV and $T\Delta S = +0.40$ eV). The authors stressed that these results show that not only the kinetics but also the thermodynamics of electron transfer reactions in PS I are significantly affected by the recruitment of the foreign plastoquinone-9 into the A_1 site.

The electronic structure of the PS I donor and acceptor centers is investigated by the whole set of modern physical methods, including ESR, ENDOR, FT-IS, and so on [150–152]. The orientation of the primary donor cation radical ($P_{700}^{\bullet +}$) in the single crystals of photosystem I from the thermophilic cyanobacterium, *S. elongatus*, was investigated by ESR and ENDOR techniques [152]. The orientation was found to be similar to those in the purple bacteria. The similarity of direction of the principal axes of the $P_{700}^{\bullet +}$ g-tensor in single crystals of PS I and in bacterial reaction centers was demonstrated by X$^-$, Q$^-$, and W-band ESR spectroscopy.

In a review [15], the detection of the magnetic resonance parameters of $P_{700}^{\bullet +}$ (cation radical) and $3P_{700}$ (triplet state) of the primary donor by EPR techniques was described. Conclusions about the electronic structure, in particular about the spin and charge distribution in this species, were drawn. The results are corroborated by studies of model systems and of the primary donor in genetically modified photosystem I preparations, which give information on the effect of the protein surroundings. Emphasis is placed on a theoretical description of P_{700} in its various states, which is based on a comparison with MO calculation.

2.4.2
Reaction Center of Photosystem II

Photosystem II, which occurs in cyanobacteria and plants, consists of about 20 different protein subunits and 14 integrally bound lipids [130, 153–163]. It includes six redox cofactors that are able to trap electrons (or holes) in minima of Gibbs energy. These cofactors are the oxygen-evolving complex (OEC), the amino acid residue tyrosine (Tyr), the reaction center chlorophyll, pheophytin, and plastoquinone molecules, Q_A and Q_B. All these cofactors except Q_B are bonded to a twisted pair of hydrophobic proteins known as D1 and D2. The D1 and D2 proteins form the scaffolding of the photosystem II complex. The protein comprises five transmembrane helices (A–E) organized in a manner almost identical to that of the L and M subunits of the reaction center of photosynthetic bacteria.

The first crystal structure of PS II from a thermophilic cyanobacterium *Thermosynechococcus elongatus* at a resolution below 4.0 Å was reported in 2001 [163, 164]. The structure of photosystem II, obtained by Kamiya and Shen [164], was built on the basis of the sequences of PS II large subunits D1, D2, CP47, and CP43; extrinsic 33 and 12 kDa proteins and cytochrome c_{550}; and several low molecular mass subunits. The arrangement of chlorophylls and cofactors, including two β-carotenes identified in a region close to the reaction center, which provided important clues to the secondary electron transfer pathways around the reaction center, was shown. Possible ligands for the Mn cluster were determined. In particular, the C terminus of D1 polypeptide was found to be connected to the Mn cluster directly. The crystal structure of PS II from *T. elongatus* at 2.9 Å resolution, reported in Ref. [155], allowed the unambiguous assignment of all 20 protein subunits and complete modeling of all 35 chlorophyll a molecules and 12 carotenoid molecules, 25 integral lipids and 1 chloride ion per monomer (Figure 2.16). Taking into account the presence of the third plastoquinone Q_C and the second plastoquinone transfer channel, the authors

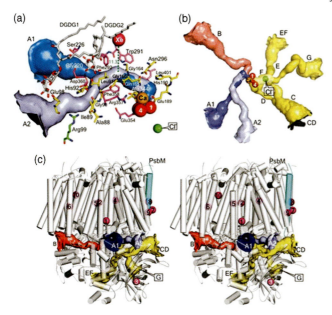

Figure 2.16 Possible substrate and product channels to the lumen and Xe positions in PSII, (a) Detailed view of two of the putative water and oxygen channels (see text for details) connecting the Mn$_4$Ca cluster with the lumen. The walls of the channels are shown in blue (channel A1) and light blue (channel A2), respectively. Residues forming the walls of the channels are shown in stick presentation, along with D1 (yellow), CP43 (magenta), PsbO (green), lipids DGDG1 and DGDG2 (gray). Also shown as spheres are the Mn$_4$Ca cluster (red and orange, Mn2 and Mn3 are not seen), Cl$^-$ (green labeled Cl$^-$ and the closest Xe Xe5) (purple), with distances given in angstroms, (b) View from the lumenal side onto the membrane plane showing the Mn$_4$Ca cluster and possible water and oxygen channels in blue (A1), light blue (A2) and pink (B), and possible proton channels (C to G) in yellow (exits of merging channels labeled as CD and EF). Position of Cl$^-$ indicated by a green sphere. Protein omitted for clarity. (c) Stereo view of channels and Xe positions in PSII. One monomer is shown with the cytoplasm above and the lumen below. All 20 protein subunits are shown as gray cylinders, except PsbM, which is colored cyan. Xe positions are shown as purple spheres (labeled 1 to 10). Openings of the channels are labeled A1, A2, B, CD, EF, G.

suggest mechanisms for plastoquinol–plastoquinone exchange, and we calculated other possible water or dioxygen and proton channels. Putative oxygen positions obtained from a Xenon derivative suggested a role for lipids in oxygen diffusion to the cytoplasmic side of PS II and chloride position suggests a role in proton transfer reactions because it is bound through a putative water molecule to the Mn$_4$Ca cluster at a distance of 6.5 Å and is close to two possible proton channels.

Time-resolved photovoltage measurements on destacked photosystem II membranes from spinach with the primary quinone electron acceptor Q(A) either singly or doubly reduced have been performed to monitor the time evolution of the primary radical pair P_{680}^+Pheo$^-$ [166]. On the basis of a simple reversible reaction scheme, the measured apparent rate constants and relative amplitudes allowed determination of sets of molecular rate constants and energetic parameters for primary reactions in the reaction centers with doubly reduced Q_A as well as with oxidized or singly reduced Q_A. The standard free energy difference between the charge-separated state $P_{680}^+ Pheo^-$ and the equilibrated excited state $(Chl(N)P_{680})^*$ was found (-50 meV). In contrast, single reduction of Q_A led to a large change in ΔG_0 degrees (approximately $+40$ meV).

It was shown [167] that the charge separation photoelectron pathway across the membrane, $P_{680}^* \rightarrow Ph_A \rightarrow Q_{AP} \rightarrow Q_{BP}$, is similar to those in the bacterial RCs. The Q_{BP}, after receiving two electrons and two protons, is replaced by a plasma plastoquinone. The electron trapping states on the reducing side of PS II were denoted T_0, T_1, T_2, T_3, and T_4. These correspond to the redox cofactors P_{680}, Chl^*, $Pheo^-$, QA^-, and QB^-, respectively.

Transfer of electrons between artificial electron donors diphenylcarbazide (DPC) and hydroxylamine (NH$_2$OH) and reaction center of manganese-depleted photosystem II complexes was studied using the direct electrometrical method [167]. It was shown that reduction of redox-active amino acid tyrosine Y_z by DPC is coupled with generation of transmembrane electric potential difference ($\delta\Psi$). The amplitude of this phase comprised <17% of that of the $\delta\Psi$ phase due to electron transfer between Y_z and the primary quinone acceptor Q_A. This phase was associated with vectorial intraprotein electron transfer between the DPC binding site on the protein–water interface and the tyrosine Y_z. It is suggested that NH$_2$OH is able to diffuse through channels with a diameter of 2.0–3.0 Å visible in PS II structure and leading from the protein–water interface to the Mn(4)Ca cluster binding site with the concomitant electron donation to Y(Z) (Figure 2.17). The slowing of $\Delta\Psi$ decay in the presence of

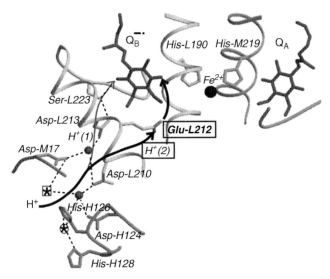

Figure 2.17 Proton transfer pathway in photosynthetic reaction centers from R. sphaeroides [167]. Reproduced with permission from Elsevier.

NH$_2$OH indicates effective electron transfer between the artificial electron donor and the reaction center of PS II.

Rapid-scan FTIR difference spectroscopy was used to investigate proton and electron transfer reactions in photosynthetic reaction centers from *R. sphaeroides*. Experiments at different temperatures and in the presence of D$_2$O have provided indication that a transient band at 1707 cm^{-1} is given by a transient protonation of the side chain of a Asp or Glu residue situated on the proton transfer pathway from the cytoplasm to the Q$_B$ site. Experiments in D$_2$O on a Asp-M17 $\rightarrow$ Asn mutant reaction center, where the proton and electron transfer reactions are slowed down compared to the wild type, showed that the kinetic isotope effect induced by H/D exchange slows down the electron transfer reaction after the first flash, confirming a strong coupling between proton and electron transfer. Rapid-scan FTIR experiments on Cd^{2+}-treated reaction centers showed that upon addition of Cd^{2+}, which inhibits proton uptake, the Q$_A^-$Q$_B$ $\rightarrow$ Q$_A$Q$_B^-$ reaction is slowed down. Interestingly, the transient 1707 cm^{-1} band is not visible in the first spectrum recorded early after the flash. This strongly suggests its identification with a residue situated on the proton transfer pathway, which is perturbed upon metal cation binding.

It was reported [158] that the crystal structure of PS II from *T. elongatus* obtained with 2.9 Å resolution allowed the assignment of all 20 protein subunits and complete modeling of all 35 chlorophyll a molecules and 12 carotenoid molecules, 25 integral lipids, and 1 chloride ion per monomer. Owing to the presence of a third plastoquinone Q$_C$ and a second plastoquinone transfer channel, the authors suggested mechanisms for plastoquinol–plastoquinone exchange, and other possible water or dioxygen and proton channels. Putative oxygen positions obtained from a Xenon derivative indicated a role for lipids in oxygen diffusion to the cytoplasmic side of PS II. The chloride position suggests a role in proton transfer reactions because it is bound through a putative water molecule to the Mn$_4$Ca cluster at a distance of 6.5 Å and is close to two possible proton channels.

It was shown [162] that the tetrameric (bacterio)chlorophyll structures in reaction centers of photosystem II of green plants and in bacterial reaction centers (BRCs) are similar and play a key role in the primary charge separation. The Stark effect measurements on PS II reaction centers have revealed an increased dipole moment for the transition at approximately 730 nm. The far-red absorption and emission was interpreted as an indication of the state with charge transfer character in which the chlorophyll monomer plays the role of an electron donor. The role of bacteriochlorophyll monomers (BA and BB) in BRCs was revealed by different mutations of axial ligand for Mg central atoms. Femtosecond measurements showed the electron transfer to B-branch with a time constant of approximately 2 ps. These results were discussed in terms of obligatory role of BA and ΦB molecules located near P for efficient electron transfer from P*.

In their work, Niklas *et al.* [168] characterized the electron spin density distribution of the electron transfer chain in photosystem I (A1)$^{\bullet -}$ with the aim of understanding the influence of the protein surrounding it. The light-induced spin-polarized radical pair $P_{700}{}^+A_1{}^-$ and the photoaccumulated radical anion $A_1{}^-$ was studied using advanced pulse EPR, ENDOR, and TRIPLE techniques at Q-band (34 GHz). Exchange with fully deuterated quinone in the A_1 binding site allowed differentiation between proton hyperfine couplings from the quinone and from the surrounding protein. In addition, the density function theory calculations in a model of the A_1 site were performed and provided proton hyperfine couplings. Comparison with vitamin K1, VK1, in organic solvents led to the conclusion that the single H bond present in both the radical pair $P_{700}{}^+A_1{}^-$ and the photoaccumulated radical anion $A_1{}^-$ is the crucial factor that governs the electronic properties of the system of interest.

2.5
Water Oxidation System

The oxidation of two water molecules to produce a dioxygen proceeds in a stepwise manner as described in the four S-state cycle with the following intermediate states: $S_0 \rightarrow S_1 \rightarrow S_2 \rightarrow S_3 \rightarrow S_4$ [169]. The radical cation $P_{680}[\text{sup} +]$ has a very high oxidizing potential, recently estimated to be 1.3–1.4 V [170], which is required for the water splitting reaction. This contrasts with 0.4 V produced by the bacterial equivalent consisting of a "special pair" of bacteriochlorophylls, which is reduced by a cytochrome after photooxidation. The cationic radical $P_{680}[\text{sup} +]$ is reduced by a redox-active tyrosine, known as Tyr[subz] (Tyr[sup161] of D1 subunit), to generate a neutral tyrosine radical Tyr[subz], which acts as an oxidant for the water oxidation process at the water oxidation system (WOS) or, by another definition, the OEC.

A schematic view of the photosystem II (PS II) complex in the thylakoid membrane from S. elongatus, obtained in the work [171], is shown in Figure 2.18.

Loll et al. [172] described cyanobacterial photosystem II structure showing locations of and interactions between 20 protein subunits and 77 cofactors per monomer. Assignment of 11 β-carotenes yields insights into electron and energy transfer and photoprotection mechanisms in the reaction center and antenna subunits. The high number of 14 integrally bound lipids reflects the structural and functional importance of these molecules for flexibility within and assembly of photosystem II. A lipophilic pathway was proposed for the diffusion of secondary plastoquinone that transfers redox equivalents from photosystem II to the photosynthetic chain. The structure provides information about the Mn_4Ca cluster.

The advance versions of X-ray techniques such as absorption near-edge structure (XANES) and EXAFS were employed for investigation of Mn clusters of OEC [173–177]. A breakthrough was recently achieved by combining crystallography and EXAFS measurements. This approach allowed collecting polarized XANES and EXAFS spectra along the three crystallographic axes of PS II. The structure of the cluster presented in the review [175] is unlike either the 3.0 or the 3.5 Å resolution X-ray structures or other previously proposed models. The authors suggested that the

2.5 Water Oxidation System | 77

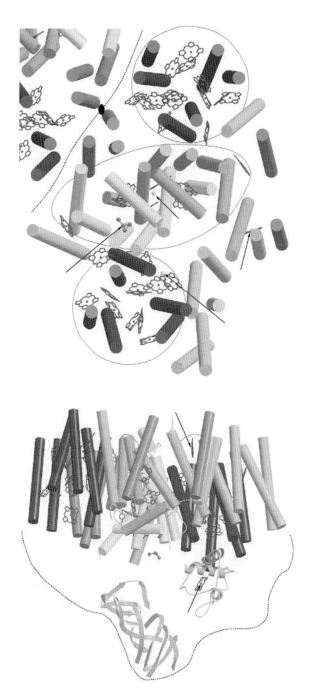

Figure 2.18 A schematic view of the photosystem II (PS II) complex in the thylakoid membrane from *S. elongatus* [171]. Reproduced with permission from *Nature*.

differences between the models derived from X-ray spectroscopy and crystallography are predominantly because of the damage to the Mn_4Ca cluster by X-rays under conditions used for the structure detection by X-ray crystallography. They concluded that a more accurate description should consider the charge density on the Mn atoms, which includes the covalency of the bonds and delocalization of the charge over the cluster.

Direct detection of oxygen ligation to the Mn_4Ca cluster of photosystem II by X-ray emission spectroscopy was reported and the outlook for Kβ″ spectroscopy was demonstrated (Figures 2.19 and 2.20) [174]. This spectroscopic technique was used to study the O ligands of the Mn_4Ca cluster that catalyzes photosynthetic water splitting and allows direct detection of the bridging oxo groups of manganese atoms. The authors suggested the involvement of bridging oxo groups or the high-valent MnIV=O or MnV≡O species in the mechanism of the formation of the O−O bond in the water oxidation reaction of photosystem II.

Polarized EXAFS measurements on PS II single crystals allowed to constrain the Mn_4Ca cluster geometry to a set of three similar high-resolution structures [176]. Combining polarized EXAFS and X-ray diffraction data, the cluster was placed within PS II, taking into account the overall trend of the electron density of the metal site and the putative ligands. Data on the single-crystal Mn-EXAFS of *S. elongatus* PS II in the S_1 allowed to suggest models of the Mn_4CaO_n cluster of the OEC deduced from the

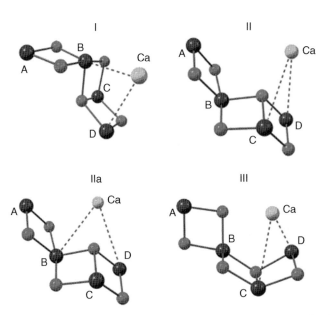

Figure 2.19 (a) Energy diagram of Mn Kb transitions in MnO. The Kb″ and Kb2,5 transitions are from valence molecular orbitals: Kb″ is the O 2s to Mn 1s "crossover" transition. (b) Logarithmic plot of the MnO Kb*n* spectrum. The O−Mn crossover Kb″ transition is highlighted [174].

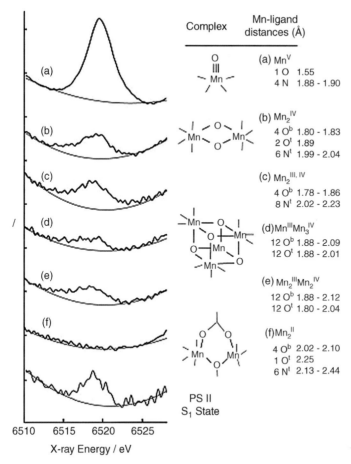

Figure 2.20 The Kb'' emission spectra from a series of multinuclear Mn complexes with oxo bridging groups, an MnV oxo complex, and PS II in the S1 state. The crossover peak from the O ligand is prominent when short bridging Mn–O distances are present [174].

single crystal. Electronic structure of the Mn_4O_xCa cluster in the S_0 and S_2 states of the oxygen-evolving complex of photosystem II was investigated using advance pulse ^{55}Mn-ENDOR and EPR spectroscopy [178]. The ^{55}Mn hyperfine coupling constants of the S_0 and S_2 states of the OEC were interpreted on the basis of Y-shaped spin-coupling schemes with up to four nonzero exchange coupling constant, J. This analysis established that the oxidation states of the manganese ions in S_0 and S_2 are, at 4 K, Mn4(III, III, III, IV) and Mn4(III, IV, IV, IV), respectively. By applying a "structure filter" that is based on the recently reported single-crystal EXAFS data on the Mn_4O_xCa cluster [179], the authors showed that (i) this new structural model is fully consistent with EPR and ^{55}Mn-ENDOR data, (ii) assign the Mn oxidation states to the individual Mn ions, and (iii) propose that the known shortening of one 2.85 Å

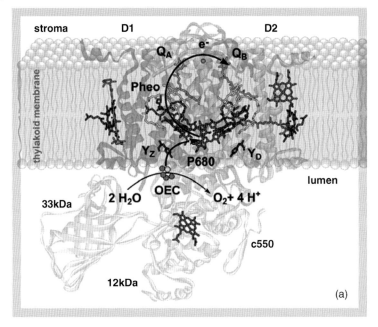

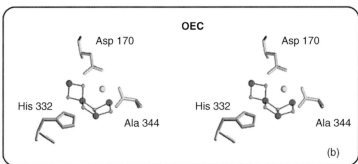

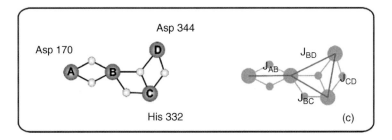

Mn–Mn distance in S_0 to 2.75 Å in S_1 corresponds to a deprotonation of a í-hydroxo bridge between MnA and MnB, that is, between the outer Mn and its neighboring Mn of the í3-oxo bridged moiety of the cluster. These results were summarized in a molecular model for EOS (Figures 2.21 and 2.22).

Taking information from various spectroscopic measurements into account, a tentative assignment of the individual Mn oxidation states, and the exchange coupling constants, Lubitz et al. [179] constructed a structure shown in Figure 2.23. This model allowed for the first time a specific description of the already known structural change during the $S_0 \rightarrow S_1$ transition to a contraction of the MnA–MnB distance due to the oxidation of one Mn(III) to Mn(IV), which is coupled to a deprotonation of a m-OH bridge (top two rows in Figure 2.23).

Twelve structural models for the S_2 state of the oxygen-evolving complex of photosystem II are evaluated in terms of their magnetic properties [180]. The set includes 10 models based on the fused twist' core topology derived by polarized EXAFS spectra and 2 related models proposed in recent mechanistic investigations. Optimized geometries and spin population analyses suggested that Mn(iii) is associated with a pentacoordinate environment, unless a chloride is directly ligated to the metal. Seven models display a doublet ground state and are considered spectroscopic models for the ground state corresponding to the multiline signal (MLS) of the S_2 state of the OEC, whereas the remaining five models display a sextet ground state and could be related to the $g = 4.1$ signal of the S_2 state. A quantum chemical method for the calculation of ^{55}Mn hyperfine coupling constants was subsequently applied to the S_2 MLS state models and the quantities that enter into the individual steps of the procedure (site spin expectation values, intrinsic site isotropic hyperfine parameters, and projected ^{55}Mn isotropic hyperfine constants) were analyzed and discussed in detail with respect to the structural and electronic features of each model.

The hyperfine sublevel correlation (HYSCORE) spectroscopy on $H_2^{17}O$-enriched PS II samples poised in the paramagnetic S2 state was employed [181]. This approach allowed authors to resolve the magnetic interaction of one solvent exchangeable ^{17}O that is directly ligated to one or more Mn ions of the Mn_4O_xCa cluster in the S_2 state of PS II (Figure 2.24). Direct coordination of ^{17}O to Mn was supported by the strong ($A \approx 10$ MHz) hyperfine coupling.

Figure 2.21 Panel a shows a schematic view of the photosystem II (PS II) complex in the thylakoid membrane that is based on the 3.0 Å crystal structure given in Ref. [172]. For clarity, the inner antenna proteins CP43 and CP47, which are involved in harvesting the light energy, and the cytochrome b559 subunits are not shown. The other core proteins of PS II are shown with corresponding labels. The cofactors of the D1, D2, cyt c550, and cyt b559 proteins are placed on top of the proteins. Panel b presents a stereoview of the Mn_4O_xCa cluster together with selected ligands that are referred to in this chapter. The structure shown was derived by Yano et al. [176] on the basis of polarized EXAFS spectroscopy on PS II single crystals. The view is also approximately along the membrane plane and from a similar angle. Panel c displays a schematic top view (approximately along the membrane normal) of the Mn_4O_5 core of the Mn_4O_xCa cluster using the same color code as that above. This schematic representation allows a clearer visualization of the bridging motifs between the four Mn ions, which are labeled A, B, C, and D. On the right-hand side of panel c, the general coupling scheme is shown that is employed in this study to derive the electronic structure of the S_0 and S_2 [178]. Reproduced with permission from American Chemical Society.

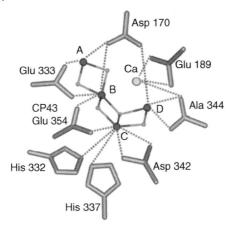

Figure 2.22 Structural model (one out of the four possible) of the Mn_4O_xCa cluster derived from polarized EXAFS spectroscopy on photosystem II single crystals: Mn [176]. The model is placed in the EXAFS-derived orientation within the protein ligands determined by crystallography [172] without any optimization [179]. Reproduced with permission from Royal Chemical Society.

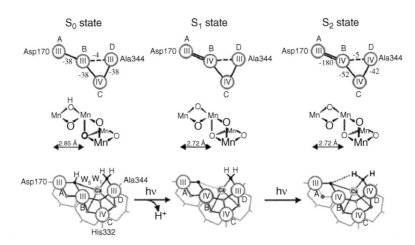

Figure 2.23 Schemes of the electronic (a) and geometric (b) structures of the Mn_4 unit within the Mn_4O_xCa cluster as derived from ^{55}Mn-ENDOR and EXAFS spectroscopies in the S_0, S_1, and S_2 states of the water-oxidizing complex of photosystem II. The exchange coupling strength (in cm^{-1}) is given in blue numbers and is also symbolized by double lines (strong), single lines (medium strength), and dashed lines (weak) antiferromagnetic coupling. (c) A possible interpretation of the available experimental data on these states, which also includes suggestions for the binding sites of the slowly (W_s) and fast (W_f) exchanging substrate water molecules (small black spheres) [179]. Reproduced with permission from Royal Chemical Society.

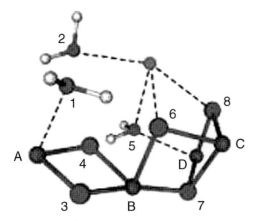

Figure 2.24 Model for the Mn$_4$O$_x$Ca cluster in photosystem II with suggested substrate water binding sites. O–O bond formation has been proposed to occur in the S$_4$ state between oxygens **1** and **2**, **3** and **4**, **4** and **5**, **1** and **6**, or **6** and **8**. All are adapted to EXAFS model. The positions of the water molecules were optimized by density function theory calculations [181]. Reproduced with permission from American Chemical Society.

The available theoretical and experimental data indicate that the photosynthetic water splitting complex consist of Mn$_4$O$_x$Ca cubane with oxygen bridges and carboxylate ligands. The dioxygen evolution occurs in the complex high-oxidation redox state, most probably by the four-electron mechanism, as it was suggested in Refs [79, 118, 182–184] (see also Section 7.3.1).

In conclusion, light energy conversion in the photosynthetic reaction centers is characterized by the high energetic efficiency and the quantum yield close to 100%. Such a result was achieved by the fulfillment of several principal conditions, which also should be realized in artificial systems of high efficiency ([80] and references therein):

1) The donor and acceptor groups should be disposed at a certain optimum distance of about 7–8 Å relative to each other. This requirement is necessary to provide a sufficiently fast direct electron transfer with the formation of the pair D$^+$A$^-$, on the one hand, and to prevent the fast pair recombination (CR) on the other hand.
2) The electron transfer driving force (ΔG_0) and redox potentials of the donor and acceptor groups should be optimal.
3) A molecular mobility in the vicinity of the donor and acceptor groups (most probably in the nanosecond temporary rank) should be provided.
4) Charge-separated pair D$^+$A$^-$ keeping strong chemical reactivity should be isolated from side reactions.
5) For effective water splitting to O$_2$ and H$_2$, the manganase or other transition metal clusters of high redox potential should be used.

References

1 Blankenship, R.E. (2002) *Molecular Mechanisms of Photosynthesis*, Blackwell Science, Oxford.
2 Dismukes, G.C. (2001) *Science*, **292** (5516), 447–448.
3 Rutherford, A.W. and Faller, P. (2001) *Trends Biochem. Sci.*, **26** (4), 341–344.
4 Witt, H.T. (1996) *Ber. Bunsensges. Phys. Chem.*, **100**, 1923–1942.
5 Hofbauer, W., Zouni, A., Bittl, R., Kern, J., Orth, P., Lendzian, F., Fromme, P., Witt, H.T., and Lubitz, W. (2001) *Proc. Natl. Acad. Sci. USA*, **98** (12), 6623–6627.
6 Hall, D.O. and Rao, K. (1999) *Photosynthesis*, 6th edn, Cambridge University Press, Cambridge.
7 Lawlor, D. (2000) *Photosynthesis*, Kluwer Academic Press, Dordrecht; Aro, E.-M. and Bertil, A.B. (2000) *Regulation of Photosynthesis*, Springer.
8 Smith, A.L. (1997) *Oxford Dictionary of Biochemistry and Molecular Biology*, Oxford University Press, Oxford, p. 508.
9 Macpherson, A.N. and Hiller, R.G. (2003) in *Light-Harvesting Antennas in Photosynthesis Advances in Photosynthesis and Respiration*, **13** (eds B.R. Green and W.W. Parson), pp. 323–352.
10 Anton, S., Marco, M., Francesco, F., Giovanni, V., and Klaus, M. (2010) *J. Phys. Chem. B*, **114** (39), 12729–12743.
11 Daviso, E., Alia, A., Prakash, S., Diller, A., Gast, P., Lugtenburg, J., Matysik, J., and Jeschke, G. (2009) *J. Phys. Chem. C*, **113** (23), 10269–10278.
12 Green, B.R., Anderson, J.M., and Parson, W.W. (2003) in *Light-Harvesting Antennas in Photosynthesis, Advances in Photosynthesis and Respiration*, vol. 13 (eds B.R. Green and W.W. Parson), Kluwer, pp. 1–28.
13 Cogdell, R.J. and van Grondelle, R. (2003) in *Light-Harvesting Antennas in Photosynthesis Advances in Photosynthesis and Respiration*, vol. 13 (eds B.R. Green, and W.W. Parson), Kluwer, pp. 169–194.
14 Blankenship, R.E. and Matsuura, K. (2003) in *Light-Harvesting Antennas in Photosynthesis, Advances in Photosynthesis and Respiration*, **13** (eds B.R. Green, and W.W. Parson), Kluwer, pp. 195–217.
15 Lubitz, W. (2006) in *Photosystem I: The Light-Driven Plastocyanin: Ferredoxin Oxidoreductase, Advances in Photosynthesis and Respiration*, vol. 24, (ed. J.H. Golbeck), Springer, pp. 245–269.
16 Liu, X.D. and Shen, Y.G. (2004) *FEBS Lett.*, **569** (1–3), 337–340.
17 Sporlein, S., Zinth, W., Meyer, M., Scheer, H., and Wachtveitl, J. (2000) *Chem. Phys. Lett.*, **322** (6), 454–464.
18 Gilmore, A.M., Hazlett, T.L., and Govindjee (1995) *Proc. Natl. Acad. Sci. USA*, **92** (6), 2273–2277.
19 Förster, T.H. (1948) 2 (1), 55–75.
20 Koepke, J., Hu, X., Muenke, C., Schulten, K., and Michel, H. (1996) *Structure*, **4** (5), 581–597.
21 Brotosudarmo, T.H.P., Kunz, R., Bohm, P., Gardiner, A.T., Moulisova, V., Cogdell, R.J., and Kohler, J. (2009) *Biophys. J.*, **97** (5), 1491–1500.
22 Gabrielsen, M., Gardiner, A.T., and Cogdell, R.J. (2009) in *Purple Phototrophic Bacteria, Advances in Photosynthesis and Respiration*, vol. 28 (eds C.N. Hunter, F. Daldal, M.C. Thurnauer, and J.T. Beatty) Springer, pp. 135–153.
23 Cogdell, R.J., Gardiner, A.T., Gabrielsen, M., Southall, J., Roszak, A.W., Isaacs, N.W., Fujii, R., and Hashimoto, H. (2008) in *Photosynthetic Protein Complexes* (ed. P. Fromme), Wiley-VCH Verlag GmbH, Weinheim, pp. 325–340.
24 Loll, B., Kern, J., Zouni, A., Saenger, W., Biesiadka, J., and Irrgang, K.-D. (2005) *Photosynth. Res.*, **86** (1–2), 175–184.
25 Melkozernov, A.N. and Blankenship, R.E. (2005) *Photosynth. Res.*, **85** (1), 33–50.
26 Prakash, J.S., Sundara, B., Masroor, A., Bhagwat, A.S., and Mohanty, P. (2003) *J. Plant Physiol.*, **160** (2), 175–184.
27 Cardoso, M.B., Smolensky, D., Heller, W.T., and O'Neill, H. (2009) *J. Phys. Chem. B*, **113** (51), 16377–16383.
28 Standfuss, J., Terwissscha van Scheltinga, A.C., Lamborghini, M., and Kuehlbrandt, W. (2005) *EMBO J.*, **24** (5), 919–928.

29 Law, C.J. and Cogdell, R.J. (2008) in *The Primary Processes of Photosynthesis, Part I, Comprehensive Series in Photochemical & Photobiological Sciences*, vol. **8** (ed. G. Renger), Royal Society of Chemistry, Cambridge, pp. 205–259.

30 Richter, M.F., Baier, J., Southall, J., Cogdell, R.J., Oellerich, S., and Koehler, J. (2007) *Proc. Natl. Acad. Sci. USA*, **104** (51), 20280–20284.

31 Ahn, T.K., Avenson, T.J., Ballottari, M., Cheng, Y.-C., Niyogi, K.K., Bassi, R., and Fleming, G.R. (2008) *Science*, **320** (5877), 794–797.

32 Moulisova, V., Luer, L., Hoseinkhani, S., Brotosudarmo, T.H.P., Collins, A.M., Lanzani, G., Blankenship, R.E., and Cogdell, R.J. (2009) *Biophys. J.*, **97** (11), 3019–3028.

33 Struempfer, J. and Schulten, K. (2009) *J. Chem. Phys.*, **131** (22), 225101/1–225101/9.

34 Cong, H., Niedzwiedzki, D.M., Gibson, G.N., LaFountain, A.M., Kelsh, R.M., Gardiner, A.T., Cogdell, R.J., and Frank, H.A. (2008) *J. Phys. Chem. B*, **112** (34), 10689–10703.

35 Liu, X.D. and Shen, Y.G. (2004) *FEBS Lett.*, **569** (1–3), 337–340.

36 Kramer, D., Bassi, R., Li, X.P., Gilmore, A.M., Caffarri, S., Golan, T., and Niyogi, K.K. (2004) *J. Biol. Chem.*, **279** (22), 22866–22874.

37 Kühlbrandt, W. (1987) *J. Mol. Biol.*, **194** (4), 757–762.

38 Liu, Z., Yan, H., Wang, K., Kuang, T., Zhang, J., Gui, L., An, X., and Chang, W. (2004) *Nature*, **428** (6980), 287–292.

39 Pascal, A.A., Liu, Z., Broess, K., van Oort, B., van Amerongen., H., Wang, C., Horton, P., Robert, B., Chang, W., and Ruban, A. (2005) *Nature*, **436** (7047), 134–137.

40 Ruban, A.V., Berera, R., Ilioaia, C., van Stokkum, I.H., Kennis, J.T., Pascal, A.A., van Amerongen, H., Robert, B., Horton, P., and van Grondelle, R. (2007) *Nature*, **450** (7169), 575–578.

41 Green, B.R. and Parson, W.W. (eds) (2003) in *Light-Harvesting Antennas in Photosynthesis, Advances in Photosynthesis and Respiration*, vol. 13 (eds B.R. Green and W.W. Parson), Kluwer, p. 544.

42 van Amerongen, H. and van Grondelle, R. (2001) *J. Phys. Chem. B*, **105** (3), 604–617.

43 Fromme, P., Schlodder, E., and Jansson, S. (2003) *Light-Harvesting Antennas in Photosynthesis, Advances in Photosynthesis and Respiration*, **13** (eds B.R. Green and W.W. Parson), Kluwer, pp. 253–279.

44 Yeremenko, N., Kouril, R., Ihalainen, J.A., D'haene, S., Van Oosterwijk, N., Andrizhiyevskaya, E.G., Keegstra, W., Dekker, H.L., Hagemann, M., Boekema, E.J., Matthijs, H.C., and Dekker, J.P. (2004) *Biochemistry*, **43** (32), 10308–10313.

45 Loll, B., Kern, J., Zouni, A., Saenger, W., Biesiadka, J., and Irrgang, K.-D. (2005) *Photosynth. Res.*, **86** (1–2), 175–184.

46 Crimi, M., Dorra, D., Bösinger, C.S., Giuffra, E., Holzwarth, A.R., and Bassi, R. (2001) *Eur. J. Biochem.*, **268** (2), 260–267.

47 Holub, O., Seufferheld, M.J., Gohlke, C., Govindjee, and Clegg, R.M. (2000) *Photosynthetica*, **38** (4), 581–599.

48 Elrad, D., Krishna, K., Niyogi, K.K., and Grossman, A.R. (2002) *Plant Cell*, **14** (8), 1801–1816.

49 Croce, R., Müller, M.G., Caffarri, S., Bassi, R., and Holzwart, A.R. (2003) *Biophys. J.*, **84** (4), 2517–2532.

50 Arteni, A.A., Ajlani, G., and Boekema, E.J. (2009) *Biochimi. Biophys. Acta Bioenergetics*, **1787** (4), 272–279.

51 DeRuyter, Y.S. and Fromme, P. (2008) in *Cyanobacteria: Molecular Biology, Genomics and Evolution* (eds A. Herrero and E. Flores), Caister Academic Press, pp. 217–269.

52 Aro, E.-M. (2000) *Regulation of Photosynthesis*, Springer.

53 Abresch, E.C., Gong, X.-M., Paddock, M.L., and Okamura, M.Y. (2009) *Biochemistry*, **48** (48), 11390–11398.

54 Marchanka, A., Paddock, M., Lubitz, W., and van Gastel, M. (2007) *Biochemistry*, **46** (51), 14782–14794.

55 Michel, H. and Deisenhofer, J. (1986) X-ray diffraction studies on a crystalline bacterial photosynthetic center. A

progress report and conclusions on the structure of the photosystem II reaction center, in *Encyclopedia of Plant Physiology, New Series*, vol. 19 (eds L.A. Stachelin and C.J. Arntzen), Springer, Berlin, pp. 371–381.
56. Feher, G. (1992) *Isr. J. Chem.*, **32** (4), 375–378.
57. Hoff, A.J. and Deisenhofer, J. (1997) *Phys. Rep.*, **287** (1), 1–247.
58. Rees, D.C., Komia, H., Yeates, T.O., Allen, J.P., and Feher, G. (1989) *Annu. Rev. Biochem.*, **38**, 607–633.
59. Shuvalov, V.A. and Krasnovsky, A.A. (1981) *Biofizika*, **26** (3), 544–556.
60. Paul, K.F., Hughes, A.V., Heathcote, P., and Jones, M.R. (2005) *Trends Plant Sci.*, **10** (6), 275–282.
61. Allen, P. and Williams, J.C. (1998) *FEBS Lett.*, **4381** (2), 5–9.
62. Wöhri, A.B., Wahlgren, W.Y., Malmerberg, E., Johansson, L.C., Neutze, R., and Katona, G. (2009) *Biochemistry*, **48**, 9831–9838.
63. Katagiri, S. and Kobori, Y. (2009) *Appl. Magn. Reson.*, **37** (1–4), 177–189.
64. Wraight, C.A. and Gunner, M.R. (2009) in *Purple Phototrophic Bacteria, Advances in Photosynthesis and Respiration*, vol. 28 (eds C.N. Hunter, F. Daldal, M.C. Thurnauer, and J.T. Beatty), Springer, pp. 379–405.
65. Krupyanskii, Yu.F., Mikhailyuk, M.G., Esin, S.V., Eshchenko, G.V., Moroz, A.P., Okisheva, E.A., Seifullina, N.Kh., Knox, P.P., and Rubin, A.B. (2006) *Biofizika*, **51** (1), 13–23.
66. Mobius, K., Savitsky, A., and Fuchs, M. (2004) in *Very High Frequency (VHF) ESR/EPR Biological Magnetic Resonance*, vol. 22 (eds O. Grinberg and L.J. Berliner), Kluwer/Plenum, pp. 45–93.
67. McAuley, K.E., Fyfe, P.K., Cogdell, R.J., Isaacs, N.W., and Jones, M.R. (2000) *FEBS Lett.*, **467** (2–3), 285–290.
68. Paddock, M.L., Flores, M., Isaacson, R., Shepherd, J.N., and Okamura, M.Y. (2009) *Appl. Magn. Reson.*, **37** (1–4), 39–48.
69. Shchepetov, D.S., Chernavskii, D.S., Gorokhov, V.V., Pashchenko, V.Z., and Rubin, A.B. (2009) *Biofizika*, **54** (6), 1026–1036.
70. Alekperov, S.D., Vasil'ev, S.I., Kononenko, A.A., Lukashov, E.P., Panov, V.I., and Semenov, A.E. (1988) *Dokl. Akad. Nauk SSSR*, **303** (2), 341–344.
71. Kulik, L. and Lubitz, W. (2009) *Photosynth. Res.*, **102** (2–3), 391–401.
72. Allen, J.P., Cordova, J.M., Jolley, C.C., Murray, T.A., Schneider, J.W., Woodbury, N.W., Williams, J.C., Niklas, J., Klihm, G., Reus, M., and Lubitz, W. (2009) *Photosynth. Res.*, **99** (1), 1–10.
73. Hofbauer, W., Zouni, A., Bittl, R., Kern, J., Orth, P., Lendzian, F., Fromme, P., Witt, H.T., and Lubitz, W. (2001) *Proc. Natl. Acad. Sci. USA*, **98** (12), 6623–6662.
74. Kulikov, A.V., Bogatyrenko, V.R., Melnikov, A.V., Syrtzova, L.A., and Likhtenshtein, G.I. (1979) *Biofizika*, **24** (2), 178–183.
75. Likhtenshtein, G.I., Kulikov, A.V., Kotelnikov, A.I., and Bogatyrenko, V.R. (1982) *Photobiochem. Photobiol.*, **3** (2), 178–182.
76. Likhtenshtein, G.I., Kotel'nikov, A.I., Kulikov, A.V., Syrtsova, L.A., Bogatyrenko, V.R., Mel'nikov, A.I., Frolov, E.N., and Berg, A.I. (1979) *Intern. J. Quant. Chem.*, **16** (3), 419–435.
77. Likhtenshtein, G.I. (1979) *Multinuclear Redox Metalloenzymes*, Nauka, Moscow.
78. Likhtenshtein, G.I. (1988) *Chemical Physics of Redox Metalloenzyme Catalysis*, Springer, Berlin.
79. Likhtenshtein, G.I. (2003) *New Trends in Enzyme Catalysis and Mimicking Chemical Reactions*, Kluwer Academic/Plenum Publishers, NY.
80. Likhtenshtein, G.I. (2008) *Pure Appl. Chem.*, **80** (10), 2125–2139.
81. Likhtenshtein, G.I., Syrtsova, L.A., Samuilov, V.D., Frolov, E.N., Borisov, A.U., and Bogatyrenko, V.R. (1975) *XII International Botanic Congress: Theses of Reports*, Nauka, Leningrad, p. 429.
82. Likhtenshtein, G.I. (1988) *J. Mol. Catalysis*, **48** (1), 129–138.
83. Chance, B. and DeVault, D. (1964) *Ber. Bunsenges. Physik. Chem.*, **68**, 722–726.

84 Fyfe, P.K., Potter, J.A., Cheng, J., Williams, C.M., Watson, A.J., and Jones, M.R. (2007) *Biochemistry*, **46** (37), 10461–10472.

85 Lancaster, C. and Roy, D. (2008) in *The Primary Processes of Photosynthesis, Part II, Comprehensive Series in Photochemical & Photobiological Sciences*, **9** (ed. G. Renger), Royal Society of Chemistry, Cambridge, pp. 5–54.

86 Ren, Y., Ke, W., Li, Y., Feng, L., Wan, J., and Xu, X.K. (2009) *J. Phys. Chem. B*, **113** (30), 10055–10058.

87 Ikegami, T., Ishida, T., Fedorov, D.G., Kitaura, K., Inadomi, Y., Umeda, H., Yokokawa, M., and Sekiguchi, S. (2010) *J. Comput. Chem.*, **31** (2), 447–454.

88 Allen, J.P., Cordova, J.M., Jolley, C.C., Murray, T.A., Schneider, J.W., Woodbury, N.W., Williams, J.C., Niklas, J., Klihm, G., Reus, M., and Lubitz, W. (2009) *Photosynth. Res.*, **99** (1), 1–10.

89 Marchanka, A., Savitsky, A., Lubitz, W., Mobius, K., and van Gastel, M. (2010) *J. Phys. Chem. B.*, **114**, 14364.

90 Flores, M., Isaacson, R., Abresch, E., Calvo, R., Lubitz, W., and Feher, G. (2007) *Biophys. J.*, **92** (2), 671–682.

91 Bixon, M. and Jortner, J. (1999) Electron transfer: from isolated molecules to biomolecules, in *Advances in Chemical Physics*, vol. 107, Part 1 (eds J. Jortner and M. Bixon), John Wiley & Sons, Inc., NY, pp. 35–202.

92 Yakovlev, A.G., Shkuropatov, A.Y., and Shuvalov, VA. (2000) *FEBS Lett.*, **466** (2–3), 209–212.

93 Ykovlev, A.G. and Shuvalov, V.A. (2001) *Biochemistry (Moscow)*, **66** (2), 211–220.

94 Khatypov, R.A., Khmelnitskiy, A.Y., Khristin, A.M., and Shuvalov, V.A. (2010) *Dokl. Biochem. Biophys.*, **430** (1), 24–28.

95 Likhtenshtein, G.I. (1996) *J. Photochem. Photobiol. A Chem.*, **96** (1), 79–92.

96 Likhtenshtein, G.I. (2000) Depth of immersion of paramagnetic centers, in *Magnetic Resonance in Biology*, vol. 18 (eds L. Berliner, S. Eaton, and G. Eaton), Kluwer Academic Publishers, Dordrecht, pp. 309–347.

97 Calvo, F R., Abresch, E.C., Bittl, R., Feher, G., Hofbauer, W., Isaacson, R.A., Lubitz, W., Okamura, M.Y., and Paddock, M.L. (2000) *J. Am. Chem. Soc.*, **122** (30), 7327–7341.

98 Xu, Q. and Gunner, M.R. (2001) *Biochemistry*, **40** (10), 3232–3241.

99 Chamorovskii, S.K., Kononenko, A.A., Petrov, E.G., Pottosin, I.I., and Rubin, A.B. (1986) *Biochimi. Biophys. Acta Bioenergetics*, **848** (3), 402–410.

100 Paddock, M.L., Flores, M., Isaacson, R., Shepherd, J.N., and Okamura, M.Y. (2009) *Appl. Magn. Reson.*, **37** (1–4), 39–48.

101 Ren, Y., Ke, W., Li, Y., Feng, L., Wan, J., and Xu, X. (2009) *J. Phys. Chem. B*, **113** (30), 10055–10058.

102 Xu, Q., Baciou, L., Sebban, P., and Gunner, M.R. (2002) *Biochemistry*, **41** (30), 10021–10025.

103 Remy, A. and Gerwert, K. (2003) *Nat. Struct. Biol.*, **10** (8), 637–644.

104 Sabine, H., Bremm, O., Garczarek, F., Derrien, V., Liebisch, P., Hermes, S., Paola, L., and Sebban, P. (2006) *Biochemistry*, **45** (2), 353–359.

105 Saclay, Gif-sur.-Yvette. (2007) *Biochemistry*, **46** (15), 4459–4465.

106 Witt, H.T. (1996) *Ber. Bunsensges. Phys. Chem.*, **100**, 1923–1942.

107 Paddock, M.L., Feher, G., and Okamura, M.Y. (2003) *FEBS Lett.*, **555** (1), 45–50.

108 Xu, Q., Axelrod, H.L., Abresch, E.C., Paddock, M.L., Okamura, M.Y., and Feher, G. (2004) *Structure*, **12** (4), 703–715.

109 Krammer, E.-M., Till, M.S., Sebban, P., and Ullmann, G.M. (2009) *J. Mol. Biol.*, **388** (3), 631–643.

110 Likhtenshtein, G.I. (1993) *Biophysical Labeling Methods in Molecular Biology*, Cambridge University Press, Cambridge, NY.

111 Likhtenshtein, G.I. (1979) Study of protein dynamics by spin-labeling, Mösbauer spectroscopy, and NMR, in *Special Colloque. AMPER on Dynamic Processes in Molecular Systems* (ed. A. Losche), Karl-Marx University, Leipzig, pp. 100–107.

112 Berg, A.I., Kononenko, A.F., Noks, P.P., Khrymova, I.N., Frolov, E.N., Rubin, A.B., Likhtenshtein, G.I., Uspenskaya, N., and Khideg, K. (1979) *Molek. Biol.*, **13**, 469–477.

113 Berg, A.I., Kononenko, A.F., Noks, P.P., Khymova, I.N., Frolov, E.N., Rubin, A.B., Likhtenshtein, G.I., Goldansky, V.I., Parak, F., Bukl, M., and Mossbauer, R.L. (1979) *Molek. Biol.*, **13** (1), 81–89.

114 Kotelnikov, A.I., Likhtenshtein, G.I., Fogel, V.R., Kochetkov, V.V., Noks, P.P., Kononenko, A.A., Grishanova, N.P., and Rubin, A.B. (1983) *Zhurnal Prikladnoi Spectroskopii*, **17**, 840–846.

115 Kochetkov, B.B., Likhtenshtein, G.I., Koltover, V.K., Knox, P.P., Kononenko, A.A., Grishanova, P.G., and Rubin, A.B. (1984) *Izv. Akademii Nauk SSSR (Seria Biologicheskaya)* **4**, 572–579.

116 Likhtenshtein, G.I., Febbrario, F., and Nucci, R. (2000) *Spectrochim. Acta A*, **56**, 2011–2031.

117 Likhtenshtein, G.I. (1990) *Pure Appl. Chem.*, **62** (2), 281–288.

118 Likhtenshtein, G.I. (1979) *Multinuclear Redox Metalloenzymes*, Nauka, Moscow.

119 Beratan, D.N. and Onuchic, J.N. (1987) *J. Chem. Phys.*, **86** (8), 4489–4498.

120 Beratan, D.N., Onuchic, J.N., Betts, J.N., Bowler, B.E., and Gray, G.H. (1990) *J. Am. Chem. Soc.*, **112** (22), 7915–7921.

121 Beratan, D.N. and Skourtis, S.S. (1998) *Biological Electron Transfer Chains: Genetics, Composition and Mode of Operation*, NATO ASI Series, Series C: Mathematical and Physical Sciences, vol. 512, NATO, pp. 9–27.

122 Balabin, I.A. and Onuchic, J.N. (2000) *Science*, **290** (5489), 114–117.

123 Muh, F., Bibikova, M., Schlodder, E., and Oesterhelt, D. (2000) *Biochim. Biophys. Acta*, **1459**, 191–201.

124 Wang, H., Lin, S., Katilius, E., Laser, C., Allen, J.P., Williams, J.A.C., and Woodbury, N.W. (2009) *J. Phys. Chem. B*, **113** (3), 818–824.

125 Kolbasov, D. and Scherz, A. (2000) *J. Phys. Chem. B*, **104** (8), 1802–1809.

126 Kirmaier, C. and Holten, D. (2009) *J. Phys. Chem. B*, **113** (4), 1132–1142.

127 Dismukes, G.C. (2001) *Science*, **292** (5516), 447–448.

128 Setif, P. (2006) in *Photosystem I: The Light-Driven Plastocyanin: Ferredoxin Oxidoreductase, Advances in Photosynthesis and Respiration*, vol. 24, (ed. J.H. Golbeck), Springer, pp., 439–454.

129 Fromme, P. and Grotjohann, I. (2006) in *Photosystem I: The Light-Driven Plastocyanin: Ferredoxin Oxidoreductase, Advances in Photosynthesis and Respiration*, vol. 24, (ed. J.H. Golbeck), Springer, pp. 47–69.

130 Klimov, V.V. (2003) *Photosynth Res.*, **76** (2), 247–253.

131 Orr, L. and Govindjee (2007) *Photosynth Res*, **91** (1), 107–131.

132 Lubitz, W. (2006) in *Photosystem I: The Light-Driven Plastocyanin: Ferredoxin Oxidoreductase, Advances in Photosynthesis and Respiration*, vol. 24 (ed. J.H. Golbeck), Springer, pp. 245–269.

133 Sener, M.K., Olsen, J.D., Hunter, C.N., and Schulten, K. (2007) *Proc. Natl. Acad. Sci. USA*, **104** (40), 15723–15728.

134 Nelson, N. and Yocum, C.F. (2006) *Annu. Rev. Plant Biol.*, **57**, 521–565.

135 Thurnauer, M.C., Poluektov, O.G., and Kothe, G. (2004) in *Very High Frequency (VHF) ESR/EPR Biological Magnetic Resonance*, vol. 22 (O. Grinberg and L.J. Berliner), Kluwer/Plenum, pp. 165–206.

136 Zech, S.G., Hofbauer, W., Kamlowski, A., Fromme, P., Stehlik, D., Lubitz, W., and Bittl, R. (2000) *J. Phys. Chem. B*, **104**, 9728–9739.

137 Jordan, P., Fromme, P., Witt, H.T., Klukas, O., Saenger, W., and Krau, N. (2001) *Nature*, **411** (6840), 909–917.

138 Ben-Shem, A., Frolow, F., and Nelson, N. (2003) *Nature*, **426** (6967), 630–635.

139 Amunts, A., Toporik, H., Borovikova, A., and Nelson, N. (2010) *J. Biol. Chem.*, **285** (5), 3478–3486.

140 Zhang, Y., Chen, M., Church, W.B., Kwok, W.L., Larkum, A.W.D., and Jermiin, L.S. (2010) *Biochimi. Biophys. Acta Bioenergetics*, **1797**, 457–465.

141 Telfer, A., Lendzian, F., Schlodder, E., Barber, J., and Lubitz, W. (1998) in Proceedings of the International

Congress on Photosynthesis, 11th, Budapest, Aug. 17–22, 2, pp. 1061–1064.

142 Kitmitto, A., Mustafa, A.O., Holzenburg, A., and Ford, R.C. (1998) *J. Biol. Chem.*, **273** (45), 29592–29599.

143 Jagannathan, B. and Golbeck, J.H. (2008) *Biochimi. Biophys. Acta Bioenergetics*, **1777** (12), 1535–1544.

144 Schmidt, K.A. and Neerken, S., *Biochim. Biophys. Acta*, **1459**, 191–201; Permentier, H.P., Hager-Braun, C., and Amesz, J. (2000) *Biochemistry*, **39** (24), 7212–7220.

145 Guergova-Kuras, M., Boudreaux, B., Joliot, A., Joliot, P., and Redding, K. (2001) *Proc. Natl. Acad. Sci. USA*, **98** (8), 4437–4442.

146 Setif, P., Seo, D., and Sakurai, H. (2001) *Biophys. J.*, **81** (3), 1208–1219.

147 Melkozernov, A.N., Lin, S., Blankenship, R.E., and Valkunas, L. (2001) *Biophys. J.*, **81** (3), 1144–1154.

148 Boichenko, V.A., Hou, J.-M., and Mauzerall, D. (2001) *Biochemistry*, **40**, 7126–7132.

149 Harvey, J.M., Hou, G., Boichenko, V.A., Golbeck, J.H., and Mauzerall, D. (2009) *Biochemistry*, **48** (8), 1829–1837.

150 Rigby, S.E.J., Evans, M.C.W., and Heathcote, P. (2000) *Biochim. Biophys. Acta*, **1507**, 247–259.

151 Kim, S., Sacksteder, C.A., Bixby, K.A., and Barry, B.A. (2001) *Biochemistry*, **40**, 15384–15395.

152 Käss, H., Fromme, P., Witt, H.T., and Lubitz, W. (2001) *J. Phys. Chem. B*, **105**, 1225–1239.

153 Rhee, K.-H. (2001) *Biophys. Biomol. Struct.*, **30** (2), 307–328.

154 Zouni, A., Witt, H.T., Kern, J., Fromme, P., Krauss, N., Saenger, W., and Orth, P. (2001) *Nature*, **409**, 739–743.

155 Ferreira, K.N., Iverson, T.M., Maghlaoui, K., Barber, J., and Iwata, S. (2004) *Science (Washington, DC, United States)*, **303** (5665), 1831–1838.

156 Biesiadka, J., Loll, B., Kern, J., Irrgang, K.-D., and Zouni, A. (2004) *Phys. Chem. Chem. Phys.*, **6**, 4733–4736.

157 Gerken, S., Brettel, K., and Witt, H.T. (1988) *FEBS Lett.*, **237** (1), 69–79.

158 Guskov, A., Kern, J., Gabdulkhakov, A., Broser, M., Zouni, A., and Saenger, W. (2009) *Nat. Struct. Mol. Biol.*, **16** (3), 334–342.

159 Mueh, F., Renger, T., and Zouni, A. (2008) *Plant Physiol. Bioch. (Issy les Moulineaux, France)*, **46** (3), 238–264.

160 Loll, B., Kern, J., Saenger, W., Zouni, A., and Biesiadka, J. (2005) *Nature*, **438** (7070), 1040–1044.

161 Gardian, Z., Bumba, L., Schrofel, A., Herbstova, M., Nebesarova, J., and Vacha, F. (2007) *Biochimi. Biophys. Acta Bioenergetics*, **1767** (6), 725–731.

162 Khatypov, R.A., Khmelnitskiy, A.Y., Leonova, M.M., Vasilieva, L.G., and Shuvalov, V.A. (2008) *Photosynth. Res.*, **98** (1–3), 81–93.

163 Zouni, A., Witt, H.T., Kern, J., Fromme, P., Krauss, N., Saenger, W., and Orth, P. (2001) *Nature*, **409**, 739–743.

164 Kamiya, N. and Shen, J.-R. (2003) *Proc. Natl. Acad. Sci. USA*, **100** (1), 98–103.

165 Guskov, A., Kern, J., Gabdulkhakov, A., Broser, M., Zouni, A., and Saenger, W. (2009) *Nat. Struct. Mol. Biol.*, **16** (3), 334–342.

166 Gibasiewicz, K., Dobek, A., Breton, J., and Leibl, W. (2001) *Biophys. J.*, **80** (4), 1617–1630.

167 Alberto, M. and Winfried, L. (2008) *Vib. Spectrosc.*, **48** (1), 126–134.

168 Niklas, J., Epel, B., Antonkine, M.L., Sinnecker, S., Pandelia, M.-E., and Lubitz, W. (2009) *J. Phys. Chem. B*, **113** (30), 10367–10379.

169 Kok, B., Forbush, B., and McGloin, M.M. (1970) *Photochem. Photobiol.*, **11** (6), 457–475.

170 Rappaport, F., Guergova-Kuras, M., Nixon, P.J., Diner, B.A., and Lavergne, J. (2002) *Biochemistry*, **41** (26), 8518–8527.

171 Zouni, A., Witt, H.-T., Kern, J., Petra Fromme, P., Krauû, N., Saenger, W., and Orth, P. (2001) *Nature*, **409**, 739–742.

172 Loll, B., Kern, J., Saenger, W., Zouni, A., and Biesiadka, J. (2005) *Nature*, **438**, 1040–1044.

173 Yano, J. and Yachandra, V.K. (2009) *Photosynth. Res.*, **102** (2–3), 241–254, 174 (172).

174 Pushkar, Y., Long, X., Glatzel, P., Brudvig, G.W., Dismukes, G.C.,

Collins, T.J., Yachandra, V.K., Yano, J., and Bergmann, U. (2010) *Angew. Chem. Int. Ed.*, **49** (4), 800–803.

175 Yano, J. and Yachandra, V.K. (2008) *Inorg. Chem.*, **47** (6), 1711–1726.

176 Yano, J., Kern, J., Sauer, K., Latimer, M.J., Pushkar, Y., Biesiadka, J., Loll, B., Saenger, W., Messinger, J., Zouni, A., and Yachandra, V.K. (2006) *Science*, **314** (5800), 821–825.

177 Sauer, K., Yano, J., and Yachandra, V.K. (2005) *Photosynth. Res.*, **85** (1), 73–86.

178 Kulik, L.V., Epel, B., Lubitz, W., and Messinger, J. (2007) *J. Am. Chem. Soc*, **129** (1), 13421–13435.

179 Lubitz, W., Reijerse, E.J., and Messinger, J. (2008) *Energy Environ. Sci.*, **1**, 15–31.

180 Pantazis, D., Orio, M., Petrenko, T., Zein, S., Lubitz, W., Messinger, J., and Neese, F. (2009) *Phys. Chem. Chem. Phys.*, **11** (31), 6788–6798.

181 Su, J.-H., Lubitz, W., and Messinger, J. (2008) *J. Am. Chem. Soc.*, **130** (3), 786–787.

182 Semenov, N.N., Shilov, A.E., and Likhtenshtein, G.I. (1975) *Dokl. Akad. Nauk SSSR*, **221** (6), 1374–1377.

183 Likhtenshtein, G.I. (1988) *Chemical Physics of Redox Metalloenzymes*, Springer, Heidelberg.

184 Likhtenshtein, G.I., Kotel'nikov, A.I., Kulikov, A.V., Syrtsova, L.A., Bogatyrenko, V.R., Mel'nikov, A.I., Frolov, E.N., and Berg, A.J. (1979) *Int. J. Quantum. Chem.*, **16** (3), 419–435.

3
Photochemical Systems of the Light Energy Conversion

3.1
Introduction

Mimicking the photosynthetic functions by using synthetic model compounds holds promise for technological advances in solar energy conversion, and building molecular optoelectronics such as photonic wires and switches. In this regard, many studies focused on synthesis of reaction center (RC) mimicking compounds and on investigating mechanisms of artificial photosynthesis [1–21].

The charge separation (CS) process includes two major steps: (i) absorption and transportation of light energy of appropriate wavelength by the antenna light-harvesting molecules to the reaction center and (ii) photoinduced electron transfer (PET) to generate charge-separated entities by using the electronic excitation energy. Such investigations form a theoretical and experimental basis for construction of effective photochemical systems including dye-sensitized solar cells (DSSC).

Without an antenna system coupled to the reaction center that can collect and deliver a sufficient number of photons in the required short period of time before charge recombination (CR) occurs, the conversion efficiency of any photocatalytic system remains very low at the ambient light intensities delivered by the sun. Thus, the requirement of an antenna coupled to a photochemical reaction center is a necessary feature for efficient photocatalytic operation in artificial solar energy conversion systems. Another principal requirement is high quantum yield of the photo separation, which can be provided by optimum combination of fast electron transfer (ET) and slow recombination in photochemical cascade systems keeping high chemical redox reactivity of the photoseparated pair. Therefore, higher overall efficiencies in an artificial self-assembled system require strong, specific binding of the various self-assembling units, as occurs in natural photosynthetic constructs.

Among the numerous potential dyes, the specific classes of compounds such as porphyrins or other cyclic tetrapyrroles, carotenoides, ruthenium and zinc complexes, fullerenes, borondipyrromethene (BDPY), melanin, and so on have attracted

Solar Energy Conversion. Chemical Aspects, First Edition. Gertz Likhtenshtein.
© 2012 Wiley-VCH Verlag GmbH & Co. KGaA. Published 2012 by Wiley-VCH Verlag GmbH & Co. KGaA.

special interest. A promising means to making more effective and durable photochemical light conversion systems is the use of templates, for example, zeolites and mesoporous silica, TiO_2, nanofibres, polymers, liposomes, metal–protein complexes, and so on.

It is necessary to stress the outstanding role of fullerenes as antennas, electron acceptors, and molecular wires. Fullerenes has extremely small reorganization energy of electron transfer because no structural change occurs before and after electron transfer. Data on fullerenes for organic electronics were reviewed [22–31].

3.2
Charge Separation in Donor–Acceptor Pairs

3.2.1
Introduction

Electron transfer reactions are important in applications ranging from photochemical systems of light conversion energy, solar cells, and light-emitting diodes to molecule-based materials for electronics and photonics. The need to transport charge over long distances in molecular systems for artificial photosynthesis and organic photovoltaics makes it important to understand the fundamental nature of wire-like molecular systems.

Two distinct mechanisms of charge transport have been reported, which are superexchange and thermally activated hopping. The superexchange mechanism of coherent charge transport from the donor to the acceptor results from quantum mechanical mixing of the donor state with bridge states that are higher in energy and energetically well separated from those of the donor. The superexchange can be quatitively characterized by the damping (attenuation) factor β (Section 1.2.5) that strongly depends on a bridge structure. Charge transfer occurs in a single step with a rate that decays exponentially with donor–acceptor distance (r_{DA}) that is described by the damping factor β. Electron transfer by hopping occurs by actual oxidation/reduction of the bridge moiety when acceptor affinity is greater than that of the bridge. Charge transport rates for the hopping mechanism are only weakly distance dependent (low value of β) allowing charge to move efficiently at long distances, so that systems exhibiting this mechanism are termed "molecular wires." Such a value is also expected in the case of conductive bridge composed of conjugated groups. Thus, to distinguish between two mechanisms, it is necessary to make a quanum mechanical calculation or to use empirical data on the attenuation.

For succesive mimicking, the following requirements, sometimes contrdictory ones, should be fulfilled:

1) Creation of light-harvesting system with broad absorbtion spectrum covering sun light radiation spectra.
2) Minimization of loss of light energy during the multistep electron transfer processes.

3) Construction of systems that have long charge separation state lifetimes.
4) Maintaining high reactivity of cation and anion segments after charge separation sufficient for subsequent chemical and photochemical reactions.
5) Prevention of side reactions.

The following means for fulfillment of these requirements can be formulated:

1) Because high thermodynamic driving force can lead to stable products, the thermodynamic gap between ET intermediates should be optimum.
2) Chromophores in excited singlet state of high energy as electron donor should be preferably used.
3) To avoid a competition with the spontaneous quenching of the primary donor exited state, the first steps of the charge separation by electron transfer should be fast. Therefore, the first acceptor should be located close to the primary donor to provide an effective superexchange mechanism.
4) The best way to prevent fast charge recombination is construct a system of intermediate acceptors separated by nonconductive bridge of 6–8 Å composed of nonsaturated chemical bonds (Section 1.4) (hopping mechanism). The number of the nonadiabatic hops should be optimum to prevent the loss of energy in each hopping.
5) A promising way to slow down CR is an organization of the spin catalysis (Section 1.2.6) that provides singlet–triplet conversion on the CS pair.
6) A redox system can be protected from side reactions either by using membranes, which separate initial and final reagents, or by incorporating donor and accaptor groups in a protein or other matrix, as it is realized in natural photosynthesis.

The charge separation process in arificial donor–acceptor pair has been discussed at length in many reviews and research articles [32–49] and will be briefly dealt with here.

3.2.2
Cyclic Tetrapyrroles

Photoinduced intramolecular ET and energy transfer (EnT) processes in two rotaxanes, one containing both zinc porphyrin and C_{60} fullerene moieties incorporated around the Cu(I) bisphenanthroline core [(ZnP)2-Cu(I)(phen)2-C_{60}] and a second complex lacking the fullerene [(ZnP)$_2$-Cu(I)(phen)$_2$] (Figure 3.1), were studied by time-resolved electron paramagnetic resonance (TREPR) spectroscopy at 9.5 GHz (X-band) combined with a selective photoexcitation of the rotaxane moieties [43]. The experiments were carried out in isotropic toluene and ethanol and in anisotropic nematic liquid-crystal (E-7) media over a wide range of temperatures corresponding to the different states of the solvents. The TREPR spectra in E-7 at 170 K for $L \mid B$ and $L \perp B$ orientations were recorded 0.96 µs after laser-pulse photoexcitation at 460 nm. Presentation of the possible photoinduced processes occurring in (ZnP)$_2$-Cu(I)(phen)$_2$ in the soft crystalline and nematic phases of E-7 is given in Figure 3.2. The complementary results revealed by the optical and TREPR techniques were

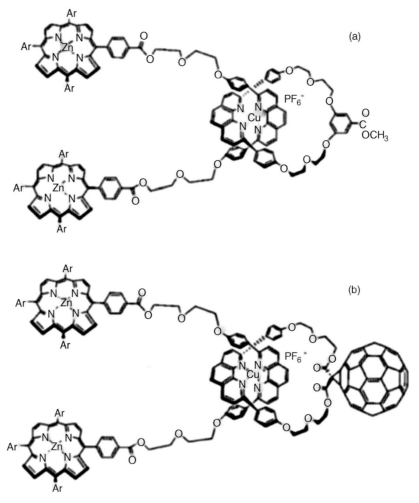

Figure 3.1 Chemical structures of the rotaxane molecules [43].

attributed to the relatively high conformational mobility of the mechanically interlocked rotaxane systems because of the solute–solvent interactions.

Light-driven ET and EnT via axial coordination in a self-assembled Zn-porphyrin–pyridylfullerene (ZnP-PyrF) complex were studied by TREPR spectroscopy at 9.5 (X-band) and 95 GHz (W-band) [42]. The studies over a wide temperature range were carried out in media of different polarity, including isotropic toluene and tetrahydrofuran (THF), and anisotropic nematic liquid crystals (LCs), E-7 and ZLI-4389. It was shown that at frozen matrices photoexcitation of the ZnP donor results mainly in singlet–singlet EnT to the pyridine-appended fullerene acceptor, while in fluid phases ET is the dominant process. In isotropic solvents, the generated radical pairs (RPs) were found to be long-lived. The authors concluded that in liquid phases of both polar and nonpolar solvents, the separation of the tightly bound complex into the

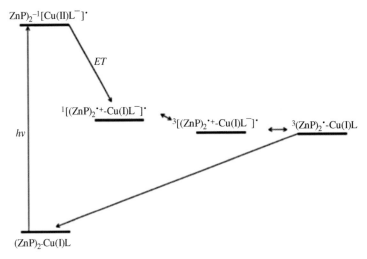

Figure 3.2 General presentation of the possible photoinduced processes occurring in (ZnP)$_2$-Cu(I) (phen)$_2$ in the soft crystalline and nematic phases of E-7 [43].

more loosely bound structure slows down the back ET (BET) process. A simple electron donor–acceptor dyad and a directly linked zinc chlorinfullerene dyad (Figure 3.3) have been developed to attain a long-lived charge separated state, where the donor and acceptor molecules were linked to a short spacer [50]. In this case, the lifetime of the CS state at 150 °C is as long as 120 s. This value is the longest CS lifetime ever reported for porphyrin-based donor–acceptor-linked systems.

The dynamics of photoinduced electron transfer from the triplet excited states of cofacial porphyrin dimers (CPDs) to a series of electron acceptors (Figure 3.4) were investigated by using laser flash photolysis measurements and compared with the porphyrin monomer [44]. CPDs linked to different spacers and the reference porphyrin monomer as electron donors and nitrobenzene derivatives and a series of benzoquinone compounds as electron acceptors were employed. According to the results of experiments, variation in the acceptors' redox potential (E_{red} versus SCE) from −0.50 to −1.30 led to variation in the electron transfer rate constants k_{ex} (M^{-1} s^{-1}) from 4×10^9 to 8×10^8, while ET from the dimer porphyrins was two times faster than that from monomer. The driving force dependence of the rate constants of photoinduced electron transfer reactions was analyzed in light of the Marcus theory of electron transfer to afford the reorganization energies of electron transfer (λ). The λ values of CPDs were found to be significantly smaller than those of the porphyrin monomer when compared at the same reorganization of the photoinduced electron transfer. The λ values increase linearly with an increase in the driving force of the photoinduced electron transfer and were accompanied by an increase in the distance between electron donor and acceptor molecules, where the electron transfer occurs.

The biomimetic model of the bacterial "special pair" donor, a cofacial zinc phthalocyanine dimer, was formed via potassium ion-induced dimerization of 4,5,4′,5′,4″, 5″,4 − ,5 − -zinc tetrakis(1,4,7,10,13-pentaoxatridecamethylene)phthalo-

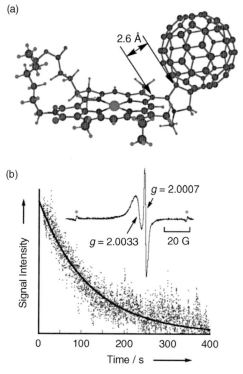

Figure 3.3 (a) Structure of ZnChC6 dyad calculated by PM3. (b) ESR spectrum and the decay time profile of the ESR signal intensity of ZnCh·$^{+}$C$_{60}$·^{1} in frozen PhCN at 150 °C [50].

cyanine [51]. The dimer was subsequently self-assembled. On the basis of detailed analysis of the kinetic data from the time-resolved emission and transient absorption spectroscopy of different timescales, it was found that the charge separation occurs from the triplet excited state of the ZnPc moiety to the C$_{60}$ moiety. The longer CS lifetime of the phthalocyanine dimer-C$_{60}$ supramolecular complex, K$_4$[ZnTCPc]$_2$: (pyC$_{60}$NH$_3^+$)$_2$ (6.7 μs), has been attained compared to that of the monomer complex, ZnTCPc:pyC$_{60}$NH^{3+} (4.8 μs). Photoinduced electron transfer in a self-assembled supramolecular ladder structure (Figure 3.5) was characterized by transient absorption spectroscopy and transient direct current photoconductivity mainly from an oligomer (rail) to the center of a terminal tetrazine, with the remaining hole being delocalized on the oligomer and subsequent charge [52]. It was observed that in the photoexcited ladder, an electron generally transfers from the trimer to a single tetrazine at one end of the ladder, with the remaining hole delocalized over three porphyrins and the electron and hole recombine in 0.19 ns.

A series of zinc porphyrin arrays comprising a meso–meso linked porphyrin dimer, a meta-phenylene linked dimer, gable-like tetramers consisting of meso–meso-linked dimers bridged via a meta-phenylene linker, and a dodecameric ring composed of this alternating dimeric pattern was prepared and intramolecular hole hopping between the porphyrin moieties was probed using ESR and electron nuclear

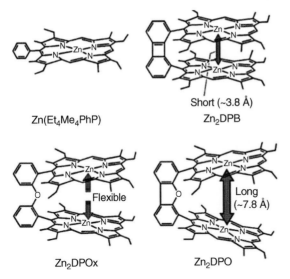

Figure 3.4 Structure of cofacial dimer porphyrins linked to different spacers and the reference porphyrin monomer [44].

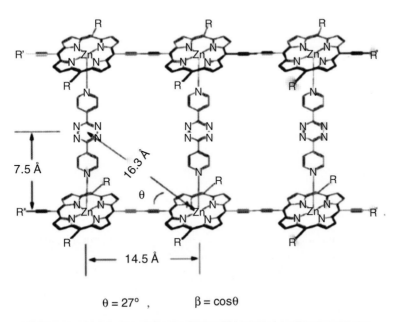

$\theta = 27°$, $\quad \beta = \cos\theta$

Figure 3.5 A self-assembled supramolecular ladder structure used in the work [52].

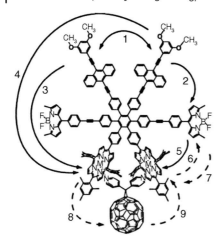

Figure 3.6 Singlet excitation energy (s) and electron (−) transfer pathways observed in heptad 1 and model compounds. Time constants for photochemical processes (ps) for different steps are shown in brackets: step 1 (0.4), step 2 (5–14), step 3 (7), step 4 (6), step 5 (2–24), step 6 (230–570), step 7 (1500–4800), step 8 (3), and step 9 (230) [49].

double resonance (ENDOR spectroscopy [49]). Rapid hole hopping was found to occur between both porphyrins within both dimers and among three porphyrins of the tetramers with rates $>10^7$ s^{-1} at 290 K. The hole hops among 8–12 porphyrins in the dodecameric ring with a rate >107 s^{-1} at 290 K, but hopping is slow at 180 K. A fullerene electron acceptor self-assembles to both porhyrins via dative bonds. It was shown that excitation energy is transferred very efficiently from all four antennas to the porphyrins (Figure 3.6). Singlet–singlet energy transfer occurs both directly and by a stepwise funnel-like pathway wherein excitation moves down a thermodynamic gradient. The porphyrin excited states donate an electron to the fullerene with a time constant of 3 ps to generate a charge-separated state with a lifetime of 230 ps. The overall quantum yield was found to be close to unity. This molecule demonstrates that by incorporating antennas, it is possible for a molecular system to harvest efficiently light throughout the visible range from ultraviolet wavelengths to <650 nm. Singlet excitation energy (s) and electron (−) transfer pathways observed in heptad and time constants for photochemical processes for different steps are shown in Figure 3.6

A series of Zn(II) porphyrin (ZnP) compounds covalently linked to different electron acceptor units, for example, naphthaleneimide (NI) and naphthalenediimide (NDI), were reported [53]. The aim of the study was to demonstrate a state-selective direction of electron transfer, where excitation to the lowest excited S_1 state of the porphyrin (Q-band excitation) would result in electron transfer to the NDI unit, while excitation to the higher S_2 state (Soret-band excitation) would lead to electron transfer to the NI unit. It was shown that electron transfer from the S_1 state to NDI occurred in solvents of both high and low polarity, whereas no electron transfer to NDI was observed from the S_2 state. With NI as acceptor, very rapid ($\tau = 200$–400 fs) electron transfer from the S_2 state occurred in all solvents. This was followed by an ultrafast ($\tau \approx 100$ fs) recombination to populate the porphyrin S1 state in nearly quantum yield. The evidence was obtained that recombination occurred from a vibrationally excited ("hot") ZnP^+NI state in the more polar solvents. The communication [20] described the design and characterization of an artificial reaction center protein that closely resembles the function of the natural photosynthetic RC. It was demonstrated that the synthetic protein, constructed using the protein design program CORE, participates in multiple redox cycles with exogenous acceptors/donors following photoexcitation. The designed metalloprotein, aRC, consists of a tetrahelical bundle functionalized with two bis-histidine-bound metal cofactors: a Ru(bpy)$_2$ moiety and a heme group. According to the calculation, photoexcitation of RC results in rapid electron transfer from the RuII complex to the heme group ($k_{ET} \geq 5 \times 10^{10}$ s^{-1}) yielding a long-lived (70 ns) charge-separated pair.

A synthetic molecular heptad that features two bis(phenylethynyl)anthracene (BPEA) and two borondipyrromethene antennas linked to a hexaphenylbenzene core that also bears two zinc porphyrins was investigated [54]. The time-dependent spectroscopic properties of the hexads and models were studied in 2-methyltetrahydrofuran at ambient temperatures using pump–probe transient absorption, fluorescence upconversion, and single-photon timing emission experiments. The two zinc porphyrin moieties attached to the hexaphenylbenzene coordinate a fullerene electron acceptor that bears two pyridine ligation units.

The hexaphenylbenzene core also supports two BPEA antennas that absorb light in the 400–500 nm region and two BDPY chromophores with absorption in the 450–550 and 330–430 nm ranges, respectively.

Photoinduced charge separation and recombination were investigated in a triad consisting of a carotenoid (C), a tetraarylporphyrin (P), and a tris(heptafluoropropyl) porphyrin (PF), C-P-PF, by means of time-resolved ESR [55]. The electron transfer process was studied in a glass of 2-methyltetrahydrofuran at 10 K, in the crystal phase at 150 K, in the liquid nematic phase of the uniaxial LC E-7 at 295 K, and in the nematic phase of the LC ZLI-1167 at 300 K. It was shown that the molecular triad undergoes a two-step photoinduced electron transfer, with the generation of a long-lived charge-separated state ($C^{\cdot +}$–P–$P^{\cdot -}$F), and charge recombination to the triplet state, localized in the carotene moiety 3C–P–PF. The large delocalized π-electron system of the porphyrin electron acceptor leads to low total reorganization energy and low sensitivity to solvent stabilization of the radical ions in a similar way as for fullerene systems.

Data on fullerene for organic electronics were revewed [28]. Rate constants of charge separation and charge recombination processes were measured. Figure 3.7 illustrates a few leading examples, in which a ZnP donor and C_{60} are connected through diverse bridges.

A simple way to obtain linked donor–acceptor entities involving metallomacrocycle complexes with fixed distance and orientation was shown by the use of coordination of axial ligands to metallomacrocycle complexes [56]. A series of electron acceptor bearing silicon phthalocyanine (SiPc) triads with fixed distance and orientation have been synthesized, using the six-coordinated nature of the central silicon atom.

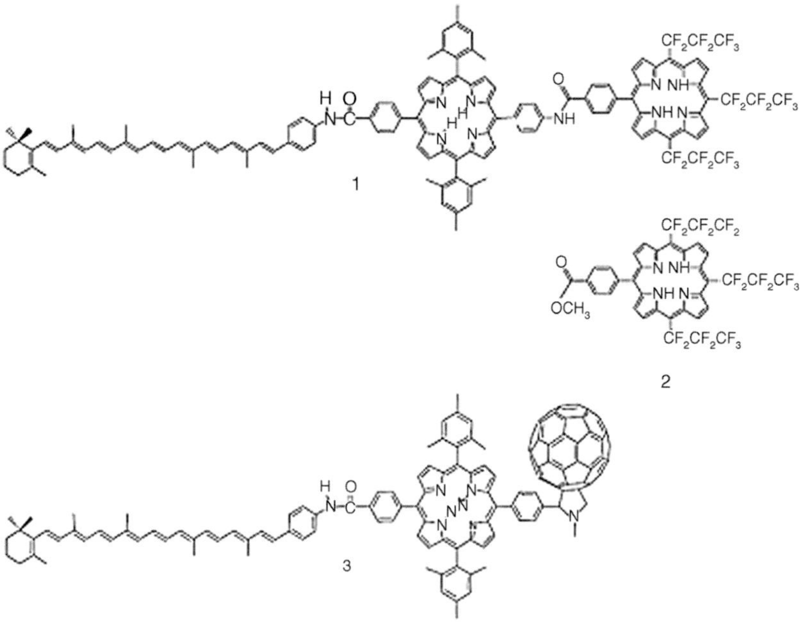

Figure 3.7 Structures of Fc–ZnP–H2P–C$_{60}$ and Fc–ZnP–ZnP–C$_{60}$ conjugates [28].

Charge separation in the triad was investigated. A panchromatic 4,4-difluoro-4-bora-3a,4a-diaza-s-indacene–zinc-phthalocyanine conjugate (Bodipy–ZnPc) 1 was synthesized and investigated [46]. Electrochemical and optical measurements provided evidence for strong electronic interactions between the Bodipy and the ZnPc constituents in the ground state. Excitation spectral analysis confirmed that the photoexcited Bodipy and the tethered ZnPc subunits interact and that intraconjugate singlet energy transfer occurs with an efficiency of 25%. Treatment of conjugate with N-pyridylfulleropyrrolidine, an electron acceptor system containing a nitrogen ligand, gave rise to the novel electron donor–acceptor hybrid through ligation to the ZnPc center. Irradiation of the resulting supramolecular ensemble within the visible range led to a charge-separated Bodipy–ZnPc$^{\bullet +}$–C$_{60}^{\bullet -}$ radical ion-pair state, through a sequence of excited state and charge transfers, characterized by a long lifetime of 39.9 ns in toluene.

The convergent synthesis, electrochemical characterization, and photophysical studies of phthalocyanine–fullerene hybrids 3–5 bearing an orthogonal geometry (the hybrid [Ru(CO)(C$_{60}$Py)Pc] and the triad [Ru$_2$(CO)$_2$(C$_{60}$Py$_2$)Pc$_2$]) were reported [57]. The characterization of the phthalocyanine–fullerene hybrids was conducted using a number of methods including femtosecond transient absorption studies and nanosecond laser flash photolysis. Electronic coupling between the two electroactive components in the ground state, which is modulated by the axial CO and 4-pyridylfulleropyrrolidine ligands, was detected. The use of ruthenium(II) phthalocyanines allowed the suppression of energy wasting and unwanted charge recombination, affording radical ion pair state lifetimes on the order of hundreds of nanoseconds for the C_{60} monoadduct-based complexes.

Turning to recent developments of molecular wires, several milestones in the development of molecular-scale electronic devices have already been passed regarding the transduction of electrons through fullerene containing compounds [55–61].

The review [62] combines the most important results of studies performed by the authors during the past decade on photoinduced electron transfer reactions of pheophytin-, phthalocyanine-, and porphyrin-fullerene dyads, in which donor and acceptor moieties are covalently linked to each other. When the center-to-center distance of the donor and acceptor pair was short (7–10 Å), both the exciplex formation and the primary electron transfer were found to be extremely fast with rate constants of $7–23 \times 10^{12}$ s^{-1} and $40–1400 \times 10^9$ s^{-1}, respectively. Rates become slower when the distance and orientational fluctuation increase. Electron donor–acceptor conjugates combining exTTF and/or TTF as donors and C_{60} as acceptor have been synthesized [63]. Fluorescence and transient absorption measurements confirm the generation of charge-separated radical ion pairs with lifetimes in the μs timescale.

In the work [64], a series of new electron acceptor bearing SiPc triads were synthesized, using the six-coordinated nature of the central silicon atom, by attaching two electron–acceptor units, fullerene Sipc-(C_{60}) and trinitrofluorenone SiPc-(TNF), and trinitrodicyanomethylenefluorene SiPc-(TNDCF)(2). The redox and photophysical properties of SiPc triads in benzonitrile were determined to evaluate the energy of the CS states and driving force of photoinduced electron transfer in SiPc triads. Photoexcitation of SiPc triads in benzonitrile results in efficient formation of the CS states, which were detected by femtosecond laser flash photolysis measurements.

3.2.3
Miscellaneous Donor–Acceptor Systems

In the review [65], the authors discussed the design of ruthenium complexes for efficient photoinitiation of electron transfer and their use for studying intracomplex electron transfer c. Direct measurement of photoinduced charge separation distances (R_{DA}) in donor–acceptor systems (Figure 3.8) for artificial photosynthetic systems was performed using OOP-ESEEM [66]. The obtained R_{DA} values (in Å) were found to be 25.0, 28.1, and 38.0 for compounds 1, 2, and 3, respectively.

New molecular triads, composed of closely spaced ferrocene–boron dipyrrin–fullerene and triphenylamine–boron dipyrrin–fullerene, were synthesized, and photoinduced electron transfer leading to charge stabilization was demonstrated using a femtosecond transient spectroscopic technique [67]. Photochemical electron donor–acceptor triads having an aminopyrene primary donor (APy) and a p-diaminobenzene secondary donor (DAB) attached to either one or both imide nitrogen atoms of a perylene-3,4:9,10-bis(dicarboximide) (PDI) electron acceptor were prepared to give DAB-APy-PDI and DABAPy-PDI-APy-DAB [68]. In toluene, both triads are monomeric, but in methylcyclohexane, they self-assemble into ordered helical heptamers and hexamers, respectively, in which the PDI molecules are π-stacked in a columnar fashion. Photoexcitation of these supramolecular assemblies resulted in

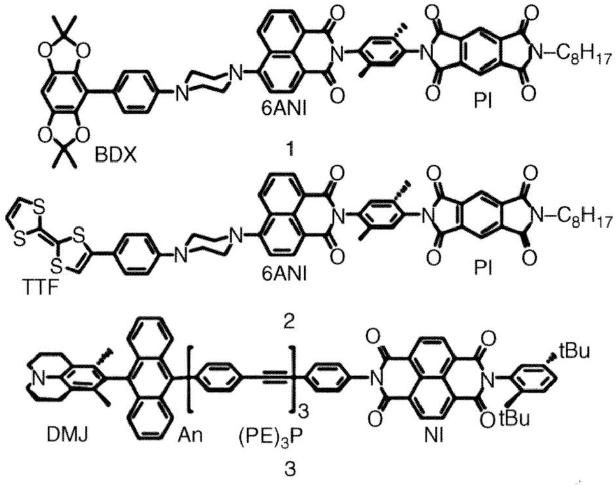

Figure 3.8 Compounds with different distances between donor and acceptor groups [66].

rapid formation of DAB$^{+\bullet}$-PDI$^{-\bullet}$ spin polarized radical ion pairs having spin–spin dipolar interactions, which show that the average distance between the two radical ions is much larger in the assemblies (31 Å) than it is in their monomeric building blocks (23 Å). This work demonstrates that electron hopping through the π-stacked PDI molecules is fast enough to compete effectively with charge recombination (40 ns) in these systems.

Spin-selective charge transport pathways through *p*-oligophenylene-inked donor–bridge–acceptor molecules were investigated using transient absorption spectroscopy, magnetic field effects (MFEs) on radical pair and triplet yields, and TREPR spectroscopy [69]. A series of donor–bridge–acceptor (D-B-A) triads have been synthesized in which the donor, 3,5-dimethyl-4-(9-anthracenyl)julolidine (DMJ-An), and the acceptor, naphthalene-1,8:4,5-bis(dicarboximide) (NI), are linked by *p*-oligophenylene (Ph$_n$) bridging units ($n = 1$–5). It was found that photoexcitation of DMJ-An produces DMJ$^{+\bullet}$-An$^{-\bullet}$ quantitatively, so that An$^{-\bullet}$ acts as a high-potential electron donor, which rapidly transfers an electron to NI yielding a long-lived spin-coherent radical ion pair (DMJ$^{+\bullet}$-An-Ph$_n$-NI$^{-\bullet}$). The charge separation and recombination reactions exhibit exponential distance dependencies with damping coefficients of 0.35 and 0.34 Å^{-1}, respectively. Two isomeric [5,6]-pyrrolidine-Ih-Sc3N@C80 electron donor–acceptor conjugates containing triphenylamine (TPA) as the donor system were synthesized [70]. It was found that when the donor is connected to the pyrrolidine nitrogen atom, the resulting dyad produces a significantly longer lived radical pair than the corresponding 2-substituted isomer for both the C$_{60}$ and the Ih-Sc3N@C80 dyad. The Ih-Sc3N@C80 dyads have considerably longer lived charge-separated states than their C$_{60}$ analogues.

Reversible switching of electron transfer paths in supramolecular donor–acceptor dyads formed by complexing benzo-18-crown-6 appended porphyrins and fulleropyrrolidine appended with an alkyl ammonium ion was demonstrated [71]. The intrasupramolecular to intermolecular switching was achieved by the addition of K$^+$, which dissociates the crown ether-alkyl ammonium complex, while intermololecular to intrasupramolecular switching was obtained by the addition of 18-crown-6 to the potassium ions of the porphyrin-crown entity. A variety of supramolecular complexes were formed by associating an electron donor-substituted flavin dyad (10-[4'-(N,N-dimethylamino)phenyl]isoalloxazine, DMA–Fl) and a family of fullerene derivatives that contains single and double hydrogen bond receptors SRC_{60} and DRC_{60} [72]. To complement these studies, the dynamics of electron transfer were investigated by femtosecond laser flash photolysis in the supramolecular clusters DMA–Fl–SRC_{60n} and [(DMA–Fl)$_2$–$DRC_{60}]_n$. In the paper [73], it was shown that donor–acceptor nanohybrid composed of single-walled carbon nanotubes (SWNTs) and coenzyme Q_{10} (CoQ_{10}) undergoes efficient photoinduced electron transfer from SWNT to CoQ_{10} to produce the charge-separated state as indicated by femtosecond laser flash photolysis and ESR measurements.

Light-driven multistep intramolecular electron transfer (IET) in a rod-like triad (Figure 3.9), in which two of the three redox components are linked by three hydrogen bonds, was studied by time-resolved electron paramagnetic resonance and optical spectroscopies [74]. According to the experiments, the electronic coupling between the oxidized donor and the reduced acceptor in the hydrogen-bonded radical ion pair MeOAn$^+$-6ANI/MELNI is similar to that of MeOAn$^+$-6ANI-Ph-NI.

The temperature dependence of intramolecular charge separation in a series of donor–bridge–acceptor molecules having phenothiazine (PTZ) donors, 2,7-oligofluorene FL_n ($n = 1$–4) bridges, and perylene-3,4:9,10-bis(dicarboximide) acceptors was studied [75]. Photoexcitation of PDI to its lowest excited singlet state resulted in the oxidation of PTZ via the FL_n bridge. In toluene, the temperature dependence of the charge separation rate constants for PTZ–FL_n–PDI ($n = 1$–4) was relatively weak and was described by the semiclassical Marcus equation. The activation energies for

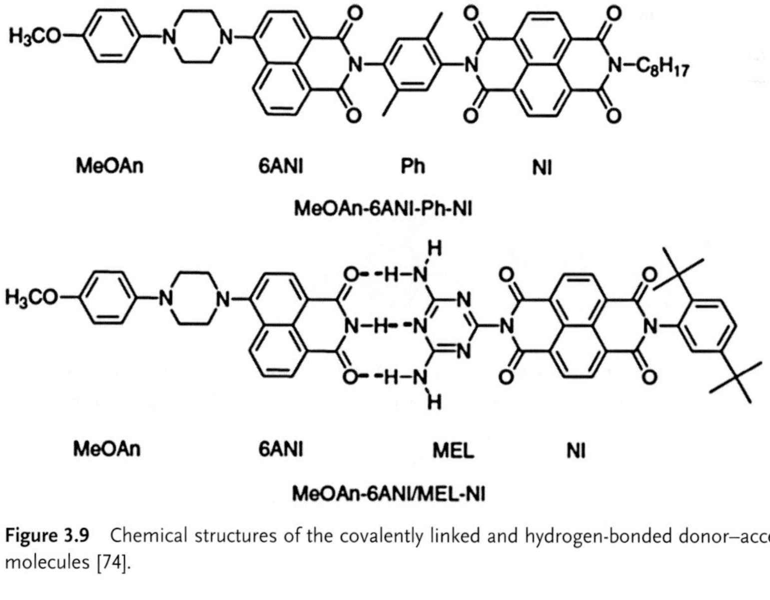

Figure 3.9 Chemical structures of the covalently linked and hydrogen-bonded donor–acceptor molecules [74].

charge separation suggest that bridge charge carrier injection was not found to be a rate-limiting step. Self-assembly strategies for integrating light harvesting and charge separation in artificial photosynthetic systems were recently discussed [16]. The authors used covalent building blocks based on chemically robust arylene imide and diimide dyes, biomimetic porphyrins, and chlorophylls. Employing small- and wide-angle X-ray scattering (SAXS/WAXS) from a synchrotron source, an advantage of the shapes, sizes, and intermolecular interactions – such as $\pi-\pi$ and/or metal–ligand interactions – of these molecules to direct the formation of supramolecular structures having enhanced energy capture and charge transport properties was demonstrated.

Spin-selective charge recombination of photogenerated radical ion pairs within a series of donor–bridge–acceptor molecules, where D = phenothiazine, B = oligo(2,7-fluorenyl), and A = perylene-3,4:9,10-bis(dicarboximide), PTZ–FL$_n$–PDI, where n = 1–4 (compounds 1–4), was studied using time-resolved electron paramagnetic resonance spectroscopy in which the microwave source is either a continuous wave or a pulsed wave [76]. The investigation results suggested that the dominant mechanism of charge recombination for $n \geq 3$ is incoherent thermal hopping, which results in wire-like charge transfer. A series of D-B-A molecules (Figure 3.10), which uses a photogenerated charge transfer state as a high-potential photoreductant to rapidly and nearly quantitatively transfer an electron across an oligo-p-phenylene bridge to produce a long-lived radical pair (RP), was investigated [77]. Time-resolved EPR spectroscopy showed directly that charge recombination of the RP initially produces a spin-polarized triplet state that can be produced *only* by hole transfer involving the HOMOs of D, B, and A within the D-B-A system. The rate constants for charge separation (k_{CS}) and recombination (k_{CR}) in 1–4 in toluene were obtained by transient absorption spectroscopy following selective photoexcitation of the CT band in 1–4 with 414 nm, 120 fs (k_{CS}) or 7 ns (k_{CR}) laser pulses. In the plots of k_{CS} and k_{CR} versus r_{DA}, $\hat{a}_{CS}$ is 0.35 Å^{-1} and $\hat{a}_{CR}$ is 0.38 Å^{-1}, respectively. Since the CS and CR reactions within 1–4 are ET and HT reactions, respectively, the measured values of $\hat{a}_{CS}$ and $\hat{a}_{CR}$ predict that $\hat{a}$TEnT is 0.73 Å^{-1} at 293 K.

Self-assembly of robust perylenediimide chromophores was used to produce an artificial light-harvesting antenna structure that in turn induces self-assembly of a

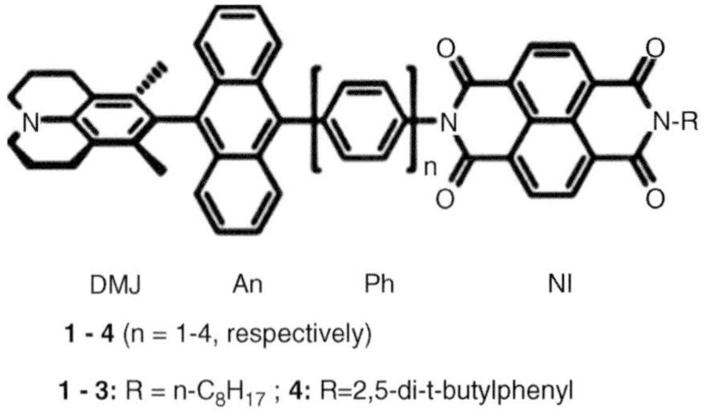

DMJ   An   Ph   NI

1 - 4 (n = 1-4, respectively)

1 - 3: R = n-C$_8$H$_{17}$; 4: R=2,5-di-t-butylphenyl

Figure 3.10 The D-B-A system consists of a 3,5-dimethyl-4-(9-anthracenyl) julolidine electron donor linked to a naphthalene-1,8:4,5-bis(dicarboximide) acceptor via a series of Ph$_n$ oligomers, where $n =$ 1–4, to give DMJ-An-Ph$_n$-NI, 1–4 [77].

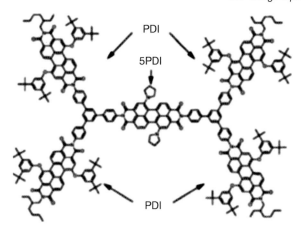

Figure 3.11 Structure of compound 1, 5PDI–PDI4 [78].

functional special pair that undergoes ultrafast, quantitative charge separation [78]. The structure (Figure 3.11) consists of four 1,7-(3′,5′-di-tert-butylphenoxy)perylene-3,4:9,10-perylene-3,4:9,10-bis(carboximide) molecules attached to a single 1,7-bis(pyrrolidin-1-yl)perylene-3,4:9,10-perylene-3,4:9,10-bis(carboximide) (5PDI) core, which self-assembles to form (5PDI-PDI4)2 in toluene. The system was characterized using both structural methods (NMR, SAXS, mass spectroscopy, and GPC) and photophysical methods (UV–vis, time-resolved fluorescence, and femtosecond transient absorption spectroscopy). It was found that energy transfer from (PDI)2 to (5PDI)2 occurred with $\tau = 21$ ps, followed by excited state symmetry breaking of 1* (5PDI)2 to produce 5PDI* + –5PDI* quantitatively with $\tau = 7$ ps. The ion pair recombines with $\tau = 420$ ps. Electron transfer occurs only in the dimeric system and does not occur in the disassembled monomer, thus mimicking both antenna and special pair function in photosynthesis.

An approach to creating noncovalent charge transfer ensembles based on two components that are linked through anion–receptor interactions was described [79]. The first component was sapphyrin, a pentapyrrolic expanded porphyrin, which is capable of carboxylate anion recognition and can act as a photodonor when irradiated with 387 nm light in the presence of an electron acceptor. The second component was the electron acceptor and consists of one of the two different C_{60} fullerene cores functionalized with multiple carboxylate anion groups arranged in a dendritic fashion. The irradiation gave rise to charge-separated states with lifetimes of 470 and 600 ps in the case of the 1:1 and 1:2 sapphyrin-fullerene ensembles, respectively. A series of double-decker lanthanide(III) bis(phthalocyaninato)-C_{60} dyads [LnIII(Pc)(Pc′)]-C_{60} (Ln = Sm, Eu, Lu; Pc = phthalocyanine) (1a–c) have been synthesized [80]. Photophysical studies on the dyads indicated that only after irradiation at 387 nm, which excites both C_{60} and [LnIII(Pc)(Pc′)] components, a photoinduced electron transfer from the [LnIII(Pc)(Pc′)] to C_{60} occurs. Four new Pt(II) terpyridyl acetylide complexes that possess a covalently linked nitrophenyl moiety were prepared and studied [81]. Specifically, the chromophore–acceptor (C–A) dyads included

[Pt(ptpy-ph-p-NO$_2$)(C–C–C$_6$H$_5$)](PF$_6$)$_3$ (1), where ptpy-ph-p-NO$_2$ = 4'-{4-(4-nitrophenyl)-phenyl}-[2,2';6',2"]terpyridine, and C–C–C$_6$H$_5$ = phenylacetylide and [Pt(ptpy-ph-m-NO$_2$)(C–C–C$_6$H$_5$)](PF$_6$)$_2$ (2), where ptpy-ph-m-NO$_2$ = 4'-(4-m-nitrophenyl-phenyl)-2,2';6',2"-terpyridine, as well as the related donor–chromophore–acceptor (D–C–A) triads [Pt(ptpy-ph-p-NO$_2$)(C–C–C$_6$H$_4$CH$_2$–PTZ)]PF$_6$ (3), where C–C–C$_6$H$_4$CH$_2$–PTZ = 4-ethynylbenzyl-N-phenothiazine, and [Pt(ptpy-ph-m-NO$_2$)(C–C–C$_6$H$_4$CH$_2$–PTZ)]PF$_6$ (4). The luminescence and transient absorption properties of the C–A dyads were found virtually identical to those of the parent chromophore, [Pt(ttpy)(C–C–C$_6$H$_5$)]PF$_6$ (5), where ttpy = 4'-p-tolyl-[2,2';6',2"] terpyridine.

Photoinduced intramolecular events of newly synthesized bis(carbazole trimer)-C$_{60}$ adducts have been studied by laser flash photolysis techniques in polar and nonpolar solvents [82]. For bis($tert$-butyl-substituted carbazole trimer)-C$_{60}$ adduct, charge separation takes place via the excited singlet state of the C$_{60}$ moiety in polar solvents as revealed by the combination of the C$_{60}$-fluorescence quenching and transient absorptions of the radical ion pair. For bis(nonsubstituted carbazole trimer)-C$_{60}$ adduct, although charge separation takes place, the charge recombination is fast because of the lower electron donor ability.

3.2.4
Photophysical and Photochemical Processes in Dual Flourophore–Nitroxide Molecules (FNO)

Dual FNO$^\bullet$ as model systems for conversion of light energy to chemical energy were proposed and developed [26, 30, 83–93]. An idea to combine a chromophore and a nitroxide in one molecule for the study of molecular dynamics of media, intramolecular fluorescence quenching (IFQ), and nitroxide fragment photoreduction was designed in the early 1980s [83].

Two types of photoseparation systems with different roles of the nitroxide fragment were implemented. In system 1 [26, 30, 83–91] (fluorophore) – (spacer) – [nitroxide] (FNO$^\bullet$), the fluorophore (F) in an excited state is an electron donor and nitroxide fragment (NO$^\bullet$) serves as an electron acceptor. In this system, photochemical and photophysical processes follow a scheme

$$FNO^\bullet + h\nu \to F^*NO^\bullet$$
$$F^*NO^\bullet \to FNO^\bullet \quad (Q[I])$$
$$F^*NO^\bullet \to F^+ NO^- \quad (CS[I])$$
$$F^+ NO^- \to FNO^\bullet \quad (CR[I])$$
$$F^+ NO^- + RH \to D-FNOH + products \quad (ChR)$$

where RH is a solvent or an ingredient; Q[I], CS[I], CR[I], and ChR designate fluorescence quenching, charge separation, charge recombination, and chemical reaction, respectively.

It was first demonstrated that nitroxide photoreduction in the dual probes occurs without a violation of the fluorophore structure [83]. This process is, in fact, the photoinduced electron transfer from molecules of the medium, which are very weak reducing agents (glycerol, ethanol, ethylene glycol, and so on), to nitroxide with the formation of hydroxylamine derivatives (**FNOH**) with a moderate reducing power. Therefore, the photochemical reactions of the dual molecules in system 1 can be considered as processes of light energy transfer.

Fluorophore–nitroxide compounds of system 2 consist of triad tethered to nitroxide [92, 93]: (fluorophore)–bridge–(acceptor)$_n$–[nitroxide] (FBA$_n$ NO$^{\cdot}$). In these systems, the nitroxide segment affects the separated charge recombination by the mechanism of spin catalysis. The following intramolecular photochemical processes take place in compounds of system 2 with two acceptors A1 and A2:

$$(D-A_1-A_2-\mathbf{FNO^{\cdot}}) + h\psi \rightarrow (D^*-A_1-A_2-\mathbf{FNO^{\cdot}})$$
$$(D^*-A_1-A_2-\mathbf{FNO^{\cdot}}) \rightarrow (D-A_1-A_2-\mathbf{FNO^{\cdot}}) \quad (Q[II])$$
$$(D^*-A_1-A_2-\mathbf{FNO^{\cdot}}) \rightarrow (D^+-A_1^--A_2-\mathbf{FNO^{\cdot}}) \quad (CS_1[II])$$
$$(D^+-A_1^--A_2-\mathbf{FNO^{\cdot}}) \rightarrow (D^+-A_1-A_2^--\mathbf{FNO^{\cdot}}) \quad (CS_2[II])$$
$$(D^+-A_1^--A_2-\mathbf{FNO^{\cdot}}) \rightarrow (D-A_1-A_2-\mathbf{FNO^{\cdot}}) \quad (CR[II]a)$$

$$(D^+-A_1-A_2^--\mathbf{FNO^{\cdot}}) \rightarrow (D-A_1-A_2-\mathbf{FNO^{\cdot}}) \quad (CR[II]b)$$

The presence of nitroxide allows one to control the lifetimes of photoinduced radical ion pairs, which is important for developing an energy conversion system, as well as molecular materials for electronics, photonics, and spintronics. Owing to the spin catalysis effect (Section 1.5), the charge recombination can be locked up and the ability of system 2 to retain the photoseparated state long enough for secondary chemical reactions of these charges to occur enhances.

In this section, we review the theoretical principles and experimental data on the photochemical and photophysical processes in the dual fluorophore–nitroxide molecules of systems 1 and 2.

3.2.4.1 System 1

As it has first been shown in our work [83], the **FNO**$^{\cdot}$ of system 1 (Figure 3.12), keeping all properties of the fluorescence and spin probes, possess a new advantage: the nitroxide segment is a strong fluorescent quencher.

(FN1)

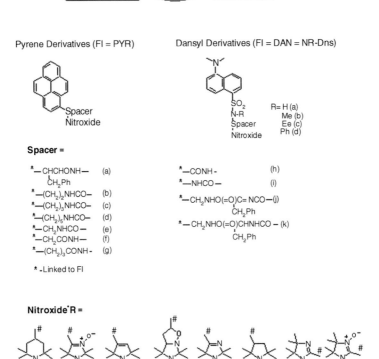

Figure 3.12 Chemical structures of dual flourophore–nitroxide molecules (private communication of Drs. V.V. Martin and A.L. Weis, Lipitek International, Inc.).

Thus, irradiation of the chromophore segment of the dual compound, dansyl-TEMPO (FN1) in a glassy liquid (glycerol 75%, water 20%, and ethanol 5%) led to the production of the hydroxylamine derivative accompanying the decay of the electron spin resonance (ESR) signal from nitroxide and a parallel eightfold increase in fluorescence [83]. Both processes occur with the same rate constant k_{red} under identical conditions.

In order to establish the mechanism of IFQ and photoreduction of the nitroxide segment in the dual molecules, a series of 17 dansyl-nitroxides of different structures were synthesized and flexibilities of the spacer group and different redox potentials of nitroxide were investigated in different media [85, 86]. Figure 3.13a and b shows the positive correlation between the rate constant of nitroxide fragment photoreduction k_{red} and the equilibrium constant K_{eq} for the chemical exchange reaction between nitroxides of different redox potential. The chemical structure of the medium,

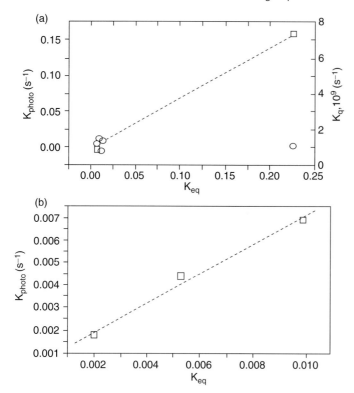

Figure 3.13 Dependence of rate constant of photoreduction, k_{red}, of fluorophore–nitroxides on redox power of nitroxide fragment, K_{eq} (squares) (a) contrary to independence of rate constant of IFQ, k_q (circles) (b) [86].

dielectric constants ($\sum_0$), and the capability of proton donating drastically affected k_{red}. Nevertheless, the rate constants of fluorescence quenching k_{qv} were found to independ on K_{eq} and the solvent nature.

On the basis of these and other available data, two mechanisms of IFQ were proposed: the major mechanism, ISC, and the minor mechanism, IET, from the excited singlet of the fluorophore (donor D*) to nitroxide (acceptor A) followed by fluorophore segment regeneration and hydroxylamine formation [26, 90]. The second mechanism is responsible for the photoreduction of nitroxide to hydroxylamine during light energy conversion.

3.2.4.2 Systems 2

In the study [92], time-resolved picosecond optical and EPR (nanosecond) spectroscopy were used to study the influence of stable free radical 2,2,6,6-tetramethylpiperidinoxyl (TEMPO, T˙) on photophysical and photochemical properties of a donor–chromophore–acceptor (D*–C–A) system, MeOAn–6ANI–Ph$_n$–A–T˙:

3 Photochemical Systems of the Light Energy Conversion

[Structures shown: MeOAn — SANI — Ph — NI — T$^\bullet$]

a: X = n-C$_8$H$_{17}$
b: X = (2,2,6,6-tetramethylpiperidinyl N-O$^\bullet$)

The distances between each component [MeOAn = p-methoxyaniline, 6ANI = 4-(N-piperidinyl)naphthalene-1,8-dicarboximide, Ph = 2,5-dimethylphenyl ($n = 0,1$), and A = naphthalene-1,8:4,5-bis(dicarboximide) or pyromellitimide (PI)] was well defined by its chemical structure. Several principal results were obtained: (1) T$^\bullet$ modulates the charge recombination rate within the triradical compared to the corresponding biradical lacking T$^\bullet$. For example, for the system 2 triad, the following values of time constants for the charge separation CS1 and CS2 and the charge recombination CR in toluene have been reported for triad without tethered nitroxide (a) and with nitroxide (b) (in brackets): CS1 = 9.8 ± 0.2 ps (7.0 ± 0.2 ps), CS2 = 430 ± 20 ps (400 ± 20 ps), and CR = 210 ± 5 ns (506 ± 10 ns). The following values of magnitudes of the resonance integral 2J were determined: 1 ± 0.5 mT (a) and <1 mT (b). The authors concluded that the nitroxide tethering does not affect markedly the charge separation and the resonance integral, but increases the charge recombination time by a factor 2.5.

The energy diagram in triad is shown in Figure 3.14.

Similar photochemical, photophysical, and spin catalysis effects were observed in triad system also having well-defined distances between the components: MeOAn-6ANI-Ph(t-butylphenylnitroxide, BPNO)-NI, where MeOAn = p-methoxyaniline,

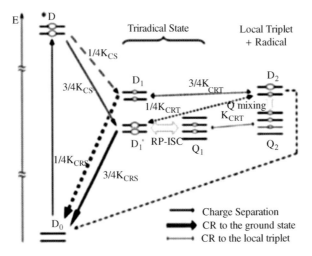

Figure 3.14 Energy level diagram showing the spin manifolds triad after charge separation and charge recombination, in a magnetic field of ~0.35 T. The size of the ellipse on each spin level represents its population qualitatively [92].

6ANI = 4-(N-piperidinyl)naphthalene-1,8-dicarboximide, Ph = phenyl, and NI = naphthalene-1,8:4,5-bis(dicarboximide) [2, 93]. MeOAn-6ANI, BPNO, and NI being attached to the 1, 3, and 5 positions of the Ph bridge, respectively, showed that BPNO influences the spin dynamics of the photogenerated triradical states 2,4(MeOAn$^+$-6ANI-Ph(BPNO)-NI$^-$. As a result, the charge recombination within the triradical is slower, compared to the corresponding biradical lacking BPNO. The above-mentioned results clearly demonstrated the possibility of controlling the charge recombination process in triads.

A nitronyl nitroxide (NN) stable radical was attached to a D-B-A system having well-defined distances between the components: MeOAn-6ANI-Ph(NN)-NI, where MeOAn = p-methoxyaniline, 6ANI = 4-(N-piperidinyl)naphthalene-1,8-dicarboximide, Ph = phenyl, and NI = naphthalene-1,8:4,5-bis(dicarboximide) (Figure 3.15) [34]. The experiments showed that NN$^{\cdot}$ influences the spin dynamics of the photogenerated triradical states 2,4(MeOAn +$^{\cdot}$-6ANI-Ph(NN$^{\cdot}$)-NI-$^{\cdot}$), resulting in slower charge recombination within the triradical compared to the corresponding biradical lacking NN. Charge recombination within the triradical results in the formation of 2,4(MeOAn-6ANI-Ph(NN$^{\cdot}$)-3*NI), in which NN$^{\cdot}$ was strongly spin-polarized. According to the authors, this effect was attributed to antiferromagnetic coupling between NN$^{\cdot}$ and the local triplet state 3*NI, which is populated following charge recombination.

A series of donor–bridge–acceptor triads have been synthesized in which the donor, 3,5-dimethyl-4-(9-anthracenyl)julolidine, and the acceptor, naphthalene-1,8:4,5-bis(dicarboximide), were linked by p-oligophenylene (Ph$_n$) bridging units (n = 1–5) [94]. It was shown that the photoexcitation of DMJ-An produces DMJ$^{+\cdot}$-An$^{-\cdot}$ quantitatively, so that An$^{-\cdot}$ acts as a high potential electron donor, which rapidly transfers an electron to NI yielding a long-lived spin-coherent radical ion pair (DMJ$^{+\cdot}$–An–Ph$_n$–NI$^{-\cdot}$). The charge transfer properties of 1–5 have been studied

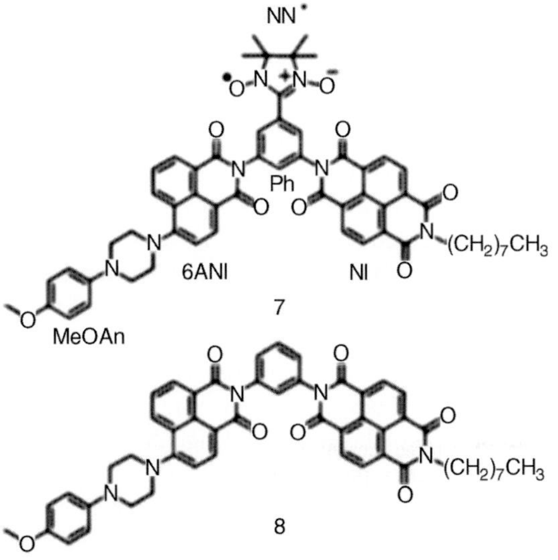

Figure 3.15 Compounds used in the work [34].

using transient absorption spectroscopy, magnetic field effect on radical pair and triplet yields, and TREPR spectroscopy. The charge separation and recombination reactions exhibit exponential distance dependencies with damping coefficients of $\beta = 0.35\,\text{Å}^{-1}$ and $0.34\,\text{Å}^{-1}$, respectively.

A simple model for analyzing the spin dynamics of a three-spin system representing a photoexcited chromophore coupled to a stable radical species was considered [9]. Perturbation theory yields a Fermi's golden rule-type rate expression that describes the formation of a local triplet on the chromophore through spin exchange with the radical. The error introduced by perturbation theory was evaluated for a number of parameters. The effect of different energetic and coupling parameters on the rate of triplet formation was explored and it was suggested how this model can be used to tune the enhanced intersystem crossing in three spin systems.

A stable 2,2,6,6-tetramethyl-1-piperidinyloxyl radical was covalently attached at its 4-position to the imide nitrogen atom of a perylene-3,4:9,10-bis(dicarboximide) to produce TEMPO-PDI 1, having a well-defined distance and orientation between TEMPO and PDI [95]. Transient optical absorption experiments in toluene following selective photoexcitation of the PDI chromophore in TEMPO-PDI showed that enhanced intersystem crossing occurs with $\tau = 45 \pm 1$ ps, resulting in the formation of TEMPO-3*PDI, while the same experiment in THF showed that the electron transfer reaction TEMPO-1*PDI $\rightarrow$ TEMPO + $^{\bullet}$-PDI-$^{\bullet}$ occurs with $\tau = 1.2 \pm 0.2$ ps and thus competes effectively with enhanced intersystem crossing. Time-resolved EPR spectroscopy of the photogenerated three-spin system TEMPO-3*PDI in toluene at 295 K indicated a broad signal assigned to spin-polarized 3*PDI, which thermalizes at longer times and is accompanied by formation of an emissively polarized TEMPO radical. The observed spin polarization of 3*PDI led the authors to propose a new spin polarization mechanism, which requires that the radical and attached triplet are in the weak exchange regime.

Future progress in the use of the **FNO$^{\bullet}$** for solving problems of light energy conversion can be associated with the developments in the following directions [26]:

- Choosing an optimum set of the donor and acceptor components of ET in system 1: donors, fluorophores with long lifetime in the excited (i.e., triplet) state and nitroxides of different redox potential, and composing a cascade systems of acceptors between the primary donor D* and the nitroxide. For example, using D* in the millisecond scale of lifetimes and increasing the distance between D and A to several angstroms to shift the competition in favor of ET compared to intersystem crossing would prolong the D^+A^- lifetime and, therefore, efficiency of the light energy conversion.
- Controlling the chemical reactivity of D^+A^- in compounds of system 2 by optimum choice of D, A, and a compound bearing electron or nuclear spin, and by playing with the capacity of molecules of the medium or ingredients to donate an electron for the reaction with D^+ and a proton for A^-.
- Incorporation of dual molecules into a nanoscale object of optimum polarity and molecular dynamics, which would be able to provide specific secondary reactions and to prevent the system from side reactions.

Building such systems appears to be the most challenging problem in the twenty-first century.

The above-mentioned new structures would allow one to prolong the lifetime of the photoseparated charged pairs and improve their specific reactivity. Construction and investigation of these structures would pave the way for the creation of efficient systems of light energy conversion that can be used for solar energy utilization.

3.3
Electron Flow through Proteins

As was shown in a large series of works by Gray and coworkers [96–101] by varying the position of the ruthenium complexes relative to metalloproteins redox-active sites, it has been possible to estimate coupling factor and its dependence on the distance between redox centers and the chemical nature and disposition of intermediate groups. Similar dependences were demonstrated in other biological systems [26, 88, 89, 102, 103].

A numerical algorithm [104, 105] was implemented to survey electron tunneling pathway in Ru-modified myoglobin and cytochrome c. The calculation results concerning the optimum pathway between ruthenated His 48 and Fe in myoglobin led to $\beta = 1.0$ Å and orientation parameter $\sigma = 0.1$ and $\Pi\varepsilon_i = 4.7 \times 10^{-6}$.

In the past two decades, several innovations in electron transfer in proteins have been reported by the Marcus group [106–110]. The artificial intelligence super-exchange methods, in which the details of the electronic structure of the protein medium were taken into account, was used for estimating the electron coupling in the matalloproteins [106, 107]. The influence of the special mutual orientation of the donor and acceptor orbitals in the Ru(bpy)$_2$ in HisX-cytochrome c on the rate electron transfer was analyzed by the transition amplitude method [108].

Recently, a new series of works on detailed mechanism of the long-distance electron transfer in proteins have been published [111–115]. Employing laser flash-quench triggering methods, Gray and Winkler [111] have shown that 20 Å coupling-limited FeII–RuIII and CuI–RuIII electron tunneling in Ru-modified cytochromes and blue copper proteins (Figure 3.16) occurs on the microsecond timescale both in solutions and in crystals. Redox equivalents can be transferred even longer distances by multistep tunneling, hopping through intervening amino acid side chains. The work has established that 20 Å hole hopping through an intervening tryptophan is two orders faster than that for direct ET.

As it is shown in Figure 3.16, the aromatic rings of the redox label and Trp122 slightly overlap, with one methyl group projecting over the indole ring and the plane of the respective p-systems making a 20.9° angle. The average separation of atoms on the overlapped six-membered rings is 3.82 Å, whereas 4.1 Å separates the edge of the Trp122 indole and the His124 imidazole. Distances between redox centers are as follows: Cu (blue) to Trp122 aromatic centroid, 11.1 Å; Trp122 aromatic centroid to Re (purple), 8.9 Å; and Cu to Re, 19.4 Å. A kinetic model of photoinduced electron transfer in ReI(CO)$_3$(dmp)-His124|(Trp122)|AzCuI is presented in Figure 3.17.

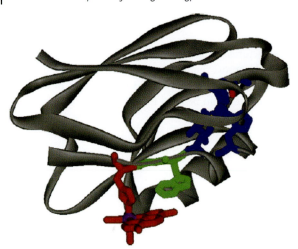

Figure 3.16 The architecture of the 1.5 Å resolution X-ray crystal structure of ReI(CO)$_3$(dmp)(His124)|(Trp122)|AzCuII [111].

Experimental data and a quantum theoretical analysis of long-range electron transfer through sensitizer wires bound in the active-site channel of cytochrome P450cam were reported. [112]. Polyfluorinated-phenyl side binding sensitizer wires to phototrigger ET from a Ru(II)-diimine to an Fe(III) (Figure 3.18) heme were

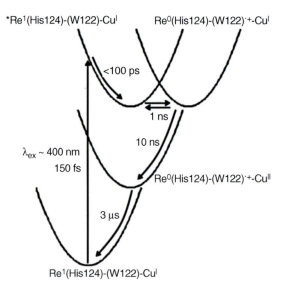

Figure 3.17 Kinetic model of photoinduced electron transfer in ReI(CO)$_3$(dmp)-His124)|(Trp122)|AzCuI. Light absorption produces electron and hole separation in the MLCT-excited ReI complex. Vibrational relaxation of *Re occurs in <100 ps, followed by migration of the hole to CuI via (Trp122)$^+$ in less than 50 ns. Charge recombination proceeds on the microsecond timescale. Elementary rate constants were extracted from fits to time-resolved luminescence, visible absorption, and infrared spectroscopic data [111].

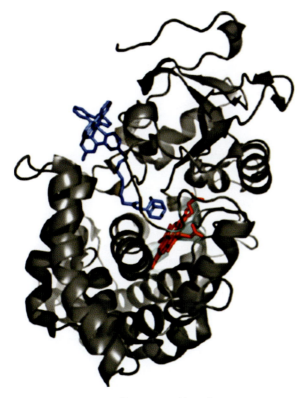

Figure 3.18 Structure of bpyRu-C9-Ad bound to P450cam [112].

employed. Each sensitizer wire consists of a substrate group with high binding affinity for the enzyme active site connected to a ruthenium-diimine through a bridging aliphatic or aromatic chain. Using combined molecular dynamics simulations and electronic coupling calculation, the authors showed that electron tunneling through perfluorinated aromatic bridges is promoted by enhanced superexchange coupling through virtual reduced states. It was found that the rate depends not only on electronic coupling of the donor and acceptor but also on the nuclear motion of the sensitizer wire.

A Ru-diimine wire, $[(4,4',5,5'\text{-tetramethylbipyridine})_2\text{Ru}(\text{F9bp})]_2^+$ (tmRu-F9bp, where F9bp is 4-methyl-4'-methylperfluorobiphenylbipyridine), was bound tightly to the oxidase domain of inducible nitric oxide synthase (iNOSoxy) and nanosecond electron tunneling was measured [113]. It was shown that the wire resides on the surface of the enzyme distant from the active-site heme. Photoreduction of an imidazole-bound active-site heme iron in the enzyme–wire conjugate ($k_{ET} = 2$ (1) $\times 10^7$ s^{-1}) was seven orders of magnitude faster than the *in vivo* process. Employing laser flash-quench triggering methods, Gray and Winkler [114] have shown that 20 Å coupling-limited Fe(II) to Ru(III) and Cu(I) to Ru(III) electron tunneling in Ru-modified cytochromes and blue copper proteins can occur on the

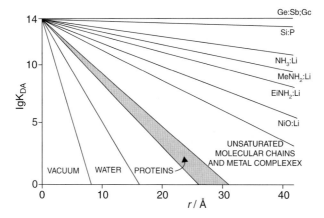

Figure 3.19 Summary plots of logarithms of the tunneling-electron rate constants in D/A systems against the distance of separation of D and/or A for a range of systems and materials. The theoretical slope of the line, $\beta = 2.303$ ln kDA/r) is zero for a .metal. (excellent conductor) while that for a vacuum is 3.5. The plot has an arbitrary value at a limiting rate of 10^{14} s^{-1} where the D/A distance is taken as zero [115].

microsecond timescale both in solution and in crystals. A unified approach to ET processes in biological and synthetic materials (such as proteins and semiconductors) was described [115]. Summary plots of logarithms of the tunneling-electron rate constants in D/A systems against the distance of separation of D and/or A for a range of systems and materials are shown in Figure 3.19.

3.3.1
Factors Affecting Light Energy Conversion in Dual Fluorophore–Nitroxide Molecules in a Protein

In order to model the effects of protein on ET in fluorophore–nitroxides, the dual probes FN1 were incorporated into the bovine serum albumin (BSA) (Figure 3.20) [84–91]. The photoreduction of the nitroxide fragment was monitored by ESR, and fluorescence quenching was measured by steady-state and picosecond time-resolved techniques. The same groups investigated and measured the factors affecting ET, namely, the molecular dynamics and micropolarity of the medium in the vicinity of the donor (by fluorescence technique) and acceptor (by ESR) moieties.

It has been shown [84] that below 240 K the photoreduction rate constant (k_{red}) for FN1 incorporated into BSA was close to zero. Above this temperature, the photoreduction rate constant k_{red} drastically increased with temperature. This increase is accompanied by a decrease in the apparent correlation time of probe nitroxide fragment rotation and the apparent correlation time of the media polar relaxation in the vicinity of the excited dansyl segment. At ambient temperature, both values reach the nanosecond scale. This conclusion was supported by the measurements of

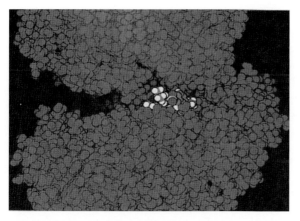

Figure 3.20 BSA modified with the dual fluorophore–nitroxide probe. Arrow indicates position of the G.I. Likhtenshtein private communication.

fluorescence polarization and ESR at ambient temperature and by direct monitoring of relaxation dynamics of the protein binding site around the dansyl moiety of the dual fluorophore–nitroxide probe FN2 using picosecond fluorescence time-resolved technique (Figure 3.21).

Hence, the nanosecond dynamics of the protein medium is one of the decisive factors affecting the photoreduction and the light energy conversion in biological and model systems. Such an intramolecular flexibility of the protein in a nanosecond range makes it possible to stabilize products of the ET reactions, oxidized donor D^+

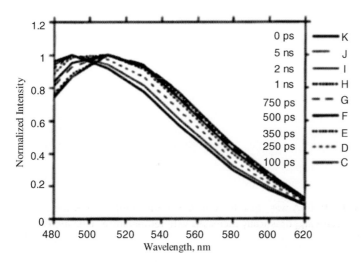

Figure 3.21 Time-resolved fluorescence spectra of dansyl-TEMPO dual probe in human serum albumin [87].

and reduced acceptor A^-, due to interactions with dipoles surrounding the protein, thus providing favorable thermodynamics for these reactions. Based on our experimental data on the local apparent dielectric constant ε_0 in the vicinity of the donor dansyl groups (fluorescence technique) and around the acceptor nitroxide segments (ESR), the following parameters of the Marcus–Levich theory were estimated for ET at $T = 300$ K [84, 90]: Gibbs energy $\Delta G_0 = -1.7$ eV, reorganization energy $\lambda = 0.9$ eV, and activation energy $\Delta G^{\#} \approx 0.25$ eV.

3.3.2
Photoinduced Interlayer Electron Transfer in Lipid Films

The photosynthetic (thylakoid) membranes support lipid–protein complexes containing pigments that participate in light absorption and charge separation and act as a barrier to inhibit the undesirable recombination and side reactions of photogenerated oxidizing and reducing species. Model vesicles are spherical, multimolecular aggregates formed by self-organization of natural or synthetic amphiphiles, in which a hydrophobic lipid bilayer separates an inner waterpool from the bulk aqueous phase. Works from several laboratories were devoted to the problems of charge separation through the lipid bilayer [116–128].

As was shown by Anderson *et al.* [116], continuous redox processes have been effected between aqueous reactants, separated by liquid membranes in which coenzyme Q, or vitamin K, is present as a carrier molecule. These membranes showed chemical specificity and function that relate closely to certain biological electron transfer processes. Light-driven transmembrane ion transport by spiropyran-crown ether supramolecular assemblies was studied [122]. According to the authors, the rates of K^+ ion leakage from phosphatidylcholine unilamellar vesicles containing a K^+-selective crown ether attached to an amphiphilic spiropyran, which undergoes reversible photoisomerization between their ring-closed (spiro) and ring-open (merocyanine) forms when exposed alternately to ultraviolet and visible light, increased markedly when the vesicle assembly was illuminated with UV light. This effect was fully reversible, with K^+ leak rates returning to their basal levels upon illumination with visible light. Transient spectrophotometry revealed that immediately following photoisomerization the merocyanine form of the dye was in a moderately polar environment, consistent with its location in the glycerophosphate backbone region of the vesicle.

Pyrene-sensitized electron transport across vesicle bilayers was investigated [123]. Endoergic electron transport across vesicle bilayers from ascorbate (Asc^-) in the inner waterpool to methylviologen (MV^{2+}) in the outer aqueous solution was driven by the irradiation of pyrene derivatives embedded in the vesicle bilayers. It was found that the initial rate of MV^{2+} reduction depends on the substituent group of the pyrenyl ring; a hydrophilic functional group linked to the pyrenyl ring by a short methylene chain acts as a sensitizer for the electron transport. Mechanistic studies using (1-pyrenyl)alkanoic acids as sensitizers suggest that the electron transport is mainly initiated by the reductive quenching of the singlet excited state of the pyrene by Asc^- and proceeds by a mechanism involving electron exchange between the

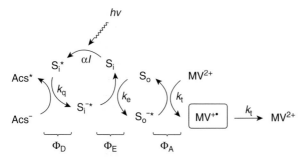

Figure 3.22 Proposed mechanism for the formation of MV^{2+} initiated by the quenching of the singlet exited state of the pyrene sensitizer by Asc^- entrapped in the inner waterpool [123].

pyrenes located at the inner and the pyrenes located at the outer interface across the vesicle bilayer. A series of unsymmetrically substituted pyrenes having both a hydrophilic group linked by a short methylene chain and a hydrophobic long alkyl group, which acted as sensitizers for the electron transport across vesicle bilayers, were synthesized. Proposed mechanism for the photochemical formation of $MV^{+\bullet}$ is shown in Figure 3.22.

Photoregulated transmembrane charge separation by linked piropyran-anthraquinone molecules was reported [124]. Amide-linked spiropyran–anthraquinone (SP-AQ) conjugates (Figure 3.23) were shown to mediate $ZnTPPS_4^-$ photosensitized transmembrane reduction of occluded $Co(bpy)_3^{3+}$ within unilamellar phosphatidylcholine vesicles by external EDTA. Experiment showed 30–35% photoconversion of the closed ring spiropyran moiety to the open-ring merocyanine (this form caused the quantum yield to decrease by sixfold in the simple conjugate and threefold for an analogue containing a lipophilic 4-dodecylphenoxy substituent on the anthraquinone moiety).

Transient spectroscopic and fluorescence quenching measurements revealed that two factors contributed to these photoisomerization-induced changes in quantum yields: increased efficiencies of fluorescence quenching of $1ZnTPPS_4^-$ by the

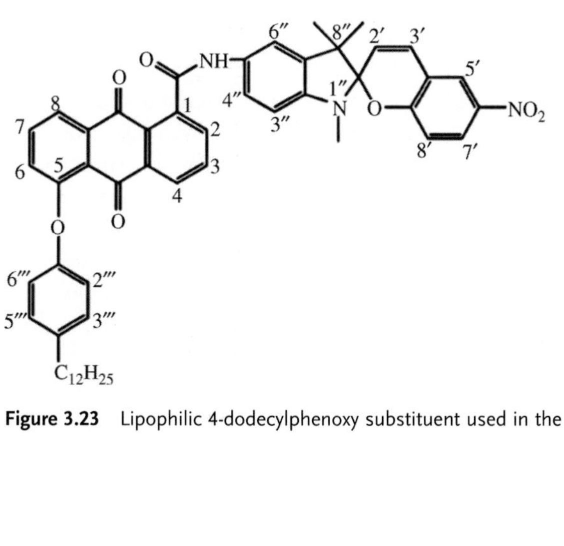

Figure 3.23 Lipophilic 4-dodecylphenoxy substituent used in the work [124].

merocyanine group and lowered transmembrane diffusion rates of the merocyanine-containing redox carriers. The minimal reaction mechanism suggested from the combined studies is oxidative quenching of vesicle-bound 3ZnTPPS$_4^-$ by the anthraquinone unit, followed by either H$^+$/e$^-$ cotransport by transmembrane diffusion of SP-AQH or, for the other redox mediators, by semiquinone anion–quinone electron exchange leading to net transmembrane electron transfer, with subsequent one-electron reduction of the internal Co(bpy)$_3^{3+}$. For the reaction ZnTPPS$_4$ + EDTA → ZnTPPS$_4^-$ + EDTA$^+$, the apparent rate constant was found to be 10^5 M^{-1} s^{-1}.

The photophysical properties of multichromophoric systems consisting of eight red or blue naphthalene diimides (NDIs) covalently attached to a p-octiphenyl scaffold, as well as a blue bichromophoric system with a biphenyl scaffold, have been investigated in detail using femtosecond time-resolved spectroscopy [125]. The blue octachromophoric systems have been shown to self-assemble as supramolecular tetramers in lipid bilayer membranes and to enable generation of a transmembrane proton gradient upon photoexcitation. The lifetime of the charge-separated state was found to increase from 22 to 45 ps by going from the bi- to the octachromophoric blue systems in methanol, while a 400 ps decay component was observed in the lipid membrane. This increase in the lifetime was explained in terms of charge migration that is most efficient when the octachromophoric systems are assembled as supramolecular tetramers in the lipid membrane. Voltammetric studies on the electron transport through a bilayer lipid membrane, BLM, containing neutral or ionic redox molecules were reported [126]. Voltammograms for the electron transport through a bilayer lipid membrane from one aqueous phase, W1, to another, W2, were recorded with the BLM containing 7,7,8,8-tetracyanoquinodimethane, TCNQ, or decamethyl ferrocene, DMFc, exposed to W1 containing [Fe(CN)$_6$]$_4^-$ and W2 containing [Fe(CN)$_6$]$^{3-}$. It was shown that the formation of an anion radical of TCNQ or a cation of DMFc in the BLM was followed by the interfacial redox reactions at both W1 | BLM and BLM | W2.

Photoinduced electron transfer from chlorophyll a through the interface of dipalmitoylphosphatidylcholine head group of the lipid bilayers was studied [127]. The photoproduced radicals were identified with ESR and radical yields of chlorophyll a were detected. The formation of vesicles was identified by changes in measured λ_{max} values from diethyl ether solutions to vesicle solution indirectly and observed directly with SEM and TEM images. Vectorial photoinduced electron transfer was studied in Langmuir–Blodgett films [128]. The monolayers of polythiophene and phthalocyanine were used as secondary electron donors in multilayers together with the monolayer of porphyrin–fullerene donor–acceptor dyad molecules mixed with octadecylamine (1: 9). Simultaneous utilization of all the three compounds in the preparation of three-layer samples allowed us to widen the range of the visible light absorption by the structure. The relative sensitivity of the sample to the excitation laser light increased by about 350 times compared to that of a dyad monolayer.

Theory of electron transfer rates across liquid–liquid interfaces was developed [129]. The theory related to the geometry of the encounter complex, the reorganization energy, and the electron transfer rate constant at a liquid–liquid interface. To treat cyclic voltammetric (CV) studies of electron transfer across the interface, the nature of the encounters was examined and a bimolecular-type rate treatment was used. When one redox pair is in large excess, it has been pointed out that a single-phase CV analysis for diffusion/reaction can be utilized. The experimental result deduced in this way for the true exchange current electron transfer rate constant at the interface was compared with that established from the present theory of the rate constant, using metal–liquid electrochemical exchange rate constants.

In conclusion, a great variety of compounds of different donor and acceptor segments and spaces were synthesized and their photochemical properties were characterized in detail. Using modern physical techniques including fast time-resolved spectroscopy and theoretical approaches, structural and molecular dynamics factors affecting electron transfer in systems under investigation were established. Nevertheless, the main strategic goal to invent and implement effective donor–acceptor supermolecules with charge-separated pairs of sufficient lifetimes and chemical reactivity and being prevented from side reactions, as it is realized in Nature, still is strongly challenged.

References

1 Gratzel, M. (2005) *Inorg. Chem.*, **44** (20), 6841–6851.
2 Gratzel, M. (2001) *Nature*, **414**, 338–344.
3 Berera, R., Herrero, C., van Stokkum, I.H.M., Vengris, M., Kodis, G., Palacios, R.E., van Amerongen, H., van Grondelle, R., Gust, D., Moore, T.A., Moore, A.L., and Kennis, J.T.M. (2007) *J. Phys. Chem. B*, **111** (24), 6868–6877.
4 Collings, A.F. and Christa Critchley, C. (2005) *Artificial Photosynthesis: From Basic Biology to Industrial Application*, Wiley-VCH Verlag GmbH, Weinheim.
5 Terazono, Y., Kodis, G., Liddell, P.A., Garg, V., Moore, T.A., Moore, A.L., and Gust, D. (2009) *J. Phys. Chem. B*, **113** (20), 7147–7155.
6 Calzaferri, G. (2010) *Top. Catal.*, **53** (3–4), 130–140.
7 Nemec, H., Nienhuys, H.-K., Zhang, F., Inganaes, O., Yartsev, A., and Sundstroem, V. (2008) *J. Phys. Chem. C*, **112** (16), 6558–6563.
8 Moule, A.J., Allard, S., Kronenberg, N.M., Tsami, A., Scherf, U., and Meerholz, K. (2008) *J. Phys. Chem. C*, **112** (33), 12583–12589.
9 Yeganeh, S., Wasielewski, M.R., and Ratner, M.A. (2009) *J. Am. Chem. Soc.*, **131** (6), 2268–2273.
10 Sauvage, F., Fischer, M.K.R., Mishra, A., Zakeeruddin, S.M., Nazeeruddin, M.K., Buerle, P., and Grätzel, M. (2009) *ChemSusChem*, **2** (8), 761–768.
11 An, B.-K., Mulherin, R., Langley, B., Burn, P., and Meredith, P. (2009) *Organic Electronics*, **10** (7), 1356–1363.
12 Mozer, A.J., Griffith, M.J., Tsekouras, G., Wagner, P., Wallace, G.G., Mori, S., Sunahara, K., Miyashita, M., Earles, J.C., Gordon, K.C., Du, L., Katoh, R., Furube, A., and Officer, D.L. (2009) *J. Am. Chem. Soc.*, **131** (43), 15621–15623.
13 Gust, D., Moore, T.A., and Moore, A.L. (2009) *Acc. Chem. Res.*, **42** (12), 1890–1898.
14 Rybtchinski, B., Sinks, L.E., and Wasielewski, M.R. (2004) *J. Am. Chem. Soc.*, **126** (39), 12268–12269.

15 Louise, E., Gusev, A.V., Ratner, M.A., and Wasielewski, M.R. (2004) *J. Am. Chem. Soc.*, **126** (17), 5577–5584.

16 Wasielewski, M.R. (2009) *Acc. Chem. Res.*, **42** (12), 1910–1921.

17 Likhtenshtein, G.I. (1993) *Biophysical Labeling Methods in Molecular Biology*, Cambridge University Press, NY.

18 McConnell, I., Li, G., and Brudvig, G.W. (2100) *Chem. Biol.*, **17** (5), 434–447.

19 Wydrzynski, T., Hillier, W., and Conlan, B. (2007) *Photosynth. Res.*, **94** (2–3), 225–233.

20 Cristian, L., Piotrowiak, P., and Farid, R.S. (2003) *J. Am. Chem. Soc.*, **125** (39), 11814–11815.

21 Giardi, M.T. and Pace, E. (2005) *Trends Biotechnol.*, **23** (5), 257–263.

22 Guldi, D.M., Illescas, B.M., Atienza, C.M., Wielopolskia, M., and Martın, N. (2009) *Chem. Soc. Rev.*, **38**, 1587–1597.

23 Seitz, W., Kahnt, A., Guldi, D.M., and Torres, T. (2009) *J. Porphyr. Phthalocyanines*, **13** (10), 1034–1039.

24 Guldi, D.M. and Prato, M. (2000) *Acc. Chem. Res.*, **33** (9), 695–703.

25 Fukuzumi, S. (2007) in *Functional Organic Materials* (eds T.J.J. Mueller and U.H.F. Bunz), Wiley-VCH Verrlag GmbH, Weinheim, 465– pp.-510.,

26 Likhtenshtein, G.I. (2008) *Pure Appl. Chem.*, **80** (10), 2125–2139.

27 Kentaro, T., Azumi, S., Tetsuro, Y., Takuzo, A., Ken-Ichi, Y., Mamoru, F., and Osamu, I. (2006) *Chem. Lett.*, **35**, 518–519.

28 Guldi, D.M., Illescas, B.M., Atienza, C.M., Wielopolskia, M., and Martın, N. (2009) *Chem. Soc. Rev.*, **38** (6), 1587–1597.

29 Brovelli, S., Meinardi, F., Winroth, G., Fenwick, O., Sforazzini, G., Frampton, M.J., Zalewski, L., Levitt, J.A., Marinello, F., Schiavuta, P., Suhling, K., Anderson, H.L., and Cacialli, F. (2010) *Adv. Funct. Mater.*, **20**, 272–280.

30 Likhtenshtein, G.I. (2003) *New Trends in Enzyme Catalysis and Mimicking Chemical Reactions*, Kluwer Academic/Plenum Publishers, NY.

31 Konno, T., El-Khouly, M.E., Nakamura, Y., Kinoshita, K., Araki, Y., Ito, O., Yoshihara, T., Tobita, S., and Nishimura, J. (2008) *J. Phys. Chem. C*, **112** (4), 1244–1249.

32 Ricks, A.B., Solomon, G.C., Colvin, M.T., Scott, A.M., Chen, K., Ratner, M.A., and Wasielewski, M.R. (2010) *J. Am. Chem. Soc.*, **132** (43), 15427–15434.

33 Berg, A., Shuali, Z., Asano-Someda, M., Levanon, H., Fuhs, M., Moebius, K., Wang, R., Brown, C., and Sessler, J.L. (2002) *J. Am. Chem. Soc.*, **121** (32), 7433–7434.

34 Chernick, E.T., Mi, Q., Kelley, R.F., Weiss, E.A., Jones, B.A., Marks, T.J., Ratner, M.A., and Wasielewski, M.R. (2006) *J. Am. Chem. Soc.*, **128** (13), 4356–4364;Chernick, E.T., Mi, Q., Vega, A.M., Lockar, J.V., Ratner, M.A., and Wasielewski, M.R. (2007) *J. Phys. Chem. B*, **111** (24), 6728–6737.

35 Verhoeven, J.W. (2006) *J. Photochem. Photobiol.*, **7** (1), 40–60.

36 Segura, J.L., Martin, N., and Guldi, D.M. (2005) *Chem. Soc. Rev.*, **34** (1), 31–41.

37 Vázquez, E. and Prato, M. (2010) *Pure Appl. Chem.*, **82** (4), 853–861.

38 Holten, D., Bocian, D.F., and Lindsey, J.S. (2002) *Acc. Chem. Res.*, **35** (1), 57–59.

39 Lukas, A.S. and Wasielewski, M.R. (2001) Approaches to a molecular switch using photoinduced electron and energy transfer, in *Molecular Switches* (ed. B.L. Feringa), Wiley-VCH Verlag GmbH, Weinheim, p. 1.

40 Ramamurthy, V. and Schanze, K.S. (1998) *Oganic and Inorganic Photochemistry*, Marcel Dekker.

41 Colvin, M.T., Giacobbe, E.M., Cohen, B., Miura, T., Scott, A.M., and Wasielewski, M.R. (2010) *J. Phys. Chem. A*, **114** (4), 1741–1748.

42 Galili, T., Regev, A., Berg, A., Levanon, H., Schuster, D.I., Moebius, K., and Savitsky, A. (2005) *J. Phys. Chem. A*, **109** (38), 8451–8458.

43 Jakob, M., Berg, A., Rubin, R., Levanon, H., Li, K., and Schuster, D.I. (2009) *J. Phys. Chem. A*, **113** (20), 5846–5854.

44 Takai, A., Gros, C.P., Barbe, J.-M., Guilard, R., and Fukuzumi, S. (2009) *Chem. Eur. J.*, **15** (12), 3110–3122.

45 Ohkubo, K. and Fukuzumi, S. (2009) *Bull. Chem. Soc. Jpn.*, **82** (3), 303–331.

46 Rio, Y., Seitz, W., Gouloumis, A., Vazquez, P., Sessler, J.L., Guldi, D.M., and Torres, T. (2010) *Chem. Eur. J.*, **16** (6), 1929–1940.

47 Heinen, U., Berthold, T., Kothe, G., Stavitski, E., Galili, T., Levanon, H., Wiederrecht, G., and Wasielewski, M.R. (2002) *J. Phys. Chem. A*, **106** (10), 1933–1937.

48 Berg, A., Shuali, Z., Asano-Someda, M., and Levanon, H. (1999) *J. Am. Chem. Soc.*, **12** (32), 7433–7434.

49 Wilson, T.M., Hori, T., Yoon, M.-C., Aratani, N., Osuka, A., Kim, D., and Wasielewski, M.R. (2010) *J. Am. Chem. Soc.*, **132** (4), 1383–1388.

50 Ohkubo, K. and Fukuzumi, S. (2009) *Bull. Chem. Soc. Jpn.*, **82** (3), 303–310.

51 D'Souza, F., Maligaspe, E., Ohkubo, K., Zandler, M.E., Subbaiyan, N.K., and Fukuzumi, S. (2009) *J. Am. Chem. Soc.*, **131** (25), 8787–8797.

52 She, C., Lee, S.J., McGarrah, J.E., Vura-Weis, J., Wasielewski, M.R., Chen, H., Schatz, G.C., Ratnerab, M.A., and Hupp, J.T. (2010) *Chem. Commun.*, **46** (4), 547–549.

53 Wallin, S., Monnereau, C., Blart, E., Gankou, J.-R., Odobel, F., and Hammarstrom, L. (2010) *J. Phys. Chem. A*, **114** (4), 1709–1721.

54 Terazono, Y., Kodis, G., Liddell, P.A., Garg, V., Moore, T.A., Moore, A.L., and Devens Gust, D. (2009) *J. Phys. Chem. B*, **113** (20), 7147–7155.

55 Di Valentin, M., Bisol, A., Agostini, G., Moore, A.L., Moore, T.A., Gust, D., Palacios, R.E., Gould, S.L., and Carbonera, D. (2006) *Mol. Phys.*, **104** (10–11), 1595–1607.

56 Fukuzumi, S., Honda, T., Ohkubo, K., and Kojima, T. (2009) *Dalton Trans.* (20), 3880–3889.

57 Rodriguez-Morgade, M.S., Plonska-Brzezinska, M.E., Athans, A.J., Carbonell, E., de Miguel, G., Guldi, DM., Echegoyen, L., and Torres, T. (2009) *J. Am. Chem. Soc.*, **131** (30), 10484–10496.

58 Nitzan, A. and Ratner, M.A. (2003) *Science*, **300** (5624), 1384–1389.

59 Carroll, R.L. and Gorman, C.B. (2002) *Angew. Chem. Int. Ed.*, **41** (23), 4378–4400.

60 Joachim, L.C., Gimzewski, J.K., and Aviram, A. (2000) *Nature*, **408**, 541–548.

61 D'Souza, F. and Ito, O. (2009) *Chem. Commun. (Camb.)* (33), 4913–4928.

62 Lemmetyinen, H., Tkachenko, N., Efimov, A., and Niemi, M. (2009) *J. Porphyr. Phthalocyanines*, **13** (10), 1090–1097.

63 Seitz, W., Kahnt, A., Guldi, D.M., and Torres, T. (2009) *J. Porphyr. Phthalocyanines*, **13** (10), 1034–1039.

64 Martin-Gomis, L., Ohkubo, K., Fernandez-Lazaro, F., Fukuzumi, S., and Sastr e-Santos, A. (2008) *J. Phys. Chem. C*, **112** (45), 17694–17701.

65 Millett, F. and Durham, B. (2002) *Biochemistry*, **41** (38), 11315–11324.

66 Carmieli, R., Mi, O., Ricks, A.B., Giacobbe, E.M., Mickley, S.M., and Wasielewski, M.R. (2009) *J. Am. Chem. Soc.*, **131** (24), 8372–8373.

67 Wijesinghe, C.A., El-Khouly, M.E., Blakemore, J.D., Zandler, M.E., Fukuzumi, S., and D'Souza, F. (2010) *Chem. Commun.*, **46** (19), 3301–3303.

68 Bullock, J.E., Carmieli, R., Mickley, S.M., Vura-Weis, J., and Wasielewski, M.R. (2009) *J. Am. Chem. Soc.*, **131** (33), 11919–11929.

69 Scott, A.M., Miura, T., Ricks, A.B., Dance, Z.E.X., Giacobbe, E.M., Colvin, M.T., and Wasielewski, M.R. (2009) *J. Am. Chem. Soc.*, **131** (48), 17655–17666.

70 Pinzon, J.R., Gasca, D.C., Sankaranarayanan, S.G., Bottari, G., Torres, T., Guldi, D.M., and Echegoyen, L. (2009) *J. Am. Chem. Soc.*, **131** (22), 7727–7734.

71 Ito, O., Sandanayaka, A.S.D., Chitta, R., and D'Souza, F. (2010) *J. Indian Chem. Soc.*, **87** (1), 13–21.

72 Murakami, M., Ohkubo, K., Hasobe, T., Sgobba, V., Guldi, D.M., Wessendorf, F., Hirsch, A., and Fukuzumi, S. (2010) *J. Mater. Chem.*, **20** (8), 1457–1466.

73 Ohtani, M. and Fukuzumi, S. (2009) *Chem. Commun.* (33), 4997–4999.

74 Jakob, M., Alexander Berg, A., Stavitski, E., Chernick, E.T., Weiss, E.A., Wasielewski, M.R., and Levanon, H. (2006) *Chem. Phys.*, **324** (1), 63–71.

75 Goldsmith, R.H., DeLeon, O., Wilson, T.A., Finkelstein-Shapiro, D., Ratner, M.A., and Wasielewski, M.R. (2008) *J. Phys. Chem. A*, **112** (19), 4410–4414.

76 Miura, T., Carmieli, R., and Wasielewski, M.R. (2010) *J. Phys. Chem. A*, **114** (18), 5769–5778.

77 Goldsmith, R.H., Sinks, L.E., Kelley, R.F., Betzen, L.J., Liu, W.H., Weiss, E.A., Ratner, M.A., and Wasielewski, M.R. (2005) *Proc. Natl. Acad. Sci. USA*, **102**, 3540–3545.

78 Rybtchinski, B., Sinks, L.E., and Wasielewski, M.R. (2004) *J. Am. Chem. Soc.*, **126** (39), 12268–12269.

79 Grimm, B., Karnas, E., Brettreich, M., Ohta, K., Hirsch, A., Guldi, D.M., Torres, T., and Sessler, J.L. (2010) *J. Phys. Chem.*, **114** (45), 14134–14139.

80 Ballesteros, B., de la Torre, G., Shearer, A., Hausmann, A., Herranz, M.A., Guldi, D.M., and Torres, T. (2010) *Chem. Eur. J.*, **16** (1), 114–125.

81 Jarosz, P., Lotito, K., Schneider, J., Kumaresan, D., Schmehl, R., and Eisenberg, R. (2009) *Inorg. Chem.*, **48** (6), 2420–2428.

82 Takashi Konno, T., El-Khouly, M.E., Nakamura, Y., Kinoshita, K., Araki, Y., Ito, O., Yoshihara, T., Tobita, S., and Nishimura, J. (2008) *J. Phys. Chem. C*, **112** (4), 1244–1249.

83 Bystryak, I.M., Likhtenshtein, G.I., Kotelnikov, A.I., Hankovsky, O.H., and Hideg, K. (1986) *Russ. J. Phys. Chem.*, **60**, 1679–1983.

84 Vogel, V.R., Rubtsova, E.T., Likhtenshtein, G.I., and Hideg, K. (1994) *J. Photochem. Photobiol. A Chem.*, **83**, 229–236.

85 Lozinsky, E., Shames, A., and Likhtenshtein, G.I. (2000) Dual fluorophore-nitroxides: models for investigation of intramolecular quenching and novel redox probes, in *Recent Research Development in Photochemistry and Photobiology*, vol. 2 (ed S.G. Pandalai), Transworld Research Network, Trivandrum, India.

86 Likhtenshtein, G.I., Nakatsuji, S., and Ishii, K. (2007) *Photochem. Photobiol.*, **83**, 871–881.

87 Likhtenshtein, G.I., Febrario, F.R., and Nucci, F. (2000) *Spectrochim. Acta Part A. Biomol. Spectrosc.*, **56**, 2011–2031.

88 Likhtenshtein, G.I., Yamauchi, J., Nakatsuji, S., Smirnov, A., and Tamura, R. (2008) *Nitroxides: Application in Chemistry, Biomedicine, and Materials Science*, Wiley-VCH Verlag GmbH, Weinheim.

89 Likhtenshtein, G.I. (2005) Dual fluorophore-ninroxides as spin-traps and redox probes. *Absracts*: A Joint Conference of the 11th *In Vivo* ESR Spectroscopy and Imaging and 8th International ESR Spin Trapping, Columbus, Ohio, September 4–8.

90 Likhtenshtein, G.I., Pines, D., Pines, E., and Khutorsky, V. (2009) *Appl. Magn. Reson.*, **35** (3), 459–472.

91 Likhtenshtein, G.I. (2009) *Appl. Biochem. Biotechnol.*, **152** (1), 135–155.

92 Qixi, M., Chernick, E.T., McCamant, D.W., Weiss, E.A., Ratner, M.A., and Wasielewski, M.R. (2006) *J. Phys. Chem. A*, **110**, 7323–7333.

93 Chernick, E.T., Mi, Q., Vega, A.M., Lockar, J.V., Ratner, M.A., and Wasielewski, M.R. (2007) *J. Phys. Chem. B*, **111** (24), 6728–6737.

94 Scott, A.M., Miura, T., Ricks, A.B., Dance, Z.E.X., Giacobbe, E.M., Colvin, M.T., and Wasielewski, M.R. (2009) *J. Am. Chem. Soc.*, **131** (48), 17655–17666.

95 Colvin, M.T., Giacobbe, E.M., Cohen, B., Miura, T., Scott, A.M., and Wasielewski, M.R. (2010) *J. Phys. Chem. A*, **114** (4), 1741–1748.

96 Winkler, J.R., Nocera, D.G., Yocom, K.M., Bordignon, E., and Gray, H.B. (1982) *J. Am. Chem. Soc.*, **104**, 5798–5800.

97 Gray, H.B. and Winkler, J.R. (2010) *Biochim. Biophys. Acta Bioenergetics*, **1797** (9), 1563–1572.

98 Gray, H.B. and Ellis, W.R. (1994) Electron transfer, in *Bioorganic Chemistry* (eds I. Bertini, H.B. Gray, S.J. Lippard,

and J.S. Valentine), University Science Books, Mill Valley, pp. 316–362.
99 Lancaster, K.M., Sproules, S., Palmer, J.H., Richards, J.H., and Gray, H.B. (2010) *J. Am. Chem. Soc.*, **132** (41), 14590–14595.
100 Tezcan, F.A., Crane, B.R., Winkler, J.R., and Gray, H.B. (2001) *Proc. Natl. Acad. Sci. U.S.A.*, **98**, 5002–5006.
101 Winkler, J.R., Dunn, A.R., Hess, C.R., and Gray, H.B. (2008) *Bioinorganic Electrochemistry* (ed. O. Hammerich and J. Ulstrup), Springer, pp. 1–23.
102 Isied, S.S., Worosila, G., and Atherton, S.J. (1982) *J. Am. Chem. Soc.*, **104**, 7659–7661.
103 Moser, C.C. and Dutton, P.L. (1992) *Biochim. Biophys. Acta*, **1101**, 171–176.
104 Beratan, D.N. and Onuchic, J.N. (1987) *J. Chem. Phys.*, **86**, 4489–4498.
105 Beratan, D.N., Onuchic, J.N., Betts, J.N., Bowler, B.E., and Gray, G.H. (1990) *J. Am. Chem. Soc.*, **112**, 7915–7921.
106 Siddarth, P. and Marcus, R.A. (1993) *J. Phys. Chem.*, **97**, 2400–2405.
107 Siddarth, P. and Marcus, R.A. (1993) *J. Phys. Chem.*, **97**, 6111–6114.
108 Stuchebrukhov, A.A. and Marcus, R.A. (1995) *J. Phys. Chem.*, **99** (19), 7581–7590.
109 Marcus, R.A. (2010) *Faraday Discuss.*, **145** (1), 9–14.
110 Marcus, R.A. (1997) *J. Electroanal. Chem.*, **438** (1–2), 251–260.
111 Gray, H.B. and Winkler, J.R. (2009) *Chem. Phys. Lett.*, **483** (1–3), 1–9.
112 Hartings, M.R., Kurnikov, I.V., Dunn, A.R., Winkler, J.R., Gray, H.B., and Ratner, M.A. (2010) *Coordin. Chem. Rev.*, **254** (3–4), 248–253.
113 Whited, C.A., Belliston-Bittner, W., Dunn, A.R., Winkler, J.R., and Gray, H.B. (2009) *J. Inorg. Biochem.*, **103** (6), 906–911.
114 Gray, H.B. and Winkler, J.R. (2010) *Biochim. Biophys. Acta Bioenergetics*, **1797** (9), 1563–1572.
115 Edwards, P.P., Gray, H.B., Lodge, M.T.J., and Williams, R.J.P. (2008) *Angewandte Chemie, International Edition*, **47** (36), 6758–6765.
116 Anderson, S.S., Lyle, I.G., and Paterson, R. (1976) *Nature*, **259**, 147–148.
117 Grimaldi, J.J., Boileau, S., and Lehn, J.M. (1977) *Nature*, **265**, 229–230.
118 Calvin, M. (1978) *Acc. Chem. Res.*, **11**, 369–374.
119 Meyer, T. (1989) *Acc. Chem. Res.*, **22**, 163–170.
120 Gust, D., Moore, T.A., and Moore, A.L. (2001) *Acc. Chem. Res.*, **34** (1), 40–48.
121 Gust, D., Moore, T.A., and Moore, A.L. (2005) in *Aritificial Photosynthesis* (eds A.F. Collings and C. Critchley), Wiley-VCH Verlag GmbH, Weinheim, pp. 187–183.
122 Khairutdinov, R.F. and Hurst, J.K. (2004) *Langmuir*, **20** (5), 1781–1785.
123 Mizushima, T., Yoshida, A., Harada, A., Yoneda, Y., Minatani, T., and Murata, S. (2006) *Org. Biomol. Chem.*, **4**, 4336–4344.
124 Zhu, L., Khairutdinov, R.F., Cape, J.L., and Hurst, J.K. (2006) *J. Am. Chem. Soc.*, **128** (3), 825–835.
125 Banerji, N., Fürstenberg, A., Bhosale, S., Sisson, A.L., Sakai, N., Matile, S., and Vauthey, E. (2008) *J. Phys. Chem. B*, **112** (30), 8912–8922.
126 Shiba, H., Maeda, K., Ichieda, N., Kasuno, M., Yoshida, Y., Shirai, O., and Kihara, S. (2003) *J. Electroanal. Chem.*, **556** (1), 1–11.
127 Lee, D.K., Seo, K.W., and Kang, Y.S. (2002) *Proc. Indian Acad. Sci. Chem. Sci.*, **114** (6), 533–538.
128 Alekseev, A.S., Efimov, A.V., Tkachenko, N.V., Saltykov, P.A., and Lemmetuinen, H. (2008) *Kratkie Soobshcheniya po Fizike*, **35** (4), 118–121.
129 Marcus, R.A. (1990) *J. Phys. Chem.*, **94** (10), 4152–4155.

4
Redox Processes on Surface of Semiconductors and Metals

4.1
Redox Processes on Semiconductors

4.1.1
Introduction

An important aspect of the efficient conversion of light energy into chemical energy is the ability of molecular catalysts to generate long-lived charge-separated states under specific conditions, which is aided by an electron injection into a semiconductor (Figure 4.1) [1–17].

The electron transfer of dye at the semiconductor surface includes several principal steps: (1) the dye molecule absorbs the energy of photon, (2) its excitation to excited singlet or triplet state, (3) injection of an electron from the excited singlet or triplet state to the conduction band of the semiconductor, (4) the oxidized dye undergoes various reactions to give stable products, (5) excitation of the semiconductor, and (6) an electron promotion of the conduction band from the valence band (VB), leaving a hole in the valence band.

4.1.2
Interfacial Electron Transfer Dynamics in Sensitized TiO_2

Spatial localization of excited state electrons in transition metal complexes used as photocatalysts or dye sensitizers in solar cells is important for an efficient electron injection into the metal oxide nanoparticles. Jakubikova *et al.* [18] used density functional theory to investigate the excited states in a prototype catalyst–chromophore assembly [(bpy)(H_2O)Ru(tpy-tpy)Ru(tpy)]$^{4+}$ ([Ru(tpy)(bpy)(H_2O)]$^{2+}$ = catalyst, [Ru(tpy)$_2$]$^{2+}$ = chromophore, tpy = 2,2′: 6′,2″-terpyridine, and bpy = 2,2′-bipyridine) and a series of related compounds. It was shown that upon excitation with visible light, [Ru(tpy)(tpy(PO_3H_2))]$^{2+}$ induces interfacial electron transfer (IET) at a rate of 1–10 ps, which is competitive with the excited state decay into the ground state. The probability of electron injection from [Ru(tpy)(bpy)(H_2O)-Ru(tpy)(tpy(PO_3H_2))]$^{4+}$ was found to be low, as the excitation with visible light localizes the

Solar Energy Conversion. Chemical Aspects, First Edition. Gertz Likhtenshtein.
© 2012 Wiley-VCH Verlag GmbH & Co. KGaA. Published 2012 by Wiley-VCH Verlag GmbH & Co. KGaA.

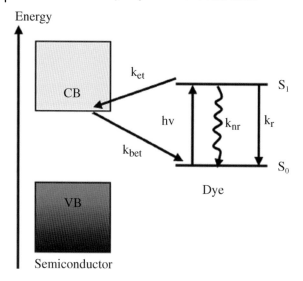

Figure 4.1 Schematic representation of photoinduced processes of dye molecules on semiconductor nanoparticles. k_r, radiative decay rate; k_{nr}, intrinsic nonradiative decay rate; k_{et}, IET rate from excited molecule to semiconductor; k_{bet}, back IET rate [3]. Reproduced with permission from Wiley.

excited electron in the tpy–tpy bridge, which does not have favorable coupling with the nanoparticle.

It was shown that chemical vapor deposition of TiO_2 combined with UV light irradiation at nanocrystal TiO_2 particle layers coated on plastic film electrodes (ITO-PET) drastically enhanced dye-sensitized photocurrent and improved photovoltage up to 750 mV, achieving an energy conversion efficiency of 3.8% [19]. The dendrimers comprised of a substituted [cis-di(thiocyanato)-bis(2,2′-bipyridyl)ruthenium(II) complex, first-generation biphenyl-based dendrons, and four, eight, or twelve 2-ethylhexyloxy surface groups were bound to the titanium dioxide of the DSSCs via carboxylate groups on one of the bipyridyl moieties [20]. The substitution improved the device performance. It was concluded that molecular volume and molar extinction coefficient are both first-order parameters in achieving high conversion efficiencies and must be taken into account when designing new dyes for efficient electron transfer and sensitizer regeneration in stable π-extended tetrathiafulvalene attached to a mesoporous TiO_2 film [21]. Using time-resolved spectroscopy, the authors observed a highly efficient regeneration of the photooxidized tetrathiafulvalene (TTF) sensitizers, even though the measured driving force for regeneration was only approximately 150 mV. Reddish purple betanin having a maximum molar absorptivity of about 65,000 M^{-1} cm^{-1} at 535 nm was immobilized on a photoanode fabricated from nanocrystalline TiO_2 on transparent conductive glass [22]. The betanin-sensitized film when employed in a dye-sensitized solar cell gave a maximum photocurrent of 2.42 mA/cm² and open-circuit photovoltage of 0.44 V in the presence of methoxypropionitrile containing I^-/I_3^- redox mediator. Interfacial processes in the dye-sensitized solar cell were

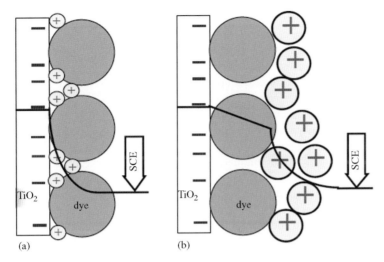

Figure 4.2 TiO$_2$/dye/solution interface shown under negative applied or photogenerated bias. The solid line shows the electrical potential drop between the TiO$_2$ and the solution. (a) In the presence of small cations, the dye oxidation potential is only slightly affected by changes in the TiO$_2$ potential. (b) In the presence of large cations, the dye potential follows, but always lags somewhat, changes in the TiO$_2$ potential [23]. Reproduced with permission from Elsevier.

reviewd [23]. The author stressed that interfacial mode of carrier generation is fundamentally different from the bulk generation occurring in conventional cells and is responsible for many of the unusual features of dye cells (and most other organic solar cells). For example, Figure 4.2 illustrates the effect of cation's size on the electrical potential drop between the TiO$_2$ and the solution.

An investigation was conducted to determine the effect of surface modification on the back electron transfer (BET) reaction of alizarin-sensitized TiO$_2$ nanoparticles using the femtosecond transient absorption technique in the visible and near-IR region [24]. It is shown that on surface modification, the rate of the BET reaction can be reduced drastically in the dye-sensitized TiO$_2$ nanoparticle system. Measurements performed in the work [25] revealed an injection time of electron injection from alizarin to colloid semiconductor TiO$_2 \tau_{inj} < 100$ fs. It was found highly multiphasic recombination dynamics with time constants from 400 fs to the nanosecond timescale. The contribution of surface trap states and their influence on the observed dynamics was investigated with alizarin adsorbed on the insulating substrate ZrO$_2$. Since the conduction band edge lies far above ($\approx$1 eV), the S1 state of alizarin, the electron injection into this band was completely suppressed. For the alizarin/ZrO2 system, the timescale for the injection into these traps was detected to be faster than 100 fs. The relaxation processes in the traps and the repopulation of the S1 state occurred within 450 fs, and the subsequent ground-state relaxation took 160 ps.

Photoinduced electron transfer in a size-quantized CdS/TiO$_2$ composite system was studied using emission and picosecond laser flash photolysis [26]. Quantum-sized CdS and TiO$_2$ particles were synthesized in reverse micelles using di-octyl sulfosuccinate. Electron transfer from photoexcited CdS to TiO_2 depended on the particle size of TiO_2 where charge transfer was observed only when TiO_2 particles were sufficiently large (>12 Å). Interactions with smaller size TiO_2 particles (≤10 Å.) with CdS instead led to enhancements in emission with an increase in quantum yield from 2.3 to 8.8%. Picosecond laser flash photolysis experiments were carried out to elucidate the interparticle electron transfer processes in the CdS/TiO_2 reverse micellar system. Electron transfer dynamics from Ru bipyridyl complexes to ZnO, Nb_2O_5, and SnO_2 nanoporous thin films were investigated using ultrafast IR spectroscopy [27]. All injection kinetics were shown to be biphasic, consisting of ultrafast (< 100 fs) and slower multiexponential increases. The comparison of amplitudes indicates a much faster injection rate to TiO_2 and Nb_2O_5 than to ZnO and SnO_2. The authors suggested that the faster electron injection to TiO_2 and Nb_2O_5 (d-conduction band) than to SnO_2 and ZnO (sp band) can be attributed to the much larger density of states in the former. Efficient electron transfer and sensitizer regeneration in stable π-extended TTF attached to a mesoporous TiO_2 film were demonstrated [28]. The measured driving force for the sensitizer regeneration was found to be only < 150 mV. The oxidation of the nonfunctionalized TTF chromophore generated the short-lived radical cation (< 500 μs) that showed a fingerprint at < 650 nm and rapidly disproportionates into the stable dication.

A series of first-generation dendrimers comprised of a substituted [cis-di(thiocyanato)-bis(2,20-bipyridyl)ruthenium(II) complex, first-generation biphenyl-based dendrons, and four, eight, or twelve 2-ethylhexyloxy surface groups were prepared [29]. The dendrimers were bound to the titanium dioxide via carboxylate groups on one of the bipyridyl moieties in a similar manner to [cis-di(thiocyanato)- bis (4,40-dicarboxylate-2,20-bipyridyl)]ruthenium(II) 1 (N3). Exchanging one pair of the carboxylate groups on one bipyridyl ligand of N3 with styryl units to give [cis-di (thiocyanato)-(4,40-dicarboxylate-2,20-bipyridyl)-(4,40-distyryl-2,20-bipyridyl]ruthenium(II) (2) resulted in an improvement in device performance (7.19% ± 0.11% for 2 versus 6.94% ± 0.12% for N3). The decrease in efficiency with increasing molecular volume (Figures 4.3 and 4.4) was explained to be due to less dye being adsorbed.

4.1.3
Electron Transfer in Miscellaneous Semiconductors

4.1.3.1 Single-Molecule Interfacial Electron Transfer in Donor–Bridge–Nanoparticle Acceptor Complexes

Photoinduced interfacial electron transfer in sulforhodamine B (SRhB), aminosilane-Tin oxide (SnO_2), and ZrO_2 nanoparticle donor–bridge–acceptor complexes (Figure 4.5) has been studied on a single-molecule and ensemble average level [30]. Shorter fluorescence lifetime on SnO_2 than on ZrO_2 was observed and attributed to IET from SRhB to SnO_2. Single-molecule lifetimes fluctuate with time and vary among different molecules, suggesting both static and dynamic IET heterogeneity in this system.

In the study [31], the electron transport through a dyad moiety from adjacent electrodes using the scanning tunneling microscope (STM) break junction method

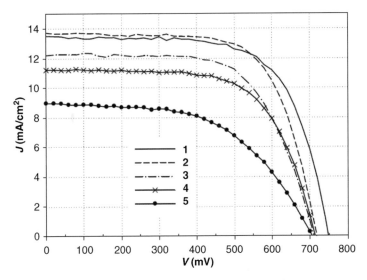

Figure 4.3 J–V curves for devices constructed from compounds 1–5 (Figure 4.4) [29]. Reproduced with permission from Elsevier.

has been characterized. The study involved the immobilization of the dyads on the surface of indium tin oxide (ITO) to form a monolayer and measure and characterize electron transport through dyad molecules in the dark and under laser illumination. It was found that the dyads form a long-lived charge separation state due to the injection of electrons into the ITO upon laser illumination. Ultrafast vibrational spectroscopy was used to examine the dynamics of interfacial electron transfer, free carrier formation, and bimolecular charge recombination and trapping in an organic photovoltaic material composed of the functionalized fullerene, PCBM, and a conjugated polymer, CN-MEH-PPV [32]. It was found that ultrafast interfacial electron transfer from CN-MEH-PPV to PCBM occurred on timescales ranging from < 100 fs to 1 ps. The frequency variation resulted in part from a vibrational Stark shift arising from an interfacial dipole formed by spontaneous charge transfer from the polymer to PCBM. The Stark shift provided a means to observe directly the formation of free carriers on the 1–10 ps timescale. It was suggested that following free carrier formation, electrons diffuse within the material and become trapped on the microsecond timescale giving a distinct peak in the vibrational spectra.

Ai et al. [33] presented a study of ultrafast photoinduced interfacial electron transfer dynamics of SnO_2 nanocrystal thin films sensitized by polythiophene derivatives (regioregular poly(3-hexylthiophene), P3HT, and regiorandom poly(3-undecyl-2,2′-bithiophene, P3UBT). ET dynamics were measured by following the dynamics of injected electrons in SnO_2 and polarons in the conjugated polymer using ultrafast mid-IR transient absorption spectroscopy. The rate of electron transfer from P3HT and P3UBT to SnO_2 films was detected to occur on subpicosecond timescale (120 ± 20 fs). Time-resolved optical second harmonic generation was used to observe hot electron transfer from colloidal lead selenide (PbSe) nanocrystals to a titanium

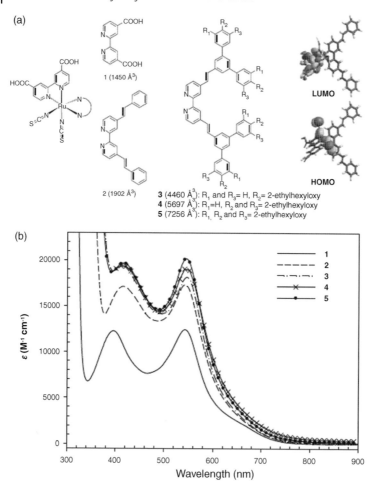

Figure 4.4 (a) The chemical structures of the materials and graphical representation of the frontier orbitals of 2 calculated by the ZINDO-1 parameter method (HyperChem 7.0). (b) The UV/visible absorption spectra of the materials (5×10^{-5} mol/l) in dimethylformamide [29]. Reproduced with permission from Elsevier.

dioxide (TiO$_2$) electron acceptor [34]. It was shown that with appropriate chemical treatment of the nanocrystal surface, this transfer occurred fast. The electric field resulting from sub-50 fs charge separation across the PbSe–TiO$_2$ interface excited coherent vibrations of the TiO$_2$ surface atoms, whose motions could be followed in real time.

The photoinduced charge injection at the interface between a fluorinated copper phthalocyanine (CuPcF$_{16}$) film deposited on a GaAs(100) wafer by means of pump–probe spectroscopy combined with ultraviolet photoemission spectroscopy (UPS) and electromodulated transmission spectroscopy was characterized [35]. The UPS characterization of the hybrid interface demonstrates that the lowest unoccupied

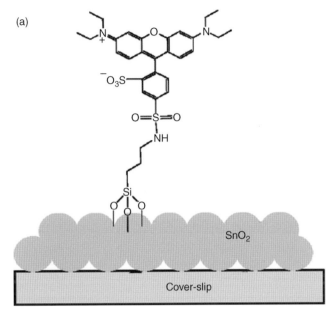

Figure 4.5 A schematic structure of SRhB–silane–SnO$_2$ [30]. Reproduced with permission from the American Chemical Society.

molecular level (LUMO) of CuPcF$_{16}$ was shown to be almost aligned with the GaAs conduction band. Upon photoexcitation of the hybrid interface with 150 fs pulses, an efficient photoinduced electron transfer from CuPcF$_{16}$ to GaAs was observed. The evolution of interfacial CuPcF$_{16}$ charge appeared to be strongly influenced by energy-level alignment at the GaAs/CuPcF$_{16}$ heterojunction. Photoinduced electron injection dynamics from Ru(dcbpy)$_2$(X)$_2$ (dcbpy = 4,4′-dicarboxy-2,2′-bipyridine; X = SCN$^-$, CN$^-$, and dcbpy, referred to as RuN3, Ru$_5$0$_5$, and Ru470, respectively) to In$_2$O$_3$ nanocrystal thin films were studied using ultrafast transient IR absorption spectroscopy [36]. After 532 nm excitation of the adsorbates, the dynamics of electron injection from their excited states to In$_2$O$_3$ were studied by monitoring the IR absorption of the injected electrons in the semiconductor. The injection kinetics were nonsingle exponentials. For samples, in pH 2 buffer, the corresponding half-time for injection from these complexes became 6 ± 1, 105 ± 20, and 18 ± 5 ps. The injection kinetics from RuN3 to In$_2$O$_3$ was found to be similar to that of SnO$_2$.

4.1.4
Redox Processes on Carbon Materials

Nanocarbon materials (fullerene, carbon nanotube, carbon nanoparticls, and so on) attract much attention as prospective technological materials for the construction of a novel electron donor–acceptor composite system for the development of light energy conversion systems. There have been many examples of covalent and noncovalent functionalization of nanocarbons with light harvester and/or electron donor molecules such as porphyrin and phthalocyanine derivatives [37–49]. The considerable interest in carbon nanotubes (CNTs) is closely related to their properties including electrical and thermal conductivities, chemical stability, and mechanical strength. As a result, this relatively novel carbon allotrope has been proposed for numerous applications such as energy conversion and electron transfer, in particular. Phthalocyanines, fullerenes, perylenes, or related macrocyclic structures serving as donor–acceptor systems together with (CNTs) were investigated in recent years.

A system of water-soluble dendritic electron donors and electron acceptors, phthalocyanines and perylenediimides (Figure 4.6) immobilized onto single-wall carbon nanotubes (SWNTs), was investigated [44]. The complementary use of spectroscopy and microscopy shed light on mutual interactions between semiconducting SWNTs and either a strong dendritic electron acceptor perylenediimide or a strong dendritic electron donor phthalocyanine. The stability of the perylenediimide/SWNT electron donor–acceptor hybrids was found to have decreased with increasing

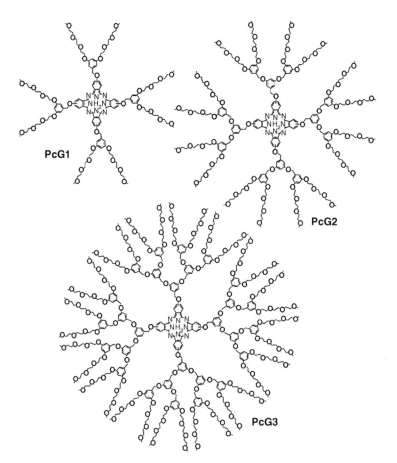

Figure 4.6 Dendric structures used in their work by Hahn et al. [44]. Reproduced with permission from the American Chemical Society.

dendrimer generation because it enhanced (i) the hydrophilicity and (ii) the bulkiness of the resulting perylenediimides. Both effects were synergetic, and, in turn, act to lower the immobilization strength on SWNT. Several spectroscopies confirmed that distinct ground- and excited state interactions prevail and that kinetically and spectroscopically well-characterized radical ion pair states are formed within a few picoseconds.

Supramolecular donor–acceptor assemblies composed of carbon nanodiamond (ND) and porphyrin (Por) were constructed through interensemble hydrogen bonding and p–p interactions [50]. The formation of the supramolecular clusters composed of ND and porphyrin has been confirmed by transmission electron microscopy, dynamic light scattering, and IR spectroscopy. The resulting supramolecular clusters have been assembled as three-dimensional arrays on nanostructured SnO_2 films using an electrophoretic deposition method for testing photoelectrochemical properties. Current–voltage characteristics in standard three-compartment cells with working electrode in dark and under illumination are shown in Figure 4.7.

Phospholipid-linked naphthoquinones separated by spacer methylene groups (C_n), PE-C_n-NQ ($n = 0, 5, 11$) (Figure 4.8) were synthesized to investigate the quinone-mediated electron transfers on a glassy carbon (GC) electrode covered with phospholipid membrane [51]. The PE-C_n-NQ was incorporated into lipid bilayer

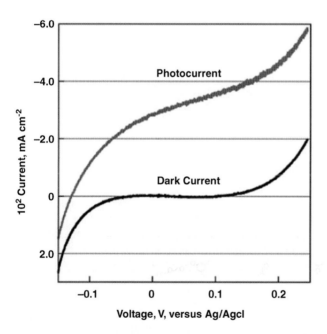

Figure 4.7 Current–voltage characteristics in standard three-compartment cells with working electrode OTE/SnO2/ND–TCPP along with a Pt wire gauze counter electrode and a reference electrode (Ag/AgCl) under white light illumination (l > 390 nm). Electrolyte: LiI 0.5 mol dm^{-3} and I_2 0.01 mol dm^{-3} in MeCN; input power $1/4$ 10 mW cm^{-2} [50]. Reproduced with permission from the Royal Chemical Society.

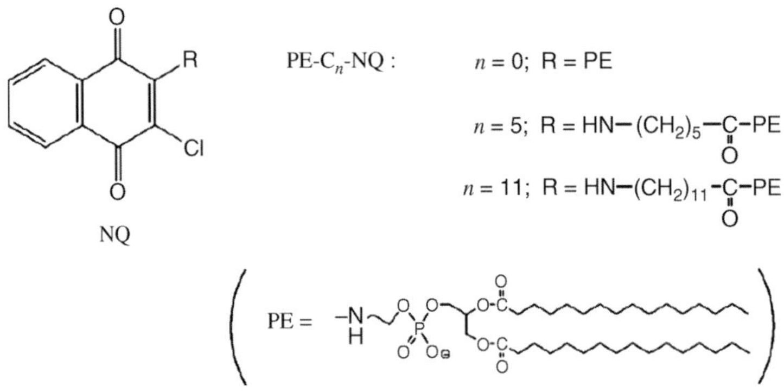

Figure 4.8 Phospholipid-linked naphthoquinones separated by spacer methylene groups (C_n), PE–C_n–NQ ($n = 0, 5, 11$) [51]. Reproduced with permission from Elsevier.

composed of phosphatidylcholine (Figure 4.9) and exhibited characteristic absorption spectral change corresponding to their redox state, quinone/hydroquinone. Results showed that the PE-C_n-NQ exhibited electron transfer associated with proton transfer in the lipid membranes, depending on the diffusivity of the redox species in the membrane and pH.

In the study [52], the adsorption of polyphenol oxidase (PPO) on SWCNTs and its electrocatalytic behavior were investigated using scanning electron microscope and atomic force microscope. The enzyme was immobilized on the SWCNTs/GC electrode. Direct electrochemistry of immobilized PPO was extensively investigated. The immobilized enzyme displayed a couple of stable redox peaks with a formal potential (E^0) of 0.102 V with respect to reference electrode in 0.05 M phosphate buffer solution (pH 7.0).

4.2
Redox Processes on Metal Surfaces

If several atoms are brought together into a molecule, their atomic orbitals split, as in a coupled oscillation (http://en.wikipedia.org/wiki/Metal). This produces a number of molecular orbitals proportional to the number of atoms. When a large number of atoms (on the order $\times 10^{20}$ or more) are brought together to form a solid, the number of orbitals becomes exceedingly large, and the difference in energy between them becomes very small, so the levels may be considered to form continuous *bands* of energy rather than the discrete energy levels of the atoms in isolation. Some intervals of energy contain no orbitals, no matter how many atoms are aggregated, forming *bandgaps*. A metal is a chemical element that is a good conductor of both electricity and heat and forms cations and ionic bonds with nonmetals. In a metal, atoms readily lose electrons to form cations. Those ions are surrounded by delocalized electrons, which are responsible for the conductivity. The solid produced is held by electrostatic interactions between the ions and the electron cloud, which are called metallic bonds.

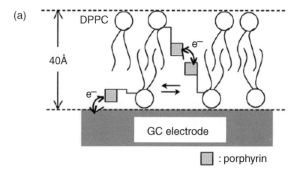

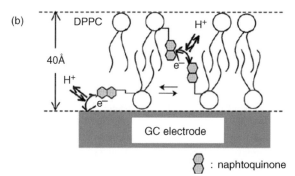

Figure 4.9 Schematic representation of electron transfer mediated by phospholipid-linked manganese porphyrins (a) and NQs (b) incorporated in DPPC lipid membranes cast on the GC electrode. The single lipid bilayers on the electrode were depicted as an interface layer between the multilamellar membrane and the electrode [51]. Reproduced with permission from Elsevier.

Vectorial multistep electron transfer at the gold electrodes modified with self-assembled monolayers (SAMs) of ferrocene–porphyrin–fullerene triads was investigated [53]. SAMs of ferrocene–porphyrin–C_{60} triads on gold electrodes were prepared. The results indicated that the triad molecules in monolayers are well packed with an almost perpendicular orientation on the gold surface. Photoelectrochemical studies were carried out in a standard three-electrode system using the gold electrodes modified with the SAMs of the triads. Stable cathodic photocurrents were observed in the presence of electron carriers such as oxygen and/or Me viologen in the electrolyte when the modified gold electrodes were illuminated with a monochromic light. According to the suggested multistep mechanism, vectorial electron transfer or partial charge transfer occurs from the excited singlet state of the porphyrin to the C_{60}, followed by the successive charge shift from the ferrocene to the porphyrin cation radical, to produce the ferrocene cation radical and the C_{60} anion radical. The C_{60} anion radical gives an electron to the counterelectrode via the electron carriers in the electrolyte solution, whereas electron transfer takes place from the gold electrode to the ferrocene cation radical, resulting in the recovery of

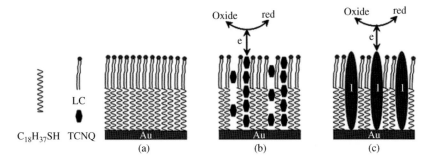

Figure 4.10 Schematic representation of mechanisms of electron transfer on the LC/SAM/Au electrode (a), TCNQ@LC/electrode (b), and 1@LC/SAM/Au electrode (c). 1 represents the TCNQ-based organometallic compound 1 with different sizes [54].

the initial state and the generation of the overall electron flow. The artificial photosynthetic cells showed a quantum efficiency of 20–25%.

A supported bilayer lipid membrane (s-BLM) containing one-dimensional compound 1, TCNQ-based (TCNQ$^1/_4$7,7,8,8-tetracyanoquinodimethane) organometallic compound {(Cu$_2$(m-Cl)(m-dppm)$_2$)(m2-TCNQ)}1, was prepared and characterized on the self-assembled monolayer of 1-octadecylmercaptan deposited on Au electrode (Figure 4.10) [54]. Cyclic voltammetry and electrochemical impedance spectroscopy (EIS) results showed that the compound 1, dotted inside s-BLM, can act as a mediator for electron transfer across the membrane. According to the suggested mechanism of the electron transfer across s-BLM by TCNQ takes place by electron hopping, while TCNQ-based organometallic compound serves as conducting material.

Mesoporous silica of the MCM-48 type as a support for fluorophore molecules was functionalized with 3-aminopropyl and 3-thiolopropyl groups. [55]. Then, molecules consisting of a fluorophore pyrene group and receptor fragments with donor N and O atoms were covalently attached by grafting carboxyl groups to amino groups on a mesoporous groups on the same surface. The vicinity of Au nanorods in a material with pyrene derivatives covalently grafted on the functionalized MCM-48 silica enhanced fluorescence of the composite material due to the surface plasmon resonance effect. The fluorescence emission of the prepared recognition material is quenched specifically owing to the photoinduced electron transfer effect after coordination reactions with Cu(II) ions. Hydrophobic magnetic nanoparticles (NPs) consisting of undecanoate-capped magnetite (Fe$_3$O$_4$, average diameter about 5 nm) were used to control quantized electron transfer to surface-confined redox units and metal NPs [56]. The attracted magnetic NPs formed a hydrophobic layer on the electrode surface changing the mechanisms of the surface-confined electrochemical processes. A quinone monolayer-modified Au electrode demonstrates an aqueous type of the electrochemical process ($2e^- + 2H^+$ redox mechanism) for the quinone units in the absence of the hydrophobic magnetic NPs, while the attraction of the magnetic NPs to the surface results in the step-wise single-electron transfer mechanism characteristic of a dry nonaqueous medium. The attraction of the hydrophobic magnetic NPs to the Au electrode surface modified with Au NPs (1.4 nm) yields a microenvironment with a low dielectric constant that results in the single-electron quantum charging of the Au NPs.

The blue copper protein from *Pseudomonas aeruginosa*, azurin, immobilized at gold electrodes through hydrophobic interaction with alkanethiol self-assembled monolayers of the general type [-S-(CH$_2$(n)-CH$_3$] ($n = 4$, 10, and 15), was employed to gain detailed insight into the physical mechanisms of short- and long-range biomolecular electron transfer [57]. Fast scan cyclic voltammetry and a Marcus equation analysis were used to determine unimolecular standard rate constants of a direct ET process and reorganization free energies for variable n (Figure 4.11), temperature (2–55 °C), and pressure (5–150 MPa) conditions. All the ET data, addressing SAMs with thickness variable over 12 Å, were described by using a single reorganization energy (0.3 eV). Logarithmic plots of unimolecular standard rate constants for Az electron exchange at Au electrodes modified by CH$_3$-terminated alkanethiol SAMs with variable methylene numbers are presented in Figure 4.11.

Determination of the transient tunneling barrier employing ultrafast electron transfer dynamics at NH$_3$/Cu(111) interfaces was the subject of the work [58]. Time-resolved two-photon photoemission of amorphous NH$_3$ layers on Cu(111) photo-injection of electrons was followed by charge solvation giving a transient potential barrier at the interface for the ET to the substrate. It was shown that the electrons are localized at the ammonia–vacuum interface and the ET rate depends exponentially on the NH$_3$ layer thickness with inverse range parameters β between 1.8 and 2.7 nm^{-1}.

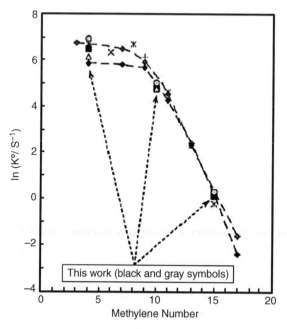

Figure 4.11 Logarithmic plots of unimolecular standard rate constants for Az electron exchange at Au electrodes modified by CH$_3$-terminated alkanethiol SAMs with variable methylene number [57]. Reproduced with permission from the American Chemical Society.

Systematic analysis of this time-resolved and layer thickness-dependent data enabled the authors to detect the temporal evolution of the interfacial potential barrier. According to experiment results, the tunneling barrier forms after $\tau E = 180$ fs and subsequently rises more than three times faster than the binding energy gain of the solvated electrons.

The rapid electron transfer reaction of the blue copper protein azurin adsorbed on different electrodes (pyrolytic graphite "edge," PGE, and gold modified with self-assembled monolayers of various 1-alkanethiols) has been studied by cyclic and square wave voltammetry [59]. By using large values (0.075–0.3 V) for the square wave amplitude, the electron transfer rate was measured over a continuously variable driving force in either direction. Values for k_{ET} at zero driving force were found to depend on the nature of the electrode; by contrast, the maximum rate constant, k_{max}, was essentially invariant with a rate of $(6 \pm 3) \times 10^3$ s^{-1} at 0 °C for both oxidation and reduction. Using Marcus theory, the potential dependence yielded an extremely low value for the reorganization energy ($\lambda < 0.25$). The temperature dependence for both PGE (-40 to 0 °C) and gold (0–50 °C) electrodes gave good fits to the Arrhenius equation, with activation energies $E_a = 19.5 \pm 1.1$ and 2.6 ± 1.7 kJ mol^{-1}, respectively. It was suggested that electron transfer is dependent upon prior formation of a highly ordered protein configuration on the electrode surface. Electron transfer processes of cytochrome c at interfaces were investigated [60]. To assess possible consequences of electric fields on the redox processes of cytochrome c, the protein was immobilized to self-assembled monolayers on electrodes and studied by surface-enhanced resonance Raman spectroscopy.

In the work [61], electrochemistry of surface-modified cytochrome c (cyt c) bound electrostatically to carboxylate-terminated alkanethiol SAMs revealed highly anisotropic electronic coupling across the protein/monolayer interface of a HOOC(CH$_2$)$_2$S/gold-bead electrode. Substitution of a lysine residue with alanine at position 13 in recombinant rat cyt c (RC9-K13A) lowered the interfacial ET rate more than five orders of magnitude, whereas ET was only slightly affected by replacement of lysine-72 or lysine-79 with alanine. The results clearly showed that lysine-13 is directly involved in coupling the protein to the SAM carboxylate terminus. A strategy for the immobilization of cytochrome c on the surface of chemically modified electrodes was demonstrated and used to investigate the protein's electron transfer kinetics [62]. Mixed monolayer films of alkanethiols and ω-terminated alkanethiols (terminated with pyridine, imidazole, or nitrile groups that are able to ligate with the *heme*) were used to adsorb cytochrome c on the surface of gold electrodes. The use of mixed films, as opposed to pure films, allows the concentration of adsorbed cytochrome to remain diluted and ensures a higher degree of homogeneity in their environment.

Robust voltammetric responses were obtained for wild-type and Y$_{72}$F/H$_{83}$Q/Q107H/Y108F azurins adsorbed on CH$_3$(CH$_2$)$_n$SH:HO(CH$_2$)$_m$SH ($n = m = 4$, 6, 8, 11; $n = 13$, 15, $m = 11$) SAM gold electrodes at pH 4.6 and high ionic strengths [63]. It was shown that electron transfer rates do not vary substantially with ionic strength, suggesting that the SAM Me head group binds to azurin by hydrophobic interactions. A binding model in which the SAM hydroxyl head group interacts with the Asn47 carboxamide accounted for the relatively strong coupling to the copper center that can be inferred from the ET rates. According to experiments, electron tunneling through $n = 8, 13$ SAMs are higher at pH 11 than those at pH 4.6, possibly owing to enhanced coupling of the SAM to Asn47 caused by deprotonation of nearby surface residues.

Grigoriev et al. [64] studied the variation in electron transmission through Au–S–benzene–S–Au junctions and related systems as a function of the structure of the Au:S contacts. For junctions with semiinfinite flat Au(111) electrodes, the highly coordinated in-hollow and bridge positions were connected with broad transmission peaks around the Fermi level due to a broad range of transmission angles from transverse motion, resulting in high conductivity and weak dependence on geometrical variations. Electron transport in molecular wire junctions was a subject of a review [65]. It was underlined that in molecular conductance, junctions are structures in which single molecules or small groups of molecules conduct electric current between two electrodes and the connection between the molecule and the electrodes greatly affects the current–voltage characteristics.

The structure–electronic structure relationship of nonmetalated *meso*-tetraphenyl porphyrin (2H-TPP) on the (111) surfaces of Ag, Cu, and Au was studied with a combination of scanning tunneling microscopy, photoelectron spectroscopy, and density functional theory [66]. The authors observed that the molecules form a 2D network on Ag(111), driven by attractive intermolecular interactions, while the surface migration barriers are comparatively small and the charge transfer to the adsorbed molecules is minimal. It was shown that the limiting factor in the formation of self-organized networks is the nature of the frontier orbital overlap and the adsorbate–interface electron transfer. Gold nanocluster systems have shown potential as components of electrochemical sensors and devices [67]. Hydrophilic gold nanoclusters were tethered to gold electrodes modified with mixed 1-octane thiol/1,9-nonane dithiol monolayers. To understand mechanism of electron transfer in these systems, the kinetics of electron transfer mediated by gold nanoclusters has been explored. Gold nanoclusters (diameter 1.8 nm) encapsulated with hydrophilic (triethylene oxide thiol, EO_3) protective layers ($AuEO_3$ clusters) were confined to gold electrodes, previously modified with mixed 1-octane thiol/1,9-nonane dithiol monolayers (Figure 4.12).

Effects of capping reagents on the electron transfer reactions on gold nanoparticles (AuNPs), attached to indium tin oxide (AuNP/ITO) electrodes, were investigated [68]. Systematic measurements of cyclic voltammograms were carried out using the AuNP/ITO electrodes prepared with different surfactants, that is, cationic CTAB and anionic SDS, for the oxidation of anionic $[Fe(CN)_6]^{4-}$ and the reduction of cationic $[Ru(NH_3)_6]^{3+}$. The results showed that the electrochemical responses were significantly improved on the AuNP/ITO electrodes, not depending on the charges of the capping surfactants and the redox species. The authors concluded that compared to thiols, the capping with surfactants would be useful in electroanalysis, in particular, of biomolecules, utilizing naked or undisturbed surfaces of AuNPs. Excited state interactions between chlorophyll a (Chla) and gold nanoparticles have been studied [69]. It was shown that the emission intensity of Chla is quenched by gold nanoparticles dominately by the process of photoinduced electron transfer from excited Chla to gold nanoparticles. Photoinduced electron transfer mechanism is

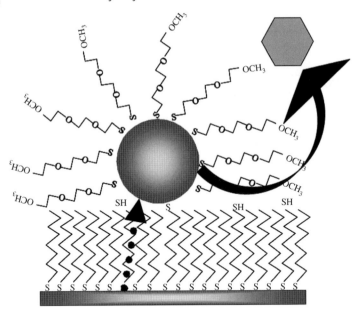

Figure 4.12 Gold nanoclusters (diameter 1.8 nm) encapsulated with hydrophilic (triethylene oxide thiol, EO_3) protective layers ($AuEO_3$ clusters) [67].

supported by the electrochemical modulation of fluorescence of Chla. Upon negatively charging the gold nanocore by external bias, an increase in fluorescence intensity was observed. It was suggested that the negatively charged gold nanoparticles create a barrier and suppress the electron transfer process from excited Chla to gold nanoparticles, resulting in an increase in radiative process. Using the nanosecond laser flash experiments of Chla in the presence of gold nanoparticles and fullerene (C_{60}), the authors demonstrated that Au nanoparticles, besides accepting electrons, can also mediate or shuttle electrons to another acceptor.

Shumyantseva et al. [70] investigated electron transfer between cytochrome P450scc (CYP11A1) and gold nanoparticles immobilized on rhodium–graphite electrodes. Thin films of gold nanoparticles were deposited on the rhodium–graphite electrodes by drop casting and cytochrome P450scc was deposited on both gold nanoparticle-modified and bare rhodium–graphite electrodes. Cyclic voltammetry indicated an enhanced activity of the enzyme at the gold nanoparticle-modified surface. The role of the nanoparticles in mediating electron transfer to the cytochrome P450scc was verified using an impedance spectroscopy.

Self-assembled monolayers of ferrocene–porphyrin–C_{60} triads (Figure 4.13) on gold electrodes were prepared to mimic photosynthetic electron transfer events where efficient conversion of light to chemical energy takes place via the long-lived charge-separated state with a high quantum yield [71]. The obtained results, together with blocking experiments using a redox probe, indicate that the triad molecules in monolayer are well packed with an almost perpendicular orientation on the gold surface. Photoelectrochemical studies under illumination with a monochromic light

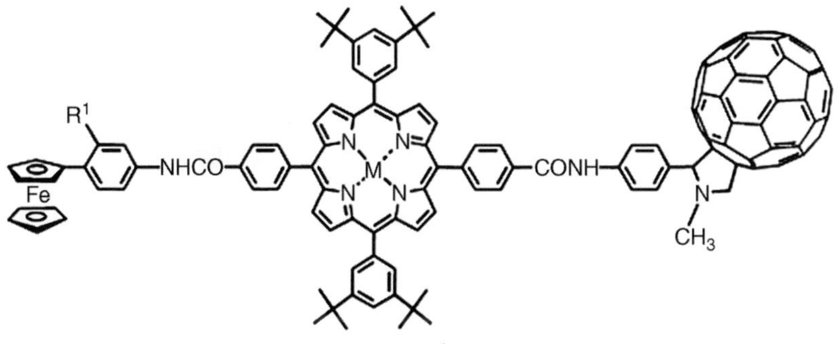

4.2 Redox Processes on Metal Surfaces | 143

1a: M=H$_2$; 1b: M=Zn, R^1=O(Ch$_2$)$_{11}$SH
2a: M=H$_2$; 2b: M=Zn, R^1=H

Figure 4.13 Ferrocene–porphyrin–C$_{60}$ triads [71] Reproduced with permission from the American Chemical Society.

were carried out in a standard three-electrode system using the gold electrodes modified with the self-assembled monolayers of the triads. A photoinduced multistep electron transfer mechanism was proposed for the photoelectrochemical cells (Figure 4.14). According to this mechanism, vectorial electron transfer or partial charge transfer occurs from the excited singlet state of the porphyrin to the C$_{60}$, followed by the successive charge shift from the ferrocene to the porphyrin cation radical, to produce the ferrocene cation radical and the C$_{60}$ anion radical. The C$_{60}$ anion radical gives an electron to the counter electrode via the electron carriers in the electrolyte solution, whereas electron transfer takes place from the gold electrode to the ferrocene cation radical, resulting in the recovery of the initial state and the generation of the overall electron flow.

Stähler *et al.* [72] discussed experiments on the ultrafast dynamics of photoinduced electron transfer and solvation processes at amorphous ice–copper

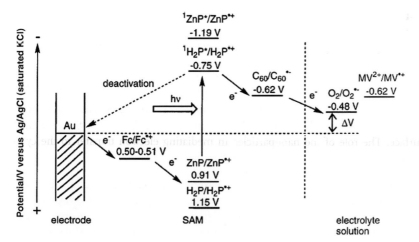

Figure 4.14 Photocurrent generation mechanism on the basis of the redox potentials [71]. Reproduced with permission from the American Chemical Society.

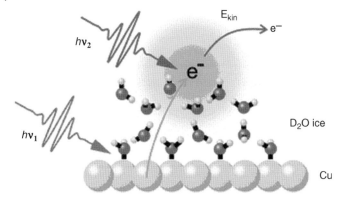

Figure 4.15 Photoelectron transfer on copper surface [72]. Reproduced with permission from the Royal Chemical Society.

interfaces (Figure 4.15). Femtosecond time-resolved two-photon photoelectron spectroscopy was employed as a direct probe of the electron dynamics, which enables the analysis of all elementary processes: the charge injection across the interface, the subsequent electron localization and solvation, and the dynamics of electron transfer back to the substrate. Using surface science techniques to grow and characterize various well-defined ice structures, a detailed insight into the correlation between adsorbate structure and electron solvation dynamics, and the location (bulk versus surface) of the solvation site, was gained.

Photoconversion properties were demonstrated for a device based on a small dye molecule, absorbing light in the near-IR region, mixed with two organic charge transport materials and together forming a dye-sensitized organic bulk heterojunction on the aluminum [73]. The organic dye molecule, phthalocyanine (1,4,8,11,15,18,22,25-octabutoxy-29H,31H-phthalocyanine), mixed with a blend of poly(3-hexylthiophene) (P3HT) and 1-(3-methoxycarbonyl)-propyl-1-phenyl-(6,6)C61 (PCBM), showed a photoconversion spectrum extending more than 150 nm toward longer wavelengths, compared to a device without such dye sensitization. Transient laser spectroscopy measurements showed that after excitation of the dye an electron transfer occurs from the dye to PCBM and a subsequent hole transfer from the dye to P3HT, which results in a long-lived ($P3HT^+$/dye/$PCBM^-$) charge-separated.

4.3
Electron Transfer in Miscellaneous Systems

The role of ZrO_2 surface states in the ultrafast photoinduced electron transfer from sensitizing dye molecules (alizarin) to semiconductor colloids was revealed [74]. Investigations on the ultrafast electron injection mechanism from the dye to wide bandgap semiconductor colloids in aqueous medium were presented. The measurements showed an injection time $\tau_{inj} < 100$ fs, suggesting that the electron transfer follows an adiabatic mechanism. The highly multiphasic recombination dynamics with time constants from 400 fs to the nanosecond timescale was found. The spectroscopic investigations showed that on ultrafast timescales the formation of an alizarin cation occurred. For the alizarin/ZrO_2 system, the timescale for the injection into these traps was detected to be faster than 100 fs. The relaxation processes in the traps and the repopulation of the S_1 state occur within 450 fs, and the subsequent ground-state relaxation takes 160 ps.

The electron transport and chemically sensing properties of individual multi-segmented Au-poly(3,4-ethylenedioxythiophene) (PEDOT)-Au nanowires were studied [75]. Temperature-dependent measurements showed that charge transport in PEDOT/poly(4-styrenesulfonic acid) (PSS) nanowires was in the insulating regime of the metal–insulator transition and dominated by hopping, while PEDOT/perchlorate (ClO_4) nanowires were slightly on the metallic side of the critical regime. Microwave reflectance studies of photoelectrochemical kinetics at low-doped Si semiconductor electrodes were reported [76]. Light- and voltage-induced changes in the microwave reflectivity of semiconductors were used to study the kinetics and mechanisms of electron transfer at semiconductor–electrolyte interfaces including steady-state, transient, and periodic responses. The results defined the range of rate constant that should be experimentally assessable using microwave reflectivity methods.

The compound $(bpy)_2MnIII(\mu\text{-}O)_2MnIV(bpy)_2$ was coupled to single CrVI charge transfer chromophores in the channels of the nanoporous oxide AlMCM-41 [77]. Observation of the 16-line EPR signal characteristic of $MnIII(\mu\text{-}O)_2MnIV$ demonstrated that the majority of the loaded complexes retained their nascent oxidation state in the presence or absence of CrVI centers. The FT-Raman spectrum upon visible light excitation of the CrVI-OII $\rightarrow$ CrV-OI ligand-to-metal charge transfer revealed electron transfer from $MnIII(\mu\text{-}O)_2MnIV$ (Mn–O stretch at 700 cm^{-1}) to CrVI, resulting in the formation of CrV and $MnIV(\mu\text{-}O)_2MnIV$ (Mn–O stretch at 645 cm^{-1}). All initial and final states were directly observed by FT-Raman or EPR spectroscopy, and the assignments were corroborated by X-ray absorption spectroscopy measurements. The endoergic charge separation products ($\Delta E_0 = -0.6$ V) remain after several minutes. Partial separation of CrV and $MnIV(\mu\text{-}O)_2MnIV$ as a consequence of hole (OI) hopping as a major contributing mechanism was poited out.

Photoinduced charge transfer events between 3 nm diameter CdSe semiconductor nanocrystals and an electron acceptor, MV^{2+}, have been probed in the subpicosecond, microsecond, and second timescale by confining the reactants to an AOT/heptane reverse micelle [78]. It was shown that the probe molecule, methyl viologen (MV^{2+}), interacts with the excited CdSe nanoparticle and quenches its emission effectively. The ultrafast electron transfer to MV^{2+}, as monitored from the exciton bleaching recovery of CdSe and the formation of $MV^{\overset{\bullet}{-}}$ radical, was completed with an average rate constant of 2.25×10^{10} s^{-1}. Under steady-state irradiation (450 nm), the accumulation of $MV^{\overset{\bullet}{-}}$ is seen with a net quantum yield of 0.1. Mediation of the electron transfer through TiO_2 nanoparticles was achieved by coupling them with the CdSe-MV^{2+} system within the reverse micelle (Figures 4.16 and 4.17). This coupling of two semiconductor nanoparticles increased the quantum yield of MV^{2+} reduction by a factor of 2.

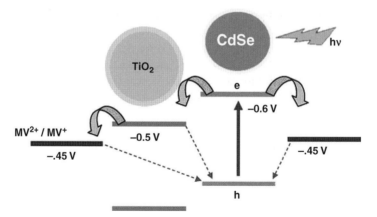

Figure 4.16 Energy level diagram (not to scale) illustrating direct and TiO$_2$-mediated electron transfer from excited CdSe to MV^{2+}. Curved arrows represent forward electron transfer, while dashed arrows represent backward electron transfer [78]. Reproduced with permission from the American Chemical Society.

Electron transfer from three conjugated amino-phenyl acid dyes to titanium and aluminum oxide nanocrystalline films was studied by using transient absorption spectroscopy with sub-20 fs time resolution over the visible spectral region [79]. The dyes attached to TiO$_2$ showed long-lived ground-state bleach signals. The transient kinetics of the dyes on TiO$_2$ revealed stimulated emission decays of about 40 fs and less than 300 fs assigned to electron injection. The same dyes on Al$_2$O$_3$ substrates displayed long stimulated emission decays. For the two of the dyes 2E,4E-2-cyano-5-(4-dimethylaminophenyl) penta-2,4-dienoic acid (NK1) and 2-[(E)-3-(4-dimethylaminophenyl)-allylidene]malonic acid (NK2), and 2E,4E-2-cyano-5-(4-diphenylamino-

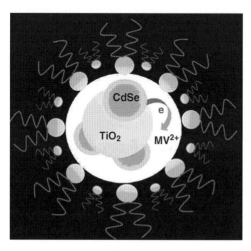

Figure 4.17 Coupling of CdSe nanoparticles with TiO$_2$ to boost the efficiency of electron transfer to MV^{2+} [78]. Reproduced with permission from the American Chemical Society.

phenyl)penta-2,4-dienoic acid (NK7) with amino methyl terminal groups, cation formation was seen at 670 nm probe wavelength. Early excited state dynamics observed in the NK7 dye bound to both TiO_2 and Al_2O_3, with pulse-limited rise and 30–40 fs decay times, was assigned to the rearrangement of charge in the amino phenyl moiety and/or possibly isomerization.

Numerous experimental and theoretical data presented in this chapter show excellent examples of effective electron injections and back reactions on the interface of semiconductors and metals. These studies promise insight into the mechanism of different intermediate reactions in existing dye solar cells and stimulation of invention of new systems of light energy conversion.

References

1 Balandin, A.A. and Wang, K.L. (2006) *Handbook of Semiconductor Nanostructures and Nanodevices* (5-vol. set), American Scientific Publishers.
2 Yu, P.Y. and Cardona, M. (2004) *Fundamentals of Semiconductors: Physics and Materials Properties*, Springer.
3 Kittel, C. (1996) *Introduction to Solid State Physics*, 7th edn, John Wiley & Sons, Inc., New York.
4 Zhang, J.Z. and Grant, C.D. (2008) *Annu. Rev. Nano Res.*, **2** (1), 1–10.
5 Gust, D., Moor, T.A., and Moor, A.L. (2009) *Acc. Chem. Res.*, **42** (12), 1890–1898.
6 Yum, J.-H., Chen, P., Gratzel, M., and Nazeruddin, M.K. (2008) *ChemSusChem*, **1** (8–9), 699–707.
7 Moon, S.-J., Yum, J.-H., Humphry-Baker, R., Karlsson, K.M., Hagberg, D.P., Marinado, T., Hagfeldt, A., Sun, L., Gratzel, M., and Nazeruddin, M.K. (2009) *J. Phys. Chem. C*, **113** (38), 16816–16820.
8 Wang, H., Nicholson, P.G., Peter, L., Zakeeruddin, S.M., and Gratzel, M. (2010) *J. Phys. Chem. C*, **114** (33), 14300–14306.
9 Jespersen, K.G., Zaushitsyn, Y., Westenhoff, S., Pullerits, T., Yartsev, A., Inganaes, O., and Sundstroem, V. (2010) in *Fluorescence of Supermolecules, Polymers, and Nanosystems, Springer Series on Fluorescence*, vol. 4, Springer, pp. 285–297.
10 Anderson, N.A., Ai, X., and Lian, T. (2003) *Ultrafast Phenomena XIII: Proceedings of the 13th International Conference*, Springer Series in Chemical Physics, **71**, pp. 325–327.
11 Anderson, N.A. and Lian, T. (2005) *Annu. Rev. Phys. Chem.*, **56**, 491–519.
12 Nazmutdinov, R.R., Bronshtein, M.D., and Schmickler, W. (2009) *Electrochim. Acta*, **55** (1), 68–77.
13 O'Regan, B. and Gratzel, M. (1991) *Nature*, **353**, 737–740.
14 Gratzel, M. (2001) *Nature*, **414**, 338–344.
15 Hossain, M.Z. (2010) *Appl. Phys. Lett.*, **96**, 053118.
16 Jin, S. and Lian, T. (2009) *Nano Lett.*, **9** (6), 2448–2454.
17 Guangzhi, X., Ciping, C., Junfeng, X., and Kittel, C. (1996) *Introduction to Solid State Physics*, 7th edn, John Wiley & Sons, Inc., New York.
18 Jakubikova, E., Martin, R.L., and Batasta, E.R. (2010) *Inorg. Chem.*, **49** (6), 2975–2982.
19 Murakami, T.N., Kijitori, Y., Kawashima, N., and Miyasaka, T. (2003) *Chem. Lett.*, **32** (11), 1076–1077.
20 Byeong-Kwan, A., Mulherin, R., Langley, B., Burn, P., Meredith, P. (2009) *Org. Electron.*, **10** (7), 1356–1363.
21 Wenger, S., Bouit, P.-A., Chen, Q., Teuscher, J., Di Censo, D., Humphry-Baker, R., Moser, J.-E., Delgado, J.L., Martin, N., Zakeeruddin, S.M., and Gratzel, M. (2010) *J. Am. Chem. Soc.*, **132** (14), 5164–5169.
22 Zhang, D., Lanier, S.M., Downing, J.A., Avent, J.L., Lumc, J., and McHalea, J.L. (2008) *J. Photochem. Photobiol. A Chem.*, **195** (1), 72–80.
23 Gregg, B.A. (2004) *Coordin. Chem. Rev.*, **248** (13–14), 1215–1224.

24 Ghosh, H.N., Ramakrishna, G., Singh, A.K., and Palit, D.K. (2004) *J. Phys. Chem. B* **108** (5), 1701–1707.

25 Huber, R., Spoerlein, S., Moser, J.E., Graetzel, M., and Wachtveitl, J. (2000) *J. Phys. Chem. B*, **104** (38), 8995–9003.

26 Sant, P.A. and Kamat, P.V. (2002) *Phys. Chem. Chem. Phys.*, **4** (2), 198–203.

27 Ai, X., Guo, J., Anderson, N.A., and Lian, T. (2006) in *Electron Transfer in Nanomaterials: Proceedings Electrochemical Society*, The Electrochemical Society, pp. 38–48.

28 Wenger, S., Bouit, P.-A., Chen, Q., Teuscher, J., Di Censo, D., Humphry-Baker, R., Moser, J.-E., Delgado, J.L., Martın, N., Zakeeruddin, S.M., and Michael Gratzel, M. (2010) *J. Am. Chem. Soc.*, **132** (14), 5164–5169.

29 An, B.-K., Mulherin, R., Langley, B., Burn, P., and Meredith, P. (2009) *Org. Electron.*, **10**, 1356–1360.

30 Jin, S., Snoeberger, R.C., III, Issac, A., Stockwell, D., Batista, V.S., and Lian, T., *J. Phys. Chem. B.* **114**, 14309–14319.

31 Bhattacharyya, S., Kibel, A., Liddell, P.A., Gust, D., and Lindsay, S. (2009) Abstracts. 65th Southwest Regional Meeting of the American Chemical Society, El Paso, TX.

32 Pensack, R.D., Banyas, K.M., Barbour, L.W., Hegadorn, M., and Asbury, J.B. (2009) *Phys. Chem. Chem. Phys.*, **11** (15), 2575–2591.

33 Ai, X., Anderson, N., Guo, J., Kowalik, J., Tolbert, L.M., and Lian, T. (2006) *J. Phys. Chem. B*, **110** (50), 25496–25503.

34 Tisdale, W.A., Williams, K.J., Timp, B.A., Norris, D.J., Aydil, E.S., and Zhu, X.-Y. (2010) *Science*, **328** (5985), 1543–1547.

35 Cabanillas-Gonzalez, J., Egelhaaf, H.-J., Brambilla, A., Sessi, P., Duò, L., Finazzi, M., Ciccacci, F., and Lanzani, G. (2008) *Nanotechnology*, **19** (42), 424010/1–424010/7.

36 Guo, J., Stockwell, D., Ai, X., She, C., Anderson, N.A., and Lian, T. (2006) *J. Phys. Chem. B*, **110** (11), 5238–5244.

37 Dai, L. (2006) *Carbon Nanotechnology: Recent Developments in Chemistry, Physics, Materials Science and Device Applications*, Elsevier Science, New York.

38 Guldi, D.M., Rahman, G.M.A., Zerbetto, F., and Prato, M. (2005) *Acc. Chem. Res.*, **38** (11), 871–878.

39 Tasis, D., Tagmatarchis, N., and Bianco, A. (2006) *Chem. Rev.*, **106** (3), 1105–1136.

40 Guldi, D.M., Rahman, A., Sgobba, V., and Ehli, C. (2006) *Chem. Soc. Rev.*, **35**, 471–487.

41 Chitta, R. and D'Souza, F. (2008) *J. Mater. Chem.*, **18** (13), 1440–1453.

42 Fukuzumi, S. and Kojima, T. (2008) *J. Mater. Chem.*, **18** (13), 1427–1439.

43 Cioffi, C.T., Palkar, A., Melin, F., Kumbhar, R., Echegoyen, L., Melle-Franco, M., Zerbetto, F., Rahman, G.M.A., Ehli, C., Sgobba, V., Guldi, D.M., and Prato, M. (2009) *Chem. Eur. J.*, **15** (17), 4419–4427.

44 Hahn, U., Engmann, S., Oelsner, C., Ehli, C., Guldi, D.M., and Torres, T. (2010) *J. Am. Chem. Soc.*, **132** (18), 6392–6401.

45 Cid, J.-J., García-Iglesias, M., Yum, J.-H., Forneli, A., Albero, J., Martínez-Ferrero, E., Vázquez, P., Grätzel, M., Nazeeruddin, M.K., Palomares, E., and Torres, T. (2009) *Chem. Eur. J.*, **15** (20), 5130–5137.

46 Bartelmess, J., Ballesteros, B., de la Torre, G., Kiessling, D., Campidelli, S., Prato, M., Torres, T., and Guldi, D.M. (2010) *J. Am. Chem. Soc.*, **132** (45), 16202–16211.

47 Gonzalez-Rodrıguez, D. and Bottari, G. (2009) *J. Porphyr. Phthalocyanines*, **13** (4–5), 624–631.

48 Rodrıguez-Morgade, M.S., Plonska-Brzezinska, M.E., Athans, A.J., Carbonell, E., de Miguel, G., Guldi, D.M., Echegoyen, L., and Torres, T. (2009) *J. Am. Chem. Soc.*, **131** (22), 7727–1734.

49 D'Souza, F. and Ito, O. (2009) *Chem. Commun.* (33), 4913–4928. De la Torre, G. (2009) *J. Porphyr. Phthalocyanines*, **13** (4–5), 637–644.

50 Ohtani, M., Kamat, P.V., and Fukuzumi, S. (2010) *J. Mater. Chem.*, **20** (3), 582–587.

51 Suemori, Y., Nagata, M., Kondo, M., Ishigure, S., Dewa, T., Ohtsuka, T., and Nango, M. (2008) *Colloids Surf. B Biointerfaces*, **61** (1), 106–112.

52 Mohammadi, A., Moghaddam, A.B.M., Dinarvand, R., and Rezaei-Zarchi, S. (2009) *Int. J. Electrochem. Sci.*, **4** (7), 895–905.

53 Imahori, H., Yamada, H., Nishimura, Y., Yamazaki, I., and Sakata, Y. (2000) *J. Phys. Chem. B*, **104** (9), 2099–2108.

54 Qu, M., Lv, Q., Yang, B., Zhang, W., Zhang, J., Zhan, S., and Yea, J. (2010) *Electroanalysis*, **22** (4), 375–378.

55 Orłowska, M., Kledzik, K., Mroczkiewicz, M., Ostaszewski, R., and Kłonkowski, A.M. (2008) *J. Non-Cryst. Solids*, **35** (35–39), 4426–4432.

56 Katz, E. and Willner, I. (2006) *Sensors*, **6** (4), 420–427.

57 Khoshtariya, D.E., Dolidze, T.D., Shushanyan, M., Davis, K.L., Waldeck, D.H., and van Eldik, R. (2010) *Proc. Natl. Acad. Sci. USA*, **107** (7), 2757–2762.

58 Staehler, J., Meyer, M., Kusmierek, D.O., Bovensiepen, U., and Wolf, M. (2008) *J. Am. Chem. Soc.*, **130** (27), 8797–8803.

59 Jeuken, L.J.C., McEvoy, J.P., and Armstrong, F. (2002) *J. Phys. Chem. B*, **106** (9), 2304–2313.

60 Murgida, D.H. and Hildebrandt, P. (2004) *Acc. Chem. Res.*, **37** (11), 854–861.

61 Niki, K., Hardy, W.R., Hill, M.G., Li, H., Sprinkle, J.R., Margoliash, E., Fujita, K., Tanimura, R., Nakamura, N., Ohno, H., Richards, J.H., and Gray, H.B. (2003) *J. Phys. Chem. B*, **107** (37), 9947–9949.

62 Wei, J., Liu, H., Dick, A.R., Yamamoto, H., He, Y., and Waldeck, D.H. (2002) *J. Am. Chem. Soc.*, **124** (32), 9591–9599.

63 Yokoyama, K., Leigh, B.S., Sheng, Y., Niki, K., Nakamura, N., Ohno, H., Winkler, J.R., Gray, H.B., and Richards, J.H. (2008) *Inorg. Chim. Acta*, **361** (4), 1095–1099.

64 Grigoriev, A., Skoldberg, J., Wendin, G., and Crljen, Z. (2006) *Phys. Rev. B Condens Matter Mater Phys.*, **74** (4), 045401/1-045401/16.

65 Nitzan, A. and Ratner, M.A. (2003) *Science*, **300** (5624), 1384–1389.

66 Rojas, G., Chen, X., Bravo, C., Kim, J.H., Kim, J.-S., Xiao, J., Dowben, P.A., Gao, Y., Zeng, X.C., Choe, W., and Enders, A. (2010) *J. Phys. Chem. C*, **114** (20), 9408–9415.

67 Lowy, D.A., Jhaveri, S.D., Foos, E.E., Tender, L.M., Ancona, M.G., and Snow, A.W. (2006) *Electrochem. Commun.*, **8** (12), 1821–1824.

68 Horibe, T., Zhang, J., and Oyama, M. (2007) *Electroanalysis*, **19** (7), 847–852.

69 Barazzouk, S., Kamat, P.V., and Hotchandani, S. (2005) *J. Phys. Chem. B*, **109** (2), 716–723.

70 Shumyantseva, V.V., Carrara, S., Bavastrello, V., Riley, D.J., Bulko, T.V., Skryabin, K.G., Archakov, A.I., and Nicolini, C. (2005) *Biosens. Bioelectron.*, **21** (1), 217–222.

71 Imahori, H., Yamada, H., Nishimura, Y., Yamazaki, I., and Sakata, Y. (2000) *J. Phys. Chem. B*, **104** (9), 2099–2108.

72 Stähler, J., Bovensiepen, U., Meyer, M., and Wolf, M. (2008) *Chem. Soc. Rev.*, **37** (10), 2180–2190.

73 Johansson, M.J., Yartsev, A., Rensmo, H., and Sundstrom, V. (2009) *J. Phys. Chem. C*, **113** (7), 3014–3020.

74 Huber, R., Spoerlein, S., Moser, J.E., Graetzel, M., and Wachtveitl, J. (2000) *J. Phys. Chem. B*, **104** (38), 8995–9003.

75 Cao, Y., Kovalev, A.E., Xiao, R., Kim, J., Mayer, T.S., and Mallouk, T.E. (2008) *Nano Lett.*, **8** (12), 4653–4658.

76 Cass, M.J., Duffy, N.W., Peter, L.M., Pennock, S.R., Ushiroda, S., and Walker, A.B. (2003) *J. Phys. Chem. B*, **107** (24), 5857–5863.

77 Weare, W.W., Pushkar, Y., Yachandra, V.K., and Frei, Y. (2008) *Am. Chem. Soc.*, **130** (34), 11355–11363.

78 Clifton Harris, C. and Kamat, P.V. (2009) *ACS Nano*, **3** (3), 682–690.

79 Myllyperkio, P., Manzoni, C., Dario Polli, D., Cerullo, G., and Korppi-Tommola, J. (2009) *J. Phys. Chem. C*, **113** (31), 13985–13992.

5
Dye-Sensitized Solar Cells I

5.1
General Information on Solar Cells

World energy consumption is about 4.7 10^{20} J (450 quadrillion Btu) and is expected to grow about 2% each year for the next 25 years [1, 2]. Nowadays, renewable sources comprise about 13% of all energy production. Photovoltaics or solar cells account for no more than 0.04% and most probably only in 2030 will that figure reach 1% [3].

A solar cell is a device that converts the energy of sunlight directly into electricity by photovoltaic effect (http://en.wikipedia.org/wiki/Solar_cell). Assemblies of cells are used to make panels, solar modules, or photovoltaic arrays. The term "photovoltaic" comes from the Greek φῶς (*phōs*) meaning "light" and the term "voltaic," meaning electric, from the name of the Italian physicist Volta. Solar cells are often electrically connected and encapsulated as a module that has a sheet of glass on the upside to the sun, allowing light to pass while protecting the semiconductor wafers. Solar cells are also usually connected in series in modules, creating an additive voltage. Connecting cells in parallel will yield a higher current. A scheme of an idealized photovoltaic converter is presented in Figure 5.1 [4].

Excitation by light causes a separation of the chemical potentials of the electrons of the system, which reflect the free energy of the electrons in the absorber (Figure 5.1). Contacts B and C maintain the Fermi levels of the two types of carriers from the interface with A up to the outer metal contacts. This model is based on the assumption of completely reversible interfaces for the carriers for the system in state H from A to M_{left} and the complete blockage of carriers for the system in state L by B. Complementary conditions hold for C and H and for L and M_{right}. The recombination in the cell is the emission process that takes the system from state H to state L, as indicated by the dotted arrow.

In appreciation of efficiency of a solar cell, the following parameters of the solar cell equivalent circuit (Figure 5.2) are used (http://en.wikipedia.org/wiki/Solar_cell).

Solar Energy Conversion. Chemical Aspects, First Edition. Gertz Likhtenshtein.
© 2012 Wiley-VCH Verlag GmbH & Co. KGaA. Published 2012 by Wiley-VCH Verlag GmbH & Co. KGaA.

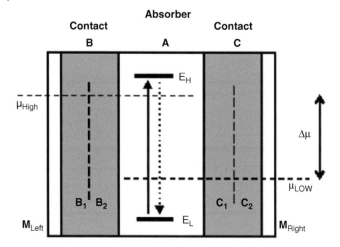

Figure 5.1 Scheme of an idealized photovoltaic converter. It consists of an absorber, A, in which photon absorption can excite electronic charge carriers, taking the system from a low-energy (L) to a high-energy (H) state with energies E_L and E_H, as indicated by the arrow (http://en.wikipedia.org/wiki/Solar_cell). [4]

In the equivalent circuit, the current produced by the solar cell (*I*) is equal to photogenerated current (I_L), minus that which flows through the diode (I_D), minus shunt current (I_{SH}):

$$I = I_L - I_D - I_{SH}$$

The current through these elements is governed by the voltage across them:

$$V_j = V + IR_S$$

where V_j is voltage across both diode and resistor R_{SH}, *V* is voltage across the output terminals, and *I* is output current.

When the cell is operated at open circuit, $I = 0$ (i.e., the circuit is broken or open), the voltage across the output terminals is defined as the *open-circuit voltage* V_{OC}.

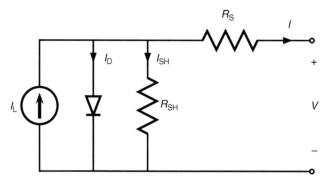

Figure 5.2 Solar cell equivalent circuit (http://en.wikipedia.org/wiki/Solar_cell).

When the cell with very low impedance is operated at *short circuit*, $V = 0$, the current I through the terminals is defined as the *short-circuit current* I_{SC}. For a high-quality solar cell (low R_S and I_0, and high R_{SH}), the short-circuit current I_{SC} *is close to* I_L. *Quantum efficiency* (QE) is the percentage of photons hitting the photoreactive surface that produce an electron–hole pair. A solar cell's *energy conversion efficiency* (η) is the percentage of power converted from absorbed light to electrical energy. A *maximum power point* is the point on the current–voltage (I–V) curve of a solar module under illumination, where the product of current and voltage is maximum. The *fill factor* (FF) is the ratio of the *maximum power point* divided by the *open-circuit voltage* (V_{oc}) and the *short-circuit current* (I_{sc}).

5.2
Dye-Sensitized Solar Cells

5.2.1
General

A dye solar cell (DSSC, DSC), the Grätzel cell that was invented by Michael Grätzel and Brian O'Regan at the École Polytechnique Fédérale de Lausanne in 1991, is a new class of low-cost solar cell, which belongs to the group of thin film solar cells [5]. It is based on a semiconductor formed between a photosensitized anode and an electrolyte, a *photoelectrochemical* system. Commercial applications, which were held up due to chemical stability problems, are now forecast in the European Union Photovoltaic Roadmap to be a potentially significant contributor to renewable electricity generation by 2020 [6]. A general scheme of DSSC is shown in Figure 5.3.

A series of reviews of recent progress in dye-sensitized solar cells (DSSCs) has been published [7–22]. The reviews introduce the structure and the principle of dye-sensitized solar cell and latest results about the development of key components, including nanoporous semiconductor films, dye sensitizers, redox electrolyte, and so on.

DSSCs' main advantages can be summarized as follow [17]: (a) good performance under standard reporting conditions; (b) stable performance at nonstandard conditions of temperature, irradiation, and solar incidence angle; (c) low cost; (d) available environmental-friendly raw materials; and (e) semitransparency and multicolor range possibilities.

There are two most important measures that are used to characterize any solar cells. One is the total amount of electrical power produced for a given amount of solar power shining on the cell. Expressed as a percentage, this is known as *the solar conversion efficiency*. The second is the QE, which is the possibility for one photon of a particular energy to create one electron. Overall QE of DSSCs is proved to be efficient reaching for green light value of about 90%.

Dye-sensitized solar cells separate the two functions provided by silicon in a traditional cell design. Normally, the silicon both acts as the source of photoelectrons and provide the electric field to separate the charges and create a current. In the dye-sensitized solar cell, the bulk of the semiconductor is used

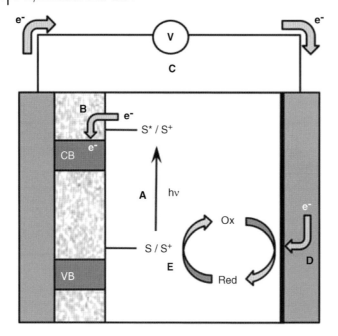

Figure 5.3 Schematic diagram of the electron transfer processes occurring in a DSSC. (a) Photon-induced dye excitation. (b) Injection of the photoexcited electron into the conduction band of the titania. (c) Movement of the electron through an external circuit. (d) Reaction of the electron with the oxidized redox couple at a platinized counter electrode. (e) Regeneration of the dye by the reduced redox couple [6]. Reproduced with permission from Elsevier.

solely for charge transport, and the photoelectrons are provided from a separate photosensitive dye. Charge separation occurs at the surfaces between the dye, semiconductor, and a media.

A dye-sensitized solar cell is consisted of the following main parts:

A photosensitive dye (D) as electron donor in excited state (D*), substrate as transparent electrode (TE), semiconductor as anode (An), counterelectrode as cathode (Cath), wire (W) connected to TE and Cath, and charge carrier bearing negative charge (ChC$^-$) in specific media (M) that separate cathode and anode.

The consecutive photochemical, chemical, and physical steps lead to the conversion of the light energy to photovoltaic current:

$$D + h\upsilon \rightarrow D^* \tag{5.1}$$

$$D^* + An \rightarrow D^+ + An^- \tag{5.2}$$

$$An^- + TE \rightarrow An + TE^- \tag{5.3}$$

$$TE^- + W \rightarrow TE + W^- \tag{5.4}$$

$$W^- + Cath \rightarrow W + Cath^- \tag{5.5}$$

$$Cath^- + ChC \rightarrow Cath + ChC^- \tag{5.6}$$

$$D^+ + ChC^- \rightarrow D + ChC \tag{5.7}$$

In all, 7263 references were found containing the concept of dye-sensitized solar cells.

5.2.2
Primary Grätzel DSSC

In the case of the original Grätzel design (Figure 5.4) [5], the cell has three primary parts. On the top is a transparent anode made of fluorine-doped tin dioxide (SnO_2:F) deposited on the back of a glass plate. On the back of the conductive plate is a thin

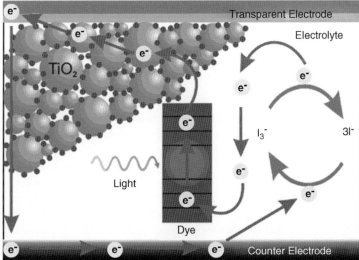

Figure 5.4 Schematic presentation of the primary Gertzel dye-sensitized solar cell.

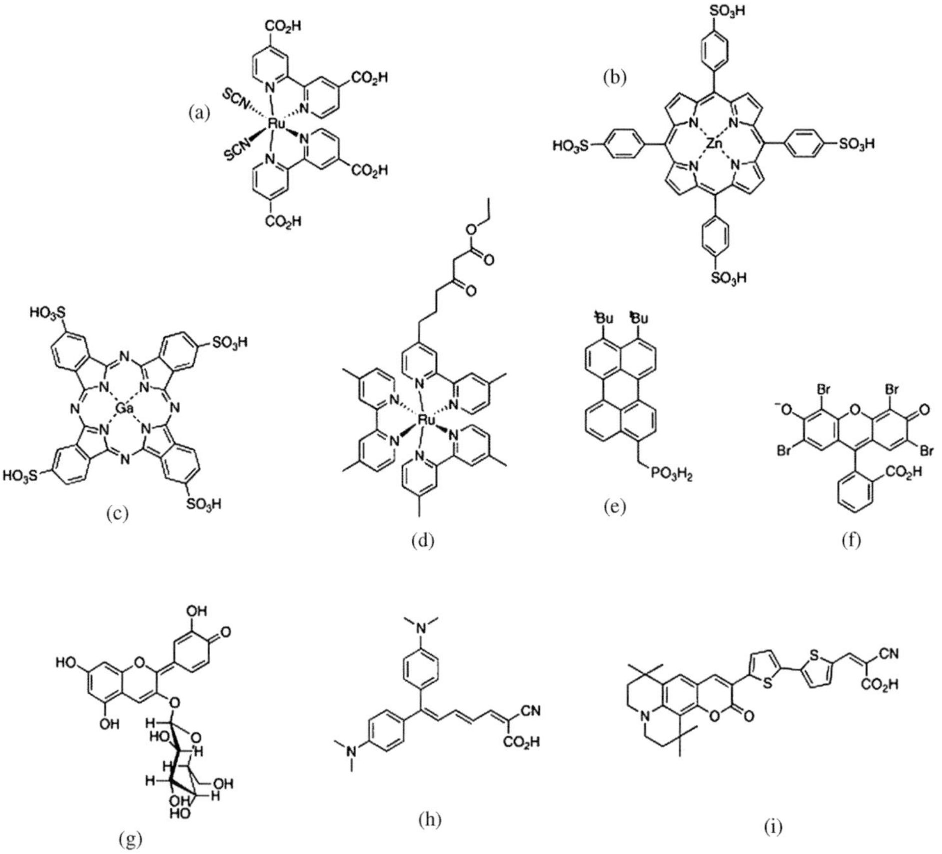

Figure 5.5 Chromophores and anchoring systems used on TiO$_2$. Ruthenium carboxypolypyridine complex **N3** (a), zinc tetrasulphonatephenylporphyrin (b) and gallium tetrasulphonatephthalocyanine (c), ruthenium acetylacetonate polypyridine complex (d), perylene dye (e), xanthene dye (Eosin Y) (f) and, natural flavonoid anthocyanin dye extracted from California blackberries (g), polyene dye NKX-2569 (h), and coumarin-based NKX-2677 (i) [38]. Reproduced with permission from Elsevier.

layer of titanium dioxide (TiO$_2$). Some chromophores and anchoring systems used on TiO2 are presented in Fig. 5.5. The plate is then immersed in a mixture of a photosensitive ruthenium-polypyridine dye (sensitizers) covalently bonded to the surface of the TiO$_2$ and a solvent. A separate backing is made of a thin layer of the iodide electrolyte spread over a conductive sheet, typically platinum metal. Sunlight enters the cell through the transparent SnO$_2$:F top contact and creates an excited state of the dye (D*). Photoelectron is injected directly into the conduction band of the TiO$_2$ and from there it moves by diffusion to the clear anode on the top. The photooxidized dye (D$^+$) strips one electron from iodide in electrolyte below the TiO$_2$, oxidizing it into iodine atom and eventually into triiodide. The triiodide is then diffused to the bottom of the cell, where the counter electrode (Cath$^-$) reintroduces the electrons after flowing through the external circuit. Thus, the charge separation occurs at the surfaces between the dye, the semiconductor, and the electrolyte.

The photochemical and chemical electron events in the primary Grätzel cell could be described by the following scheme:

$$D + h\upsilon \rightarrow D^*$$

$$D^* + TiO_2 \rightarrow D^+ + TiO_2^-$$

$$D^+ + TiO_2^- \rightarrow D + TiO_2$$

$$D^+ + I^- \rightarrow D + I$$

$$2I \rightarrow I_2$$

$$I + I^- \rightarrow I_2^-$$

$$2I_2^- \rightarrow 2I^- + I_2$$

$$I_2 + I^- \rightarrow I_3^-$$

$$D^+ + I_3^- \rightarrow D + I + I_2$$

$$I_3^- + Cath^- \rightarrow Cath + 2I^- + I$$

$$I_2 + Cath^- \rightarrow Cath + I^- + I$$

$$I + Cath^- \rightarrow Cath + I^-$$

In some specific cases, side reactions between sensitizes and reactive ion containing species should be taking into consideration.

Energy conversion efficiency is controlled by four elements: (i) light-harvesting efficiency; (ii) charge injection efficiency; (iii) electron transport and collection efficiency in the electrodes; and (iv) hole transport and collection efficiency in the electrolyte [20].

Through the efforts of Grätzel and his research group [1, 7, 10–12, 19, 22, 23], a significant improvement in the field of dye-sensitized solar cells has been achieved. In parallel, many research groups have been successfully working in this promising field to understand and improve the dye-sensitized solar cell.

5.3
DSSC Components

5.3.1
Sensitizers

5.3.1.1 Ruthenium Complexes

The following ligands of ruthenium complexes have been utilized: NCS[23], [24], dithiolates [25], β-diketonate [26], ethylenediamine [27], and 2-pyridinecarboxylate [28]. Among the numerous sensitizers developed for DSSCs, ruthenium(II) polypyridine complexes have received much attention owing to their superior performance in DSSCs [24–30]. Grätzel and coworkers reported a complex black dye, which contains ruthenium-4,4′,4″-tricarboxy-2,2′:6′,2″-terpyridine (tctpy), tetra-n-butylammonium (TBA), and monodentate isothiocyanato (NCS) functionalities; DSSCs incorporating black dye achieve 10% conversion efficiency from solar light to electricity [29]. A heteroleptic polypyridyl Ru complex, cis-Ru(4,4′-bis(5-octylthieno [3,2-b]thiophen-2-yl)-2,2′-bipyridine)(4,4′-dicarboxyl-2,2′-bipyridine)(NCS)$_2$, with a molar absorptivity of $20.5 \times 10^3 \, M^{-1} \, cm^{-1}$ at 553 nm was synthesized and demonstrated as an efficient sensitizer for a dye-sensitized solar cell, giving a power conversion efficiency of 10.53% measured under an irradiation of AM 1.5G full sunlight [30]. A new type of Ru(II) complex containing a 2-quinolinecarboxylate ligand was synthesized and its photophysical and photochemical properties were characterized [31]. Solar cells with this complex exhibited efficient panchromatic sensitization over the entire visible wavelength range extending into the near-IR region. An overall conversion efficiency of 8.2% was attained under standard air mass 1.5 irradiation (100 mW/cm^2) with a short-circuit photocurrent density of 18.2 mA/cm^2, an open-circuit photovoltage of 0.63 V, and a fill factor of 0.72.

A high molar extinction coefficient charge transfer sensitizer was developed [32]. The dye harvested visible light over a large spectral range and produced a short-circuit photocurrent density of 18.8 mA/cm^2, an open-circuit voltage of 783 mV, and a fill factor of 0.73, resulting in a solar-to-electric energy conversion efficiency (η) of 10.8, under air mass (AM) 1.5 sunlight. The time-dependent density functional theory (TDDFT) excited state calculations of the new sensitizer showed that the first three HOMOs have ruthenium t_{2g} character with sizable contribution coming from the NCS ligands. In the paper [33], Ru(II) compounds were reported that not only increase the light-harvesting efficiencies significantly, up to $40\,000\,M^{-1}\,cm^{-1}$, but also stabilize the charge-separated state by a very fast hole transfer event to regenerate Ru(II).

Stark effects after excited state interfacial electron transfer at sensitized TiO_2 nanocrystallites were revealed [34]. Photophysical studies were performed with [Ru(dtb)$_2$(dcb)](PF6)$_2$ and cis-Ru(dcb)(dnb)(NCS)$_2$, where dtb is 4,4'-(C(CH$_3$)$_3$)2-2,2'-bipyridine, dcb is 4,4'-(COOH)$_2$-2,2'-bipyridine, and dnb is 4,4'-(CH$_3$(CH$_2$)$_8$)2-2,2'-bipyridine, anchored to anatase TiO_2 particles of 15 nm in diameter interconnected in a mesoporous, 10 μm thick film immersed in Li$^+$-containing CH_3CN electrolytes with iodide or phenothiazine donors. Pulsed laser excitation resulted in rapid excited state injection and donor oxidation to yield $TiO_2(e^-)$s and oxidized donors. The spectral data were consistent with an underlying Stark effect and indicated that the surface electric field of 270 mV/m was not completely screened from the molecular sensitizer. The nonexponential screening kinetics was characterized by rate constant, τ_0^{-1}, of $1.5 \times 10^5\,s^{-1}$. The authors concluded that the electric field created by excited state injection from one sensitizer influenced the absorption spectra of other sensitizers that had not undergone photoinduced electron injection.

5.3.1.2 Metalloporphyrins

Porphyrins are important classes of potential sensitizers for highly efficient dye-sensitized solar cells owing to their photostability and potentially high light-harvesting capabilities that would allow applications in thinner, low-cost dye-sensitized solar cells. Application of metalloporphyrins in dye-sensitized solar cells for conversion of sunlight into electricity was the subject of a number of recent publications [35–42].

In the work [36], a series of Zn-porphyrins, that is, zinc metalloporphyrins for DSSC, cyano-3-(2'-(5',10',15',20'-tetraphenylporphyrinato zinc(II))yl)-acrylic acid (Zn-3), 3-(trans-2'-(5',10',15',20'-tetraphenylporphyrinato zinc(II))yl)-acrylic acid (Zn-5), 2-cyano-5-(2'-(5',10',15',20'-tetraphenylporphyrinato zinc(II))yl)-penta-2,4-dienoic acid (Zn-8), 4-(trans-2'-(2''-(5'',10'',15'',20''-tetraphenylporphyrinato zinc(II))yl)ethen-1'-yl)-1,2-benzenedicarboxylic acid (Zn-11), and 2-cyano-3-[4'-(trans-2''-(2–(~5,10-,15-,20--tetraphenylporphyrinato zinc(II))yl) ethen-1''-yl)-phenyl]-acrylic acid (Zn-13) were synthesized and characterized by using various spectroscopic techniques. Density functional theory (DFT) and TDDFT calculations showed that the key molecular orbitals (MOs) of porphyrins Zn-5 and Zn-3 are stabilized and extended to the substituent by π-conjugation, causing enhancement and redshifts of visible transitions and increasing the possibility of electron transfer from the substituent. The porphyrins were investigated for conversion of sunlight into electricity by constructing dye-sensitized TiO_2 solar cells using an I^-/I_3^- electrolyte. The cells yield close to 85% incident photon-to-current efficiencies (IPCE), and under standard AM 1.5 sunlight, the Zn-3-sensitized solar cell demonstrates a short-circuit photocurrent density of 13.0 ± 0.5 mA/cm^2, an open-circuit voltage of 610 ± 50 mV, and a fill factor of 0.70 ± 0.03. This corresponds to an overall conversion efficiency of 5.6.

The synthesis, electronic, and photovoltaic properties of six novel green porphyrin sensitizers were reported [37]. All porphyrin dyes gave solar cell efficiencies of $\geq 5\%$. Under standard global AM 1.5 solar conditions, they gave a short-circuit photocurrent density (J_{sc}) of 14.0 ± 0.20 mA/cm^2, an open-circuit voltage of 680 ± 30 mV, and a fill factor of 0.74, corresponding to an overall conversion efficiency of 7.1%. This dye gave an efficiency of 3.6% in a solid-state cell with spiro-MeOTAD as the hole transporting component. Following the development of an efficient building block approach to functionalized porphyrin arrays, Campbell et al. [38] synthesized a variety of β-carboxylic-substituted porphyrinmonomers and multiporphyrin arrays and evaluated their performance in the dye-sensitized TiO$_2$ solar cell (Fig. 5.5). The effect of porphyrin substituent, functional group position, linker conjugation, binding group, and electrolyte on the porphyrin light-harvesting efficiency was investigated. It was found that a β-substituted monoporphyrin carboxylic acid derivative with a conjugated linker shows significant advantage over any antennae-type multiporphyrin arrays. 4-trans-2′-(2″-(5″,10″,15″,20″-tetraphenylporphyrinato zinc(II)yl)ethen-1′-yl)-1-benzoic acid gives an overall efficiency of 4.2% under AM 1.5.

Several metalloporphyrins (Ru(CO)OEP, Ru(CO)TPP, and ZnTPP) have been evaluated for relative comparison and relationship to pyridyl axial binding strengths [39]. The systematic study evaluated multiple background cases using H$_2$TPP, TiO$_2$ modification with benzoic acid, or unmodified TiO$_2$ and confirmed the high affinity of Ru and Zn porphyrins for surface-anchored pyridyl sites. DSSC devices with novel mixed porphyrin assemblies were shown to give higher power performance than DSSCs utilizing sensitization with only one type of porphyrin. In a spectroscopic and DFT study of thiophene-substituted metalloporphyrins as dye-sensitized solar cell dyes, a combination of density functional theory calculations, electronic absorption, and resonance Raman spectroscopy has been applied to a series of β-substituted zinc porphyrins to elucidate how the substituent affects the electronic structure of the metalloporphyrin and assign the nature of electronic transitions in the visible region [40]. The use of conjugated β-substituents invoked a large perturbation to both the nature and the energy of the frontier molecular orbitals and resulted in the generation of additional molecular orbitals from the parent metalloporphyrin species.

Novel meso- or beta-derivatized porphyrins with a carboxyl group have been designed and synthesized for use as sensitizers in dye-sensitized solar cells [41]. Absorption spectra of porphyrins with a phenylethynyl bridge showed that both Soret and Q bands are redshifted with respect to those of porphyrin. This phenomenon was more pronounced for porphyrins, which have a pi-conjugated electron-donating group at the meso position opposite the anchoring group. Quantum chemical (DFT) results support the spectroelectrochemical data for a delocalization of charge between the porphyrin ring and the amino group in the first oxidative state of diarylamino-substituted porphyrin 5, which exhibits the best photovoltaic performance among all the porphyrins under investigation. From a comparison of the cell performance based on the same TiO_2 films, the devices made of porphyrin coadsorbed with chenodeoxycholic acid (CDCA) on TiO_2 in ratios [5]/[CDCA] = 1:1 and 1:2 have high efficiencies of power conversion making this green dye a promising candidate for colorful DSSC applications.

Typical porphyrins possess an intense Soret band at 400 nm and moderate Q bands at 600 nm, which does not match solar energy distribution on the Earth. According to Lin and Imahori [42], elongation of these compounds the could cause broadening and redshift of the absorption bands together with an increasing intensity of the Q bands compared to that of the Soret band. The efficiency of porphyrin-sensitized solar cells could be improved significantly if the dyes with larger red and near-infrared absorption could be developed.

5.3.1.3 Organic Dyes

Numerous metal-free sensitizers including cyanine, merocyanine, hemicyanine, anthocyanine, phthalocyanine, indoline, coumarin, eosin Y, perylene, anthraquinone, polyene, pentacene, triphenylamine, polyene, and other promising metal-free structures with different spectral responses have been employed over the years. Several structures have yielded efficiencies of around 8% and above ([43] and references therein). Nevertheless, to date, ruthenium complexes are the most successful dyes to achieve over 10% efficiency under standard conditions.

Pure organic dyes do not involve precious metal species, and therefore, these systems are very attractive from the viewpoint of cost reduction. A molecular design of numerous organic dyes for efficient dye-sensitized solar cells was recently reported ([44] and references therein). Among them are the following molecules: coumarin [45], merocyanine [46], cyaninea [47], hemicyanine [48], indolin [49], oligoene [50], xanten [51], triphenylamine,11dialkylaniline [52], bis (dimethylfluorenyl)amino]phenyl [53], phenothiazine [54], tetrahydroquinoline [55], squarine dyes [56, 57], cyanine [58] and styryl dyes [59], and their derivatives have been studied. However, most of these dyes cannot harvest infrared light (over 800 nm) efficiently, and the DSCs using them show only poor performance despite their ability to absorb infrared light.

An efficiency of 8–9% was obtained in the DSC by using pure organic dyes, which absorb light in the full visible light region [60–62].

Organic sensitizers comprising donor, electron-conducting, and anchoring groups were senthysized for dye-sensitized solar cell applications [63]. A solar cell employing 3-(5′-{4-[bis-(4-hexyloxy-phenyl)-amino]-phenyl}-[2,2′]bithiophenyl-5-yl)-2-cyanoacrylic acid dye spiro-OMeTAD as a hole transporting material exhibited a short-circuit photocurrent density of 9.64 mA/cm^2, an open-circuit voltage of 798 mV, and a fill factor of 0.57, corresponding to an overall conversion efficiency of 4.4% at standard AM 1.5 sunlight. Photoinduced absorption spectroscopy probes an efficient hole transfer from dyes to the spiro-OMeTAD. Chemical formulas of molecular structures of D5L6, D21L6, and D25L6 are shown in Figure 5.6.

Two organic sensitizers, MKZ-21 and MKZ-22, comprising 5,11-dioctylindolo[3,2-b]carbazole moiety as the electron donor, n-hexyl-substituted oligothiophene units as

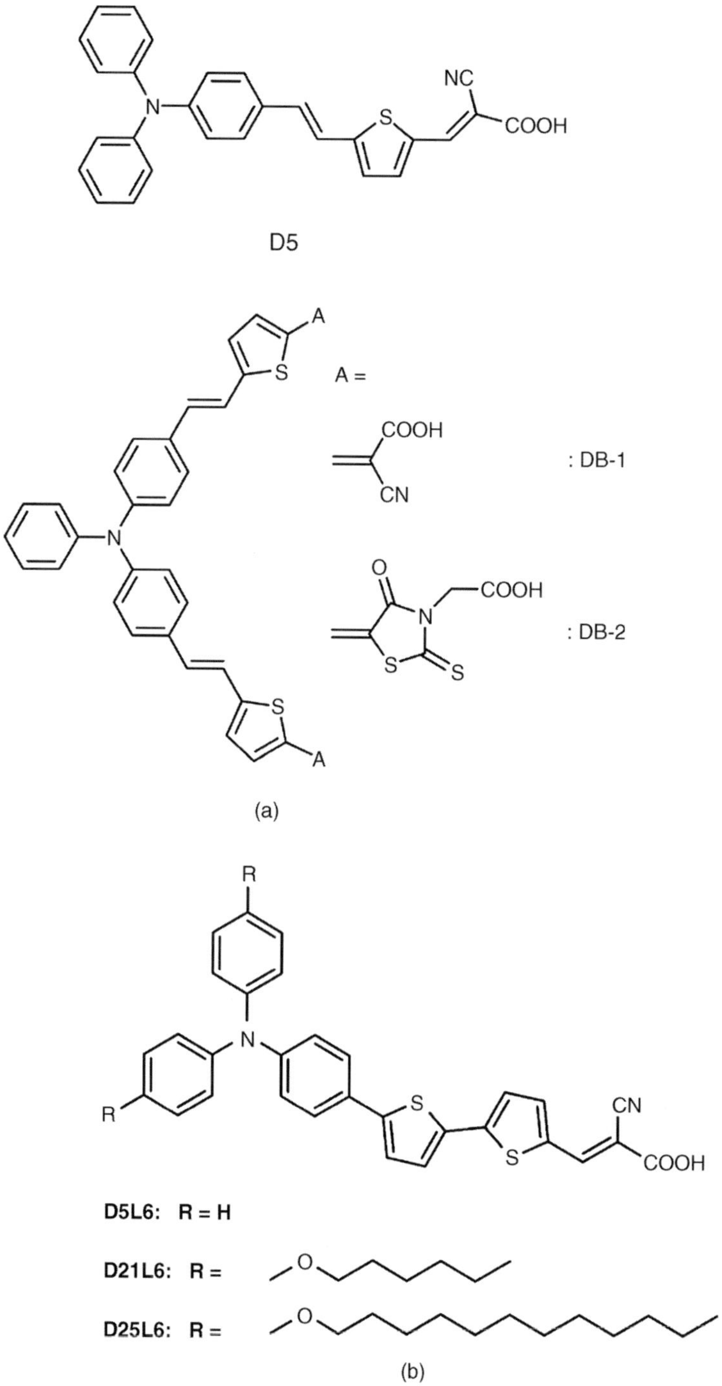

Figure 5.6 Structures of di-branched DSSC organic sensitizers and reference compound DS [63]. Reproduced with permission from the Royal Chemical Society.

the π-conjugated bridge, and cyanoacrylic acid group as the electron acceptor were designed and synthesized for application in dye-sensitized nanocrystal TiO$_2$ solar cells [64]. Upon anchoring onto TiO$_2$ film, MKZ-21 exhibited a better photovoltaic performance, that is, a monochromatic incident photon-to-current conversion efficiency (IPCE) of 83%, a short-circuit photocurrent density (J_{sc}) of 15.4 mA cm^{-2}, an open-circuit voltage (V_{oc}) of 0.71 V, and a fill factor (FF) of 0.67, corresponding to an overall conversion efficiency (η) of 7.3% under standard AM 1.5G irradiation (100 mW/cm^2).

Dibranched dianchoring organic sensitizers were prepared and used in dye-sensitized solar cells leading to redshifted IPCE maxima and increased photocurrent compared to the corresponding monobranched monoanchoring dye [65]. The sensitizers yielded a power conversion efficiency of 5.7% with enhanced stability under one-sun conditions from the dianchoring groups.

In the study [66], mesoporous titanium dioxide with nanograins of dimension in the range 16–20 nm was prepared through the soft-templating approach using various cationic surfactants such as octyl-, dodecyl-, and cetyl trimethylammonium bromide with different surfactant compositions and titania precursor concentrations. As-synthesized mesoporous titanium dioxide samples were characterized by TGA, PXRD, FESEM, HRTEM, and surface area measurements, used as photoelectrode material in DSSCs. Under global AM 1.5 solar irradiation, the best photovoltaic performance of 7.5% with a short-circuit photocurrent density of 14.2 mA/cm^2, an open-circuit voltage of 748 mV, and a fill factor of 70.83% were obtained for the DSSC using a film of mesoporous TiO2 synthesized from the cetyl trimethyl-ammonium bromide surfactant.

Among the different classes of red absorbing chromophores, squaraine dyes have received considerable attention in recent years because of their intense absorption in the red/near-IR regions [67, 68]. In the squaraine-based solar cell extinction coefficients often exceeding 300 000 mol/(cm l) showed relatively narrow band located in 600–800 nm region. Recently, three unsymmetric squaraine dyes JK-64, JK-65, and JK-64Hx (Figure 5.7), containing a bulky spirobifluorene or hexyloxyphenyl unit, were synthesized [69]. These sensitizers, when anchored onto a TiO$_2$ surface, exhibited both decreased aggregation and enhanced unidirectional flow of electrons. Under standard global AM 1.5 solar conditions, an optimized JK-64Hx sensitized cell gave a short-circuit photocurrent density (J_{sc}) of 12.82 mA/cm^2, an open-circuit voltage (V_{oc}) of 0.54 V, and a fill factor (FF) of 0.75, corresponding to an overall conversion efficiency (η) of 5.20%.

Two organic sensitizers, MKZ-21 and MKZ-22 (Figure 5.8), comprising 5,11-dioctylindolo[3,2-b]carbazole moiety as the electron donor, n-hexyl-substituted oligothiophene units as the π-conjugated bridge, and cyanoacrylic acid group as the electron acceptor, were synthesized for application in dye-sensitized nanocrystalline TiO$_2$ solar cells [70]. For the functionalized organic sensitizers MKZ-21 and MKZ-22, upon anchoring onto TiO2 film, MKZ-21 exhibited a better photovoltaic performance: a monochromatic IPCE of 83%, a short-circuit photocurrent density (J_{sc}) of 15.4 mA/cm^2, an open-circuit voltage (V_{oc}) of 0.71 V, and a fill factor (FF) of 0.67, corresponding to an overall conversion efficiency (η) of 7.3% under standard AM 1.5 G irradiation (100 mW/cm^2).

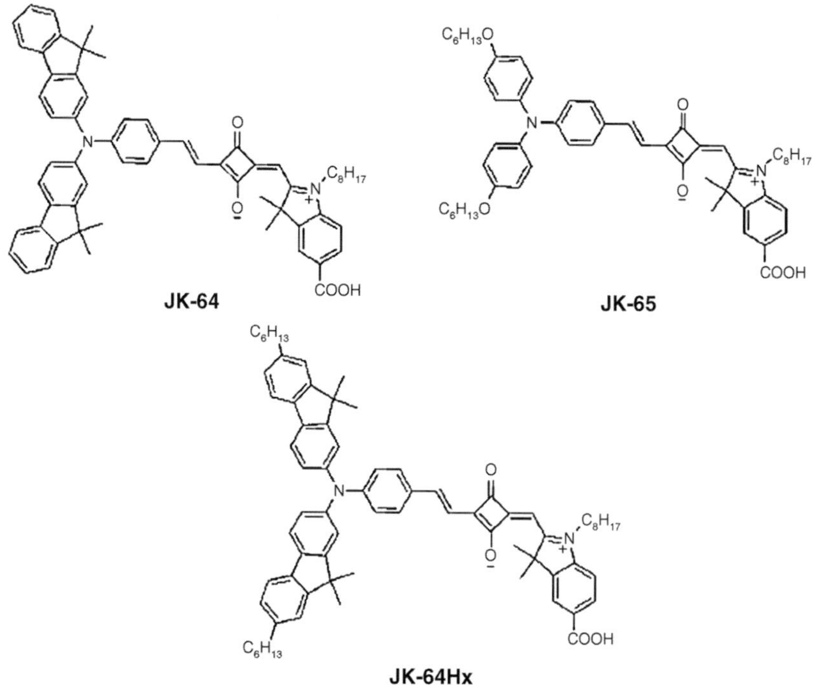

Figure 5.7 Structure of the dyes of JK-64, JK-65 and JK-64Hx [69]. Reproduced with permission from Elsevier.

New *organic* D-π-A dyes with fluorine substitution and inorganic composite electrolyte containing catalytic functional polypyrrole nanoparticles (NPs), A1, A2-H, and A2-F, possessing a remarkably high absorption extinction coefficient of $\varepsilon > 5.0 \times 10^4 \, \text{M}^{-1} \, \text{cm}^{-1}$ at peak wavelength were synthesized [71, 72]. A2-F having a key

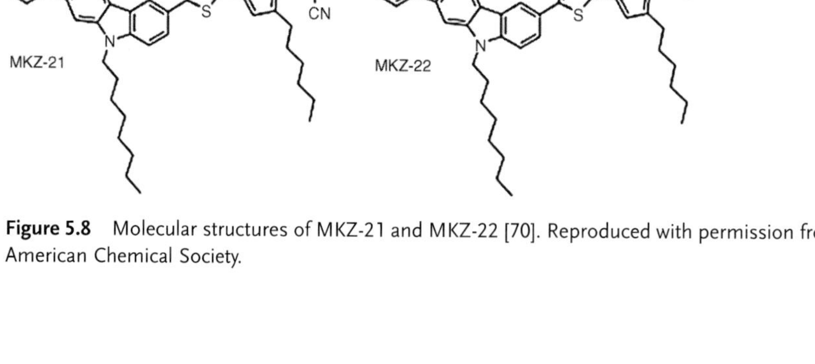

Figure 5.8 Molecular structures of MKZ-21 and MKZ-22 [70]. Reproduced with permission from American Chemical Society.

F substitution attains all solid-state DSSC performance, with optimized parameters of $\eta = 4.86\%$, $J_{SC} = 7.52\,\text{mA/cm}^2$, $V_{OC} = 0.91\,\text{V}$, and $FF = 0.71$. Dye-sensitized solar cells with cobalt-based mediators with efficiencies surpassing the record for DSSCs with iodide-free electrolytes were developed by selecting a suitable combination of a cobalt polypyridine complex and an organic sensitizer [73]. The effect of the steric properties of two triphenylamine-based organic sensitizers and a series of cobalt polypyridine redox mediators on the overall device performance in DSSCs and on transport and recombination processes in these devices was compared. The recombination and mass transport limitations were avoided by matching the properties of the dye and the cobalt redox mediator. Organic dyes with higher extinction coefficients than the standard ruthenium sensitizers were employed in DSCs in combination with outer-sphere redox mediators, enabling thinner TiO_2 films to be used. Recombination was reduced further by introducing insulating butoxyl chains on the dye. Optimization of DSSCs sensitized with a triphenylamine-based organic dye in combination with tris(2,2′-bipyridyl)cobalt(II/III) yielded solar cells with overall conversion efficiencies of 6.7% and open-circuit potentials of more than 0.9 V under 1000 W/m² AM 1.5 G illumination.

Teng et al. [74] reported the design, synthesis, and application of 10 novel organic dyes (TC201-TC602, Figure 5.9) in which triphenylamine (TPA) derivatives were used as electron donating moieties, cyanoacrylic acid as electron acceptor, and anthracene moiety as π-conjugations to bridge the donor–acceptor (D-A) systems. A double bond and triple bond were introduced into different positions of the π-conjugation systems to tune the molecular structure and their configurations. With the introduction of the anthracene moiety, together with a triple bond for the fine-tuning of molecular planar configurations and to broaden absorption spectra, the short-circuit photocurrent densities (J_{sc}) and open-circuit photovoltages (V_{oc}) of DSSCs were improved, which was attributed to much broader absorption spectra of the dyes with the anthracene moiety. Electrochemical impedance spectroscopy (EIS) analysis revealed that the introduction of the anthracene moiety suppresses the charge recombination arising from electrons in TiO_2 films with I_3^- ions in the electrolyte, which provided a prominent solar energy conversion efficiency (η) up to 7.03%, J_{sc} of 12.96 mA/cm², V_{OC} of 720 mV, and FF of 0.753 under simulated AM 1.5 irradiation (100 mW/cm²).

Different molecular layers on TiO_2 were prepared by using the p-dimethylaniline triphenylamine-based organic dye, D29, together with the coadsorbents decylphosphonic acid (DPA), dineohexyl bis(3,3-dimethylbutyl)phosphinic acid (DINHOP), and CDCA [75]. The surface molecular structure of dye and coadsorbent layers on TiO_2 was investigated by photoelectron spectroscopy (PES). Two new organic dyes adopting coplanar diphenyl-substituted dithienosilole as the central linkage have been synthesized, characterized, and used as the sensitizers for DSSCs [76]. The best DSSC exhibited a high power conversion efficiency up to 7.6% (TP6CADTS) under AM 1.5 G irradiation. Phenylethynyl-substituted porphyrin (PE1) sensitizers bearing a nitro, cyano, methoxy, or dimethylamino phenylethynyl substituent were prepared to examine the electron donating or withdrawing effects of dyes on the photovoltaic

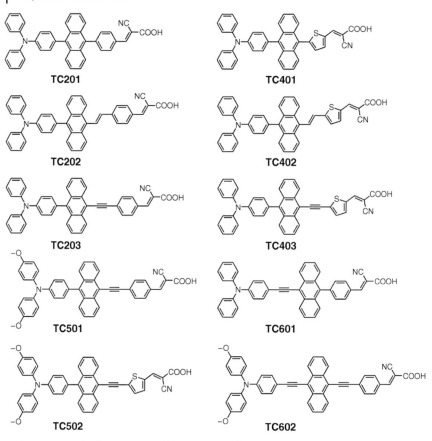

Figure 5.9 Molecular structures of 10 novel metal-free organic dyes (TC201-TC602) bridged by anthracene-containing π-conjugations [74]. Reproduced with permission from the American Chemical Society.

performance of the corresponding dye-sensitized solar cells [77]. The overall efficiencies of power conversion of the devices showed a systematic trend Me_2N-PE1 > MeO-PE1 > CN-PE1 > NO_2-PE1, for which Me_2N-PE1 has a device performance about 90% that of a N719-sensitized solar cell. The authors attributed the superior performance of Me2N-PE1 to the effective electron donating property of the dye that exhibits broadened and redshifted spectral features. Electrochemical tests indicated that both LUMO and HOMO levels show a systematic trend Me_2N-PE1 > MeO-PE1 > CN-PE1 > NO_2-PE1, consistent with the trend of variation in the short-circuit currents in this series of sensitizers.

The performances of IR dye-sensitized solar cells fabricated using novel cyanine dye NK6037 (Figure 5.10; light absorption edge: 900 nm) with interaction between the dye and the TiO_2 photoelectrode via the functional group were investigated [78]. The efficiency of electron injection was found to be 1.9% by optimizing the light

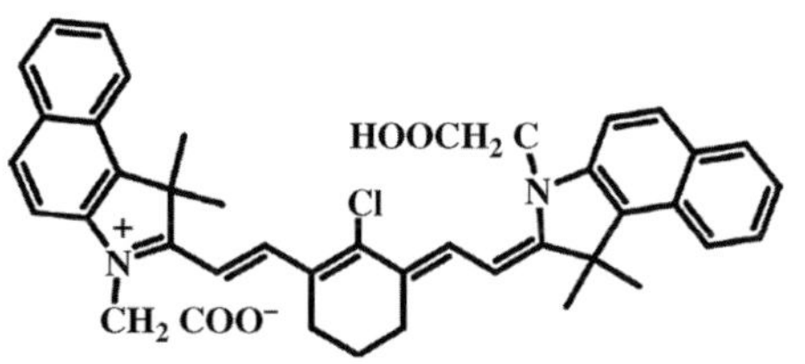

Figure 5.10 Molecular structures of infrared dyes, NK-4432 and NK-6037. NK-4432 – X_1: C_2H_5, X_2: C_2H_5, Y: BF_4; NK-6037 – X_1: CH_2COO, X_2: CH_2COOH, Y: none [78]. Reproduced with permission from Elsevier.

confining effect of the TiO_2 photoelectrode and TiO_2 film thickness, and reached 2.3% by adjusting the concentration of deoxycholic acid (DCA) in the dye solution.

The performances of infrared dye-sensitized solar cells fabricated using two novel cyanine dyes (light absorption edge: 900 nm) were investigated [79]. The efficiency reached 1.9% by optimizing the light-confining effect of the TiO_2 photoelectrode and TiO_2 film thickness, and reached 2.3% by adjusting the concentration of deoxycholic acid (DCA) in the dye solution. A method for measuring the effect of an externally applied electric field on the absorption of dye (Figure 5.11) monolayers adsorbed on flat TiO_2 substrates was developed [80]. The measured signal has the shape of the first derivative of the absorption spectra of the dyes and reverses sign along with the reversion of the direction of the change in dipole moment upon excitation relative to the TiO_2 surface. Similar signal was observed in photoinduced absorption spectra of dye-sensitized TiO_2 electrodes under solar cell conditions, demonstrating that the electric field across the dye molecules changed upon illumination.

5.3.1.4 Semiconductor Sensitizes

Semiconductors composed from PbS, CdS, CdSe, CdTe, and In_2S_3 $Cu_{2-x}S$, using as light-absorbing material in place of dye molecules, possess a number of advantages including a high light-harvesting capability, a tunable bandgap over a wide range, and a large intrinsic dipole moment [81–87]. CdSe nanocrystals as a sensitizer for TiO_2 solar cells are being widely investigated [83–87]. An improvement in the injection from CdSe nanocrystals to TiO_2 photoelectrodes efficiency has been realized by

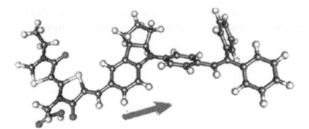

Figure 5.11 Chemical structures of D149 of the difference in dipole moment between ground and excited states ($\Delta\mu$) obtained from TD-DFT calculations indicated by an arrow. By definition, μ_f points from negative to positive charge [80]. Reproduced with permission from the American Chemical Society.

chemically attaching CdSe nanocrystals to TiO_2 via cross-linking moieties. However, the power conversion efficiencies are lagging behind those obtained for the more traditional (transition–metal complexes and organic dye-sensitized) DSSCs [88–94]. For example, the semiconductors as sensitizers suffer from a poor efficiency of about 1% [95].

In order to capture a maximum amount of the incoming light, the semiconductor nanomaterials are used as scaffold to hold large numbers of the dye molecules in a 3D matrix, increasing the number of molecules for any given surface area of the cell. The use of light scattering layers (LSLs) consisting of larger titania particles, which work as a phototrapping photovoltaic system, have been reported. Using light scattering particles, Koo et al. [96] observed efficiency increments of about 15%.

Nanostructures, namely, nanotubes [97–100], nanowires, [101], nanorods [102, 103], and inverse structures [104, 105] have appeared as the most promising materials in this field. Oriented tubule nanostructures made with carbon metal oxides such as ZnO [105, 106] and other templates were also reported [107–110].

The differences between the liquid junction dye-sensitized solar cell with solar efficiencies of 11% and the semiconductor-sensitized analogue (SSSC) with a maximum efficiency of 2.8% were discussed, considering typical charge transfer times for the various current generating and recombination processes [110]. Three main factors that could contribute to differences between the two types of cells were taken into consideration: the multiple layers of absorbing semiconductor on the oxide, the different electrolytes normally used for the two types of cells, and the charge traps in the absorbing semiconductor. Entropic effects and the irreversible nature of electron injection of the normally used Ru dye into TiO_2 were also considered. The author concluded that although the DSSC does possess some fundamental advantages, we can expect large improvements in efficiency of the SSSC, possibly reaching values comparable to the DSSC.

5.3.2
Photoanode

Since the pioneering research by Grätzel and O'Regan [5], TiO_2 has been the preferred semiconductor anode in DSSCs, despite some promising properties offered by other metal oxides such as ZnO, SnO_2, and Nb_2O_5 [111]. Several deposition techniques are generally used for preparation of TiO_2 films. The film morphology is a major variability factor in a DSSC's performance mainly because of its influence on the electron recombination rate [112, 113].

An antireflective hybrid nanostructure was fabricated using anatase TiO_2 nanobelts synthesized by a hydrothermal route and a ZnO nanowire array grown via a low-temperature solution-phase process [114]. It was shown that the replacement of TiO_2 nanoparticles with TiO_2 nanobelts improved the electron transport in the TiO_2 porous film. Rigorous coupled wave analysis and reflectance measurements indicated that the well-designed composite of TiO_2 nanobelt–ZnO nanowire array acted as an efficient antireflection coating for photoanode. It was shown that the photoanode made of this hybrid nanostructure enhances the performance of dye-sensitized solar cells by minimizing the electron–hole recombination-related and reflection-induced energy loss. A simple photoelectrochemical method was developed to measure the intrinsic electron transport resistance (R_0) of TiO_2 photoanodes [115]. A series of TiO_2/FTO photoanodes with different electron transport resistance were fabricated using conventional screen-printing technique, surface modifications using titanium organic solvent and $TiCl_4$ aqueous solution. The results suggested that these surface modifications significantly decrease R_0 values, which bestowed better photovoltaic performance than the corresponding nonmodified photoanodes.

Dye-sensitized solar cells using titanium dioxide (TiO_2) electrodes with different haze (horizontal obscuration) were investigated [116]. It was found that the incident photon-to-current conversion efficiency of DSSCs increases with increase in the haze of the TiO_2 electrodes, especially in the near infrared wavelength region. Conversion efficiency of 11.1%, measured by a public test center, was achieved using high-haze TiO_2 electrodes.

Hierarchical or one-dimensional architectures are among the most exciting developments in material science these recent years. Sauvage et al. [117] presented a nanostructured TiO_2 assembly combining these two concepts, resembling a forest composed of individual, high aspect ratio, tree-like nanostructures. These structures were used for the photoanode in dye-sensitized solar cells and a 4.9% conversion efficiency in combination with C101 dye was achieved. It was demonstrated that this morphology is beneficial to hamper the electron recombination and also mass transport control in the mesopores when solvent-free ionic liquid electrolyte is used.

Nanostructured TiO_2 hollow fibers have been prepared using natural cellulose fibers as a template [118]. This material was used to produce highly porous photoanodes incorporated into dye-sensitized solar cells and exhibited enhanced electron transport properties compared to mesoscopic films made of spherical nanoparticles. Photoinjected electron lifetime was multiplied by 3–4 in the fiber morphology, while the electron transport rate within the fibrous photoanode was doubled. A nearly quantitative absorbed photon-to-electric current conversion yield exceeding 95% was achieved upon excitation at 550 nm and a photovoltaic power conversion efficiency of 7.2% was reached under simulated air mass 1.5 (100 mW/cm^2) solar illumination.

A hollow TiO_2 nanoribbon network electrode for DSSC was fabricated by a biotemplating process combining peptide self-assembly and atomic layer deposition (ALD) [119]. A thin TiO_2 layer was deposited at the surface of an aromatic peptide of diphenylalanine nanoribbon template via the ALD process. The hollow TiO_2 nanoribbon network electrode was integrated into DSSC devices. Hollow TiO_2 nanoribbon-based DSSCs exhibited a power conversion efficiency of 3.8%. In the electrolyte containing Li^+ and tert-n-butylammonium (TBA^+), the band edge movement, trap state distribution, electron recombination, and electron transport in dye-sensitized solar cells before and after TiO_2 film surface coating with Yb_2O_3 was studied [120]. After surface coating, the band edge shifted negative in the Li^+ electrolyte. In both types of electrolytes, the Yb_2O_3-coated TiO_2 film suppressed the recombination and slowed down the electron transport. A significant efficiency improvement in the black dye-sensitized solar cell through protonation of TiO_2 films was reported [121]. The influence of acid pretreatment of TiO_2 mesoporous films prior to dye sensitization on the performance of dye-sensitized solar cells based on $[Bu_4N]_3[Ru(Htcterpy)(NCS)_3]$ (tcterpy = 4,4′,4″-tricarboxy-2,2′,2″-terpyridine) was described. It was found that the HCl pretreatment caused an increase in overall efficiency by 8%, with a major contribution from photocurrent improvement. The analysis of incident photon-to-electron conversion efficiency, UV–visible absorption spectra, redox properties of the dye and TiO_2, and the impedance spectra of the dye-sensitized solar cells indicated that photocurrent enhancement was attributable to the increases in electron injection and/or charge collection efficiency. The suppression of electron transfer from conduction band electrons to the I_3^- ions in the electrolyte upon HCl pretreatment was shown. An overall efficiency of 10.5% with the black dye was obtained under illumination of simulated AM 1.5 solar light (100 mW/cm^2) using an antireflection film on the cell surface.

A process for synthesizing vertically aligned TiO_2 nanotube arrays on fluorine-doped tin oxide, transparent conducting oxide (TCO), through a liquid-phase conversion process using ZnO nanowire arrays as a template, was described [122]. The resulting TiO_2 nanotube arrays was integrated with the fabrication process of DSSCs and led to improved photovoltaic performance compared to ZnO nanowire-based DSSCs. It was shown that the lifetime of photogenerated electrons in TiO_2 nanotubes is more than an order of magnitude larger than that in sintered TiO_2 nanoparticles. The authors stressed that this result provides opportunities to further improve DSSCs, for example, by employing solid-state electrolytes and redox mediators with faster kinetics. A contact-free method of surface potential measurement using scanning Kelvin probe microscopy (SKPM) was conducted to probe the nature of nanocrystal TiO_2/dye interface [123]. In combination with electrical measurements, an effort has been made to establish a correlation between the nature of sensitizing dye and the observed surface potential with open-circuit voltage (V_{oc}) after dye-sensitized solar cell fabrication.

Oriented tubule nanostructures made with carbon [105], metal oxides such as ZnO and TiO_2 in particular [106, 124–127], have been applied to DSSC electrodes, replacing the semiconducting mesoporous layer. For example, a 9.3% efficiency was obtained with oriented anatase nanowires and they are particularly favorable in solvent-free ionic liquid electrolytes [128].

Titanium dioxide nanotubes (TiO_2 NTs) with various sizes have been prepared by low-temperature chemical synthesis using anatase TiO_2 particles with different crystallite sizes in NaOH solution and used as a photoelectrode in a dye-sensitized solar cell [124]. The electrodes made from modified TiO_2 NTs showed a strong dependence on their surface area and resultant amount of dye adsorption; the surface area decreased with increase in the diameter of the NT from 9.8 to 23.6 nm. A flexible photoanode with zinc oxide film on titanium foil was prepared and its application in a DSSC was investigated [125]. The ZnO film with a mosaic structure, composed of densely packed ZnO nanosheets (ZnONS), was obtained by calcining a film of layered hydroxide zinc carbonate (LHZC), which was previously grown directly on a Ti foil via chemical bath deposition (CBD). The highly porous ZnONS film with a thickness of about 25 μm facilitated the preparation within 4 h under CBD conditions.

Mesoporous carbon (MC) with a surface area of 380 m²/g was prepared and employed as the carbon support of Pt catalyst for counterelectrode of dye-sensitized solar cells [126]. Pt/MC samples containing 1 wt% Pt were prepared by reducing chloroplatinic acid on MC using wet impregnation. It was found that Pt nanoparticles were uniform in size and highly dispersed on MC supports. The average size of Pt nanoparticles was about 3.4 nm. Pt/MC electrodes were fabricated by coating Pt/MC samples on fluorine-doped tin oxide glass.

5.3.3
Injection and Recombination

In the field of research on dye-sensitized solar cells, more and more attention is drawn to understanding the electron transfer kinetics in this type of cells [129–141]. The effect of changing the electron transfer distance has been investigated for a number of ruthenium complexes [135] and for organic dyes [136], showing different distance dependence up to an exponential distance dependence of the electron transfer rate constant with an attenuation factor β of 1 Å^{-1} [138, 139].

Several theoretical works took into account the effect of nonlinear recombination kinetics involving transfer of electrons in the TiO_2 to I^{3-} ions in the electrolyte [140–144]. Simulation of steady-state characteristics of dye-sensitized solar cells and the interpretation of the diffusion length were reported [142]. It was shown that the diffusion length, L_n, defined from the probability of collection, was independent of the macroscopic perturbation for $\beta \neq 1$ only for a small perturbation, and in this case, it coincided with the value $\lambda_n = (D_n \tau_n)^{1/2}$ that can be measured by impedance spectroscopy (IS) under homogeneous conditions. The increase in the diffusion length with the potential, usually observed experimentally, was attributed to the increase in the free carrier lifetime. A steady-state method for determination of the electron diffusion length in dye-sensitized solar cells was described and illustrated with data obtained using cells containing three different types of electrolyte [143]. The method was based on using near-IR absorbance techniques to establish pairs of illumination intensity for which the total number of trapped electrons is the same at open circuit as at short circuit. Electron diffusion length values obtained by this method were compared with values derived by intensity-modulated methods and by impedance measurements under illumination.

A direct computation of the diffusion length, as a crucial parameter controlling the electron collection efficiency in DSSCs, was carried out [144]. The computation was performed for a DSSC with a short diffusion length by running a random walk numerical simulation with an exponential distribution of trap states and explicit incorporation of recombination. The diffusion length and the lifetime were estimated from the average distance traveled and the average survival time of the electrons between recombination events. The results demonstrate the compensation effect between diffusion and recombination that keeps the diffusion

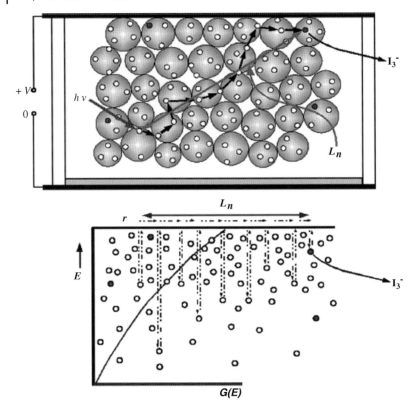

Figure 5.12 Illustration of the random walk numerical procedure utilized in this work to compute the electron diffusion length, L_n. A three-dimensional network of traps is distributed randomly and homogeneously in space. The energies of the sites are taken from an exponential distribution $G(E)$ given by Equation (5.10). A recombining character is given to an arbitrary amount of traps (solid circles) so that when an electron reaches one of these traps, it may undergo recombination (reaction with I^{3-}) and be removed from the sample [144]. Reproduced with permission from the American Chemical Society.

length approximately constant on a wide range of illumination intensities or applied biases. Further developing the model, the authors introduced a recombination probability that depends exponentially on the Fermi level, which leads to a nonconstant diffusion length. Figure 5.12 illustrates the random walk numerical procedure utilized in this work to compute the electron diffusion length.

Electron transfer from three conjugated amino-phenyl acid dyes to titanium and aluminum oxide nanocrystalline films was studied by using transient absorption spectroscopy with sub-20 fs time resolution over the visible spectral region [145]. Global analysis of the transient kinetics of the dyes on TiO_2 revealed stimulated emission decays of about 40 fs and less than 300 fs assigned to electron injection. The same dyes on Al_2O_3 substrates displayed long stimulated emission decays. For two of the dyes, 2E,4E-2-cyano-5-(4-dimethylaminophenyl) penta-2,4-dienoic acid

(NK1), 2-[(E)-3-(4-dimethylaminophenyl)-allylidene]malonic acid (NK2), and 2E,4E-2-cyano-5-(4-diphenylaminophenyl)penta-2,4-dienoic acid (NK7)) with amino methyl terminal groups, cation formation was seen at 670 nm probe wavelength. It was suggested that early excited state dynamics observed in the NK7 dye bound to both TiO_2 and Al_2O_3, with pulse-limited rise and 30–40 fs decay times, could be the reason for the low efficiency reported for the NK7 sensitized solar cell.

Zhu et al. [146] reported on the microstructure and dynamics of electron transport and recombination in dye-sensitized solar cells incorporating oriented TiO_2 nanotube arrays prepared from electrochemically anodized Ti foils. The morphology of the nanotube arrays that were characterized by scanning electron microscopy (SEM), transmission electron microscopy, and X-ray diffraction consisted of closely packed nanotubes, several micrometers in length, with typical wall thicknesses and intertube spacing of 8–10 nm and pore diameters of approximately 30 nm. Molecules of the N719 dye, [tetrabutylammonium]$_2$[Ru(4-carboxylic acid-4′-carboxylate-2,2′-bipyridyl)$_2$(NCS)$_2$], were shown to cover both the interior and the exterior walls of the nanotubes The transport and recombination properties of the nanotube and nanoparticle films used in DSSCs were studied by frequency-resolved modulated photocurrent/photovoltage spectroscopies. Recombination was found to be much slower in the nanotube films, indicating that the nanotube-based DSSCs have significantly higher charge collection efficiencies than their nanoparticle-based counterparts. Analysis of photocurrent measurements indicated that the light-harvesting efficiencies of nanotube-based DSSCs were higher than those found for DSSCs incorporating nanoparticles owing to stronger internal light-scattering effects.

Intensity-modulated photocurrent spectroscopy and intensity-modulated photovoltage spectroscopy were employed to measure the dynamics of electron transport and recombination in the ZnO nanowire (NW) array-ZnO/layered basic zinc acetate (LBZA) nanoparticle composite dye-sensitized solar cells (Figure 5.13) [147]. The roles of the vertical ZnO NWs and insulating LBZA in the electron collection and transport in DSSCs was investigated by comparing the results with those in the TiO_2-NP, horizontal TiO_2-NW, and vertical ZnO-NW array DSSCs. It was shown that the electron transport rate and electron lifetime in the ZnO NW/NP composite DSSC are superior to those in the conventional TiO_2-NP cell and the ZnO NW/NP composite anode is able to sustain efficient electron collection over much greater thickness than the TiO_2-NP cell does. Consequently, a larger effective electron diffusion length is available in the ZnO composite DSSC. Current density (J) versus voltage (V) characteristics of the cells used in this work are shown in Figure 5.14.

Several techniques have been applied to determine the extract recombination time [148–155]: BHJ devices, including modulated photo-induced absorption [149], transient absorption [150], photo-CELIV [151], double-injection currents [152], and time-of-flight methods [153].

Electron transport and recombination in dye-sensitized solar cells made of single-crystal rutile TiO_2 nanowires were investigated [154]. It was shown that the electron transport rate in dye-sensitized solar cells made of single-crystal rutile titanium

174 | 5 Dye-Sensitized Solar Cells I

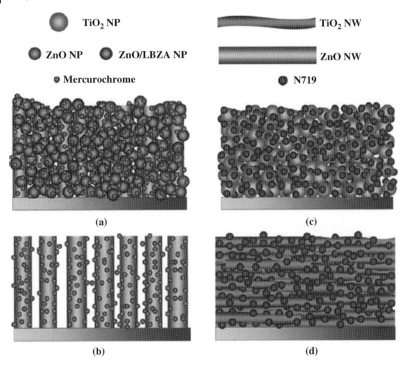

Figure 5.13 Schematics of the (a) ZnO NW array/NP composite, (b) vertical ZnO–NW array, (c) TiO$_2$–NP, and (d) horizontal TiO$_2$–NW DSSCs [147].

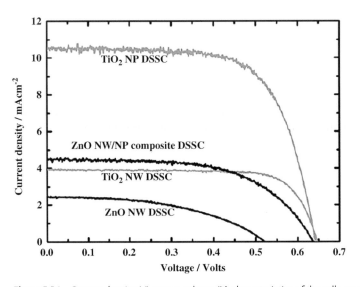

Figure 5.14 Current density (J) versus voltage (V) characteristics of the cells used in work [147].

dioxide nanowires is found to be similar to that measured in dye-sensitized solar cells made of titanium dioxide nanoparticles. Interfacial charge separation and recombination were quantified at sensitized mesoporous nanocrystal TiO_2 interfaces immersed in acetonitrile electrolyte [155]. Two sensitizers containing a phenylenethynylene spacer between a cis-Ru(NCS)$_2$ core and TiO_2 anchoring groups, and a third sensitizer not containing the spacer, cis-Ru(dcb)(bpy)(NCS)2, where bpy is 2,2′-bipyridine and dcb is 4,4′-(CO$_2$H)$_2$-bpy, were employed. It was found that excited state injection occurred with approximately the same yield for all these sensitizers and was rapid with rate constant $k_{inj} > 10^8\,s^{-1}$. Representative charge recombination rate constants from nanosecond transient absorption data were found to be three times slower for the sensitizers with the phenylenethynylene spacer.

Electrochemical impedance spectroscopy and transient voltage decay measurements are applied to compare the performance of dye-sensitized solar cells using organic electrolytes, ionic liquids, and organic hole conductors as hole transport materials (HTMs) [156]. Nanocrystalline titania films sensitized by heteroleptic ruthenium complex NaRu(4-carboxylic acid-4′-carboxylate) (4,4′-dinonyl-2,2′-bipyridyl)(NCS)(2), coded Z-907Na, were employed as working electrodes. The influence of the nature of the HTMs on the photovoltaic figures of merit, that is, the open-circuit voltage, short-circuit photocurrent, and fill factor was evaluated. In order to derive the electron lifetime, as well as the electron diffusion coefficient and charge collection efficiency, EIS measurements were performed in the dark and under illumination corresponding to realistic photovoltaic operating conditions of these mesoscopic solar cells.

5.3.4
Charge Carrier Systems

Specific interactions of the I^-/I_3^- redox mediators with the reduced and oxidized dye, Ru(4,4′-dicarboxy- 2,2′-bipyridyl)$_2$(NCS)$_2$ (N3) or Ru(dcbpy)$_2$(NCS)$_2$, have been studied by means of density functional theory with the focus on the charge transfer process involving $\{dye^+\,I^-\}$ adducts [157]. Different pathways leading to $\{dye^+\,I^-\}$ adducts have been analyzed. Mechanistic insights into the interaction of I^- with RuIII(dcbpy)$_2$(NCS)$_2$ via an SCN^- ligand directly giving rise to [RuII(dcbpy)$_2$(NCS)$_2$I]^0 have been obtained with the distinctive S–I bonding, while the binding of I^- to the N3 dye cation via I^-–dcbpy interactions has been taken into consideration. Evidence for a charge transfer process in the presence of only one I^- anion in the outer coordination sphere of the ruthenium center has been identified. Geometries and electronic structures of plausible intermediates have been computationally analyzed including a two-step regeneration reaction, [RuIII(dcbpy)$_2$(NCS)$_2$]$^+ + I^- \rightarrow$, followed by the interaction of a second I^- with the intermediate [RuIII(dcbpy)$_2$(NCS)I]$^+$ complex.

The optimized geometry of the Ru–I complex with S–I bonding is shown in Figure 5.15.

On the basis of these data, the two-step mechanism of the regeneration reaction was formulated.

176 5 *Dye-Sensitized Solar Cells I*

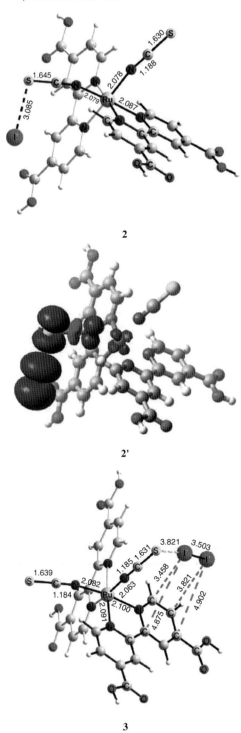

2

2'

3

Dyes' linker lengths were ranged from 17.2 to 11.0 Å. XPS studies have shown that the dye molecules are rather vertically arranged with the triphenyl amine pointing out from the surface (Figure 5.16) [158]. Energy level diagram presented in Figure 5.17 shows driving forces for injection and recombination. Three new sensitizers for photoelectrochemical solar cells were synthesized consisting of a triphenylamine donor, a rhodanine-3-acetic acid acceptor and a polyene connection [159] The photovoltaic performance of this set of dyes as sensitizers in mesoporous TiO_2 solar cells was investigated using electrolytes containing the iodide/triiodide redox couple. For all dyes, the injection rate was found to be larger than $(200\,fs)^{-1}$. The subsequent recombination reaction increased with increasing linker length, which was consistent with the concomitant decrease in driving force for this series of dyes. The lifetimes of electron transfer showed exponential distance dependence (Figure 5.18), when corrected for driving force and reorganization energy, which indicates a superexchange interaction between the electrons in TiO_2 and the radical cations of the dyes. A dependence on probe wavelength of the attenuation factor was found, giving a value of $0.38\,\text{Å}^{-1}$ at 940 nm and $0.49\,\text{Å}^{-1}$ at 1040 nm.

The influence of annealing temperature on the microstructure and dynamics of electron transport and recombination in dye-sensitized solar cells incorporating oriented titanium oxide nanotube (NT) array was investigated [160]. The morphology of the NT arrays was characterized by scanning and transmission electron microscopies and Raman and X-ray diffraction spectroscopies. Over the temperature range from 200 to 600 °C, the crystallinity, crystal phase, and structural integrity of the NT walls underwent pronounced changes whereas the overall film architecture remained intact.

A series of ruthenium phthalocyanines, a dye class with large and tunable absorption in the red, were prepared and photovoltage transients and charge density measurements were examined [161]. It was demonstrated that reduction in voltage is caused by a 100-fold increase in the rate constant for recombination at the TiO_2/electrolyte interface. By examination of the literature, the authors proposed that catalysis of the recombination reaction may be occurring for many other classes of potentially useful dyes including porphyrins, coumarins, perylenes, cyanines, merocyanines, and azulene.

An idea proposed in the work [162] was to use photochemistry to new iodide redox reactions. The reactivity of oxidized iodide species with mesoporous nanocrystal (anatase) TiO_2 thin films, related to unwanted charge recombination processes in dye-sensitized solar cells, was quantified spectroscopically on nanosecond and longer timescales in half molar iodide MeCN solution. Under forward bias conditions, TiO_2 did not react with photogenerated I radical anions, $I_2^-\cdot$, which disproportionated with a rate constant $k = 3 \times 10^9\,M^{-1}\,s^{-1}$. The authors suggested that the reduction of

Figure 5.15 The optimized geometry of the Ru-I complex with S-I bonding; charge) 0, electronic spin state) doublet; dashed black line highlights the S-I bond. **2′**: The spin density of the ground electronic state of complex **2** with one unpaired electron. **3**: The {dye $I_2 -$} complex with an outer-sphere binding of $I_2 -$; dashed lines highlight close contacts of $I_2 -$ with C atoms of the dcbpy ligand; dashed a line highlights the S-I distance. All distances are in angstroms [157]. Reproduced with permission from the American Chemical Society.

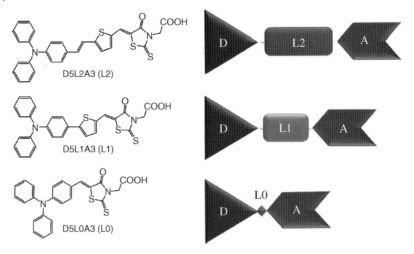

Figure 5.16 Dye structures (left) with their building block representation (right) [158]. Reproduced with permission from the American Chemical Society.

I_2^- by $TiO_2(e^-)$ does not compete kinetically with rapid $I_2^- \cdot$ disproportionation. However, $TiO_2(e^-)$ decreased the concentration of tri-iodide, I_3^-.

Shown in Figure 5.19 is a simplified scheme that displays the key features of the photoelectrochemical cell when the light absorption forms an electron–hole pair (e^-/h^+).

Bandgap excitation of TiO_2 was shown to have resulted in iodine oxidation and $I_2^- \cdot$ formation:

$$TiO_2 + hv \rightarrow TiO_2(e^-, h^+)$$

$$2I^- + TiO_2(e^-, h^+) \rightarrow TiO_2(e^-) + I_2$$

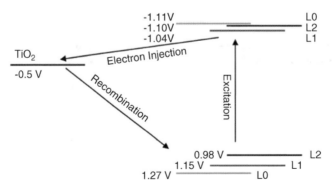

Figure 5.17 Energy level diagram showing driving forces for injection and recombination. The addition of tBP is estimated to raise the energy level of the TiO_2 CB by 0.15 V [158]. Reproduced with permission from the American Chemical Society.

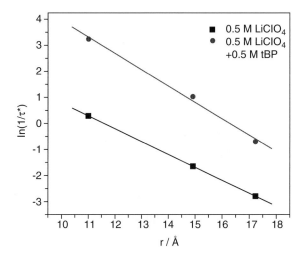

Figure 5.18 Distance dependence of the driving force-corrected averaged lifetimes of L_0–L_2 when probing at 940 nm (a) and 1040 nm (b) [158]. Reproduced with permission from the Royal Chemical Society.

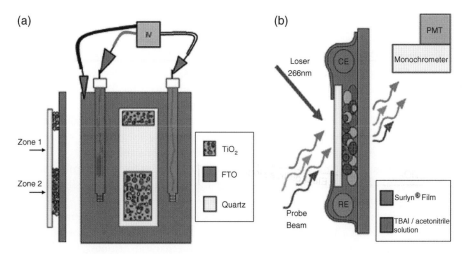

Figure 5.19 (a) Side and front view of the photoelectrochemical cell utilized that consisted of a Ag/AgNO$_3$ reference electrode (RE), a Pt counter electrode (CE), and mesoporous TiO$_2$ thin films deposited on a FTO substrate. A side view is also shown that exaggerates the 9 μm separation between the quartz coverslip and the FTO substrate. (b) Top view of the cell and the optical alignment of the 266 nm excitation, white light probe beams, dispersive element, and detector. The Surlyn films that contain the TBAI/acetonitrile solution are shown [162]. Reproduced with permission from the American Chemical Society.

New insights into recombination kinetics of poly(3 hexylthiophene):methanofullerene (P3HT:PCBM) bulk heterojunction (BHJ) solar cells, based on simultaneous determination of the density of states (DOS), internal recombination resistance, and carrier lifetime, at different steady states, by impedance spectroscopy were reported [163]. It was found that the recombination kinetics follows a bimolecular law, with the recombination time (lifetime) being inversely proportional to the density of photogenerated charges and the recombination coefficient $g = 6 \times 10^{-13}$ cm^3 s^{-1}, and the open-circuit photovoltage is governed by the carrier ability of occupying the DOS. The latter resulted in Gaussian shape and spreads in energy $s = 125$–140 meV. The energy position of the Gaussian DOS center ($E_L = 0.75$–0.80 eV), which corresponds to half occupation of the electron DOS, approximates LUMO (PCBM)–HOMO(P3HT) difference. In order to retard geminate recombination in dye-sensitized p-type NiO solar cells, surface treatment by aluminum alkoxide to form a thin insulating layer was applied [164]. Open-circuit voltage, short-circuit, and energy conversion efficiency of the *cell* were increased by the treatment. The IPCE of the light absorption wavelength of sensitizer was increased by 40%, while that at the wavelength of direct excitation of NiO was reduced.

Mechanistic details of how iodide oxidation, which yields the I—I bonds present in I_2^- and I_3^- reaction products, were subject of discussion ([165, 166] and references therein). In the work [167], the first direct evidence that electron transfer sensitized to visible light with metal-to-ligand charge transfer excited states can directly yield iodine. Visible light excitation of [Ru(bpz)$_2$(deeb)](PF$_6$)$_2$, where bpz is 2,2′-bipyrazine and deeb is 4,4′-(CO$_2$Et)2-2,2′-bipyridine, in acetonitrile solution with iodide was shown to initiate excited state electron transfer reactions that yield iodine atoms. The iodine atoms subsequently react with iodide to form the I—I bond in I_2^-. The resultant Ru(bpz$^-$)(bpz)(deeb)$^+$ + $I_2^{-\cdot}$ stores apprximately1.64 eV of free energy and returns cleanly to ground-state products with $k_{cr} = (2.1 \pm 0.3) \times 10^{10}$ M^{-1} s^{-1}. Excited state electron transfer to yield the iodine atom is <430 mV downhill. Reaction of the iodine atom with iodide to make an I—I bond lowers the free energy stored in the charge-separated state by 110 mV. Charge recombination to yield ground-state products Ru$^+$ + $I_2^{-\cdot}$ → RuII + 2I$^-$ is highly thermodynamically favored ($-\Delta G_0 = 1.64$ eV) and occurs with a rate constant of 2.1×10^{10} M^{-1} s^{-1}, almost 10 times larger than the $I_2^{-\cdot}$ disproportionation rate constant.

Impedance spectroscopy has provided marked success for the determination of energetic and kinetic factors governing the operation of dye-sensitized solar cells [168, 169].

To accelerate the I—I bond breaking reaction, the "platinum thermal cluster catalyst" (PTCC) has been developed [170]. The PTCC provided low platinum loading, superior kinetic performance, and mechanical firmness [171]. Other cheaper alternatives were used such as various forms of carbon [172], carbon black [173], graphite [174], activated carbon199, single-wall carbon nanotubes [175], polymer materials, such as PEDOT [(poly(3,4-ethylenedioxythiophene)], polypyrrole [172], polyaniline, and gold [24]. A single-plate storage battery composed of Nafion-coated polypyrrole and I_3^-,I$^-$|Pt electrodes in an interdigitated comb-like structure was set in an energy-storable dye-sensitized solar cell (ES-DSSC) [175]. The new ES-DSSC has a simple electrochemical cell structure having a single electrolyte with a typical photoanode and a dual functional electrode assembly.

New promising electrolytes and redox couples are continuously being developed in various laboratories [176–180]. The traditional I_3^-/I^- redox couple has been widely used, but with certain organic dyes the redox couples Br_3^-/Br^- are also used [181–185]. The power conversion efficiencies of the semiconductors as sensitizers are lagging behind those obtained for the more traditional (transition metal complexes and organic dye-sensitized) DSCs [186–192]. For example, the semiconductors as sensitizers suffer from a poor efficiency of about 1% [192].

An ionic liquid is a salt in the liquid state below some arbitrary temperature, such as 100 °C. While ordinary liquids such as water are predominantly made of electrically neutral molecules, ionic liquids are largely made of ions and short-lived ion pairs. Good chemical and thermal stability, negligible vapor pressure, nonflammability, high ionic conductivity, and a wide electrochemical window of liquid solvent, such as 1,3-dialkylimidalozium iodide, appear to be positive features of ionic liquids and stimulated their applications as solvents in DSSC [193, 194]. It is important to note that ionic liquids work simultaneously as iodide source and as solvent. Ionic liquid electrolytes based on a number of imidazolium, quaternary ammonium, and phosphonium cations have been developed for porphyrin dye-sensitized solar cells yielding efficiencies of up to 5.2% at 0.68 sun. Many families including those based on imidazolium, 13,14 pyrrolidinium, ammonium 15, phosphonium-based ILs [195, 196], diethylisobutylmethyl phosphoniumbis(trifluoromethanesulfonyl)amide), and other ionic liquids were investigated [197–199].

Ionic liquid electrolytes based on a number of imidazolium, quaternary ammonium, and phosphonium cations have been developed for porphyrin dye-sensitized solar cells yielding efficiencies of up to 5.2% at 0.68 sun [197]. The authors of work [197] introduced the concept of using eutectic melts to produce **solvent-free liquid** redox electrolytes. Ionic liquid electrolytes based on a number of imidazolium, quaternary ammonium and phosphonium cations have been developed for porphyrin dye sensitised solar cells yielding efficiencies of up to 5.2% at 0.68 (Sun [197]). Using a ternary melt in conjunction with a nanocrystal titania film and the amphiphilic heteroleptic ruthenium complex $Z = 907$Na as a sensitizer, the authors reached excellent stability (Figure 5.20) and an unprecedented efficiency of 8.2% under air-mass 1.5 global illumination. The results are of importance to realize large-scale outdoor applications of mesoscopic DSSCs.

Among the iodide salts that form room-temperature ionic liquids, 1-propyl-3-methylimidazolium iodide (PMII) has the lowest viscosity. The conductivity of pure iodide melts increases in the order 1-hexyl-3-methylimidazolium iodide < 1-butyl-3-methylimidazolium iodide < PMII, mirroring the behavior of their fluidity, while 1-allyl-3-methylimidazolium iodide showed a higher conductivity than PMII in the liquid state.

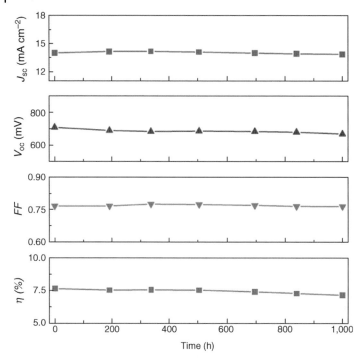

Figure 5.20 Detailed photovoltaic parameters of a cell measured under the irradiance of AM 1.5G sunlight during successive full-sun visible-light soaking at 60 C [197]. Reproduced with permission from *Nature*.

5.3.5
Cathode

The state of the art of the two main catalyst material types used in DSSC such as metal and carbon-based nanostructures for cathode preparation was reviewed [21]. One of the characteristics of the cathodes, also named counterelectrodes (CEs), is the ability to keep the overvoltage low at photocurrent densities up to 20 mA/cm^2. Platinum has been the preferred material for the CE since it is a good catalyst for the triiodide reduction. A light reflecting CE is usually employed, consisting of a conducting TCO glass onto which a 2 m thick Pt mirror is deposited by sputtering. Schematic representation of the materials that can be used in DSSC as catalyst is shown in Figure 5.21.

A review of metal nanoparticles and carbon-based nanostructures as advanced materials for cathode application in dye-sensitized solar cells was recently reported [21]. The review described methods for the fabrication of cathodes for dye-sensitized solar cells employing nanostructured materials. The attention was focused on metal nanoparticles and nanostructured carbon, among which nanotubes and graphene, whose good catalytic properties make them ideal for the development of counterelectrode substrates, transparent conducting oxide, and advanced catalyst materials.

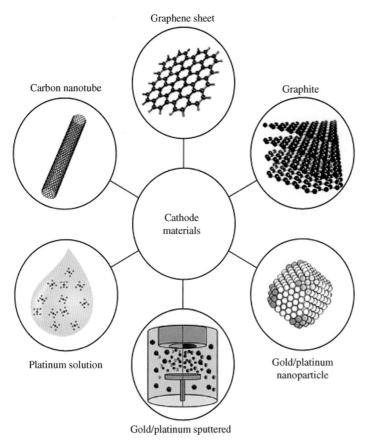

Figure 5.21 Schematic representation of the materials that can be used in DSSC as catalysts [21].

Highly efficient plastic substrate dye-sensitized solar cells (by the press method) were developed [200]. The conversion efficiency of plastic substrate DSSCs was improved by adjusting the press conditions, the thickness of the TiO_2 layer, and the surface treatment of the plastic substrate. An efficiency of 8% was achieved for these cells with a 0.25 cm^2 cell area under 100 mW/cm2 (AM 1.5, 1 sun). A 1.1 cm^2-sized plastic substrate DSC was fabricated that had an efficiency of 7.6%.

The most commonly used cathodes in DSSCs are fluorine-doped tin oxide (FTO) glass with Pt deposited on them [201]. Pt exhibits excellent catalytic activity for I_3^- reduction and high conductivity. Nevertheless, problems arise from the rigid nature of FTO glass substrates, such as relative high sheet resistance, high cost, and inconvenient transportation [202]. Alternative cheap catalysts and substrates for CEs were developed including carbon black [203], hard carbon sphere [204], carbon nanotube [205], conductive polymers, such as polypyrrole [206], polyaniline (PANI) [207], poly(3,4-methylenedioxythiophene), metals, plastic foils, and graphite paper [208].

Commercial graphene nanoplatelets in the form of optically transparent thin films on F-doped SnO_2 (FTO) exhibited, as a cathode, high electrocatalytic activity toward I_3^-/I^- redox couple, particularly in electrolyte based on ionic liquid (Z952) [209]. The charge transfer resistance, R_{CT}, was smaller by a factor of 5–6 in ionic liquid, compared to values in traditional electrolyte based on methoxypropionitrile solution (Z946). Electrocatalytic properties of graphene nanoplatelets for the I_3^-/I^- redox reaction were found to be proportional to the concentration of active sites (edge defects and oxidic groups), independent of the electrolyte medium.

A novel and low-cost counterelectrode material for DSSCs TiN-CNTs was fabricated by hydrolysis of a titanium salt on CNTs and subsequent nitridation, in which TiN nanoparticles with a size of 5–10 nm were dispersed on the surface of CNTs [210]. The material showed a comparable photovoltaic performance with the conventional Pt electrode, which was attributed to the combination of superior electrocatalytic activity and high electrical conductivity derived from highly active TiN nanoparticles supported by a CNT electron transport network. A flexible composite electrode, which is composed of conducting polyaniline (PANI) as electroactive material and flexible graphite (FG) as conducting substrate for counter electrode, has been fabricated [211]. The photovoltaic parameters of DSSCs were dependent on the oxidation state and the thickness of the PANI film. Higher photocurrent density and efficiency have been obtained by using emeraldine PANI compared to pernigraniline. A DSSC with the composite CE showed an overall conversion efficiency of 7.36%, which was comparable to 7.45% of that with Pt electrode under the same test conditions. Dye-sensitized solar cells with a mesoporous network of interconnected TiO_2 nanocrystals, promising sensitizers, novel electrolytes, and platinum (Pt) as a counterelectrode, which are attracting widespread scientific and technological interest as a high-efficiency and low-cost alternative to conventional inorganic photovoltaic devices, were reviewed ([212] and references therein). For example, counterelectrodes in DSSCs have been prepared by Pt vacuum deposition or thermal annealing of a Pt precursor on a transparent conductive oxide (TCO) substrate to reduce the overpotential for reduction of I_3^- to I^- in redox electrolytes.

Carbon black-based counterelectrodes in DSSCs were fabricated and their use provided a remarkable conversion efficiency of approximately 9% under $100\,mW/cm^2$ [203].

In the study [213], a highly conductive conjugated polymer composite was deposited by vapor phase polymerization. Poly[3,4-ethylenedioxythiophene:para-toluenesulfonate] (PEDOT:PTS) itself was used as the counterelectrode in a dye-sensitized solar cell. The maximum photocurrent of $9.7\,mA/cm^2$, open-circuit voltage of 759 mV, fill factor of 0.71 with a power conversion efficiency of 5.25% were observed for glass-based wet-type dye-sensitized solar cell, under illumination of $100\,mW/cm^2$. It was observed that the resistance, during operation of the dye-sensitized solar cells, due to the I_3^- conversion was less with PEDOT:PTS-coated cathodes than with standard platinum-coated fluorine-doped tin oxide and was confirmed by steady-state electrochemical measurements.

An approach to a Pt-free counter electrode has used several conducting polymers to obtain a maximum conversion efficiency, approximately 7.8% under $100\,mW/cm^2$ [214].

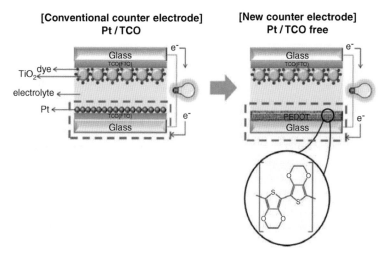

Figure 5.22 Schematic diagram of DSSC with Pt- and TCO-free counter electrode [217]. Reproduced with permission from Royal Chemical Society.

A power conversion efficiency of 7.1% has been achieved with the microporous polyaniline counterelectrode [215]. Efficiency of 7.93% by using poly(3,4-alkylenedioxythiophene) nanoporous layers prepared from electrooxidative polymerization was reported [216]. The authors concluded that conductive polymers have the potential to replace the Pt counterelectrode in DSSCs because their high electrochemical cell efficiency is comparable to cells using a Pt/FTO counterelectrode.

Highly conductive poly(3,4-alkylenedioxythiophene) films (Figure 5.22) were synthesized on a glass substrate by a modified simple presolution/*in situ* polymerization method [217]. Only films without TCO were used as a counterelectrode in the fabrication of DSSCs, resulting in a power conversion efficiency of 5.08%, whereas for a DSSC with a Pt/FTO counterelectrode it was 5.88%. The obtained efficiency was the first of its kind reported to date using only a conducting polymer as a counterelectrode. According to the authors, if the electric conductivity of the PEDOT film can be further increased, the performance of a PEDOT-only counterelectrode can be improved, which eliminates the utility of Pt and TCO.

A single-plate storage battery composed of Nafion-coated polypyrrole and $I_3^-,I^-|Pt$ electrodes in an interdigitated comb-like structure was set in an ES-DSSC [218]. The new ES-DSSC has a simple electrochemical cell structure having a single electrolyte with a typical photoanode and a dual functional electrode assembly.

5.3.6
Solid-State DSSC

Solid-state dye-sensitized solar cells are an offshoot technology of the dye-sensitized liquid junction cells that have been under active development for the past decade [219]. Solid-state dye-sensitized cells have also been referred to as dye-sensitized heterojunctions (DSHs) by virtue of the placement of the light-absorbing dye at an otherwise transparent n–p heterojunction. The promise of DSHs is the fusion of the inexpensive materials of the dye-sensitized liquid junction technology with the easier and less expensive manufacturing and packaging applicable to solid devices. One of the proposed materials for the p-type side of a DSH iCuSCN was reported in [220]. Although solid-state DSSCs have lower efficiency compared to liquid-type DSSCs, the research on the solid-state DSSCs has gained considerable attention because it is suitable for realizing flexible PV cells in a roll-to-roll production. The disadvantage of solid-state DSSCs is the problem of pore filling in thicker films because thin films of 1.5–3 μm TiO_2 have to be used. In the solid-state charge transporting systems, the regular liquid electrolyte was replaced by a p-type semiconductor layer composed of inorganic materials such CuSCN, CuI, and CuBr [221, 222] or organic molecules (cyanidine dyes, ruthenium complex, etc.) [223]. The main difficulty in this approach appeared to be optimizing the interface between the sensitized semiconductor and the electrolyte.

Various forms of carbon [224–227] such as carbon black [224], graphite [225], activated carbon [226], single-wall carbon nanotubes, polyaniline, and gold ([227] and references therein) have been used in solid-state DSSCs. The most commonly used substrates for DSSCs were made of coated glass with a transparent conducting oxide (TCO) with high electrical conductivity of a magnitude of 10^4 S cm^{-1}, and transparency better than 80% visible wavelengths. The following materials were also employed as substrates for DSSCs: SnO_2:F [228], In_2O_3:Sn and its combination [229], organic (2,2′,7,7′-tetrakis(N,N-di-p-methoxyphenyl-amine)9,9′-spirobi-fluorene spiro-MeOTAD) [222], cyanovinylene 4-nitrophenyl derivatives [230], plastic (polyethylene naphthalate [231], and polyethylene terephthalate (PET) [232]. Metals that form a conducting layer (stainless steel, W or Ti) [233] and various organic materials were also used as substrates [234–237].

Wei et al. [238] described a strategy to make a full solid-state, flexible dye-sensitized solar cell based on novel ionic liquid gel, organic dye, ZnO nanoparticles, and carbon nanotube thin film stamped onto a PET substrate. The CNTs serve both as the charge collector and as the scaffolds for the growth of ZnO nanoparticles, where the black dye molecules are anchored. It opens up the possibility of developing a continuous roll-to-roll processing for the mass production of DSSCs. Two types of polysiloxane grafted with different ratios of imidazolium iodide moieties (IL-SiO_2) have been synthesized to develop a microporous polymer electrolyte for quasi-solid-state dye-sensitized solar cell [239]. The samples were characterized by ^{1}HNMR, FT-IR spectrum, XRD, TEM, and SEM. The ionic conductivity of the electrolytes was measured by electrochemical methods. Nanostructured polysiloxane containing imidazolium iodide showed excellent compatibility with organic solvent and polymer matrix for its ionic liquid characteristics.. A dye-sensitized solar cell with gel polymer electrolyte yielded an open-circuit voltage of 0.70 V, short-circuit current of 11.19 mA/cm^2, and the conversion efficiency of 3.61% at 1 sun illumination. Solid-state dye-sensitized photovoltaic cells have been fabricated with TiO_2 as the electron conductor and CuSCN as the hole conductor (Figures 5.23 and 5.24) [240]. The substrate SnO_2

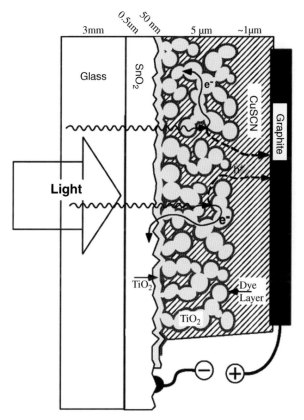

Figure 5.23 Schematic of a solid-state dye-sensitized photovoltaic cell used in Ref. [240]. Reproduced with permission from American Chemical Society.

was covered by spray pyrolysis with a thin (50 nm) solid layer of TiO_2, which prevents short circuiting between the SnO_2 and the CuSCN. It was found that the active layer consists of a 2–5 μm mesoporous TiO_2 layer, covered with a monolayer of a sensitizing dye and then impregnated with CuSCN. The CuSCN also forms a layer above the TiO_2 film whose thickness we can vary between 0.2 and 3 μm. The electrical contact to the CuSCN was made with pressed graphite or evaporated gold, both of which give similar results. The cells showed photocurrents of 8 mA/cm^2, voltages of 600 mV, and energy efficiencies of 2% at 1 sun. The final drying step after the CuSCN deposition was shown to be critical; drying in vacuum or argon is required for photocurrents above 2 mA/cm^2.

Solid-state/nanocrystalline solar cells composed of chemical-bath deposited Sb_2S_3 (antimony sulfide) as a light-absorber layer deposited on nanoporous TiO_2 and spiro-MeOTAD were fabricated [241]. The cells were used as an organic hole transporting material yielding a solar conversion efficiency of 5.2% at 0.1 sun illumination and a peak 88% of the incident monochromatic photon-to-current conversion efficiency.

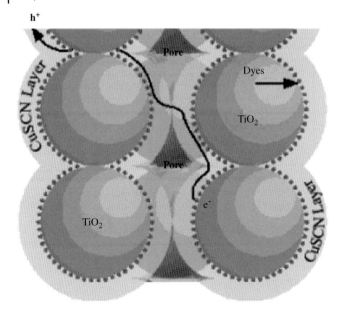

Figure 5.24 Cross section illustrating idealized pore filling by adding CuSCN layers to the surface of dye-sensitized TiO2 particles on a cubic lattice [240]. Reproduced with permission from the American Chemical Society.

Solid-state dye-sensitized solar cells have been made using nanocrystal titania, a dye sensitizer, and poly(3-hexylthiophene) [242]. Improved results were obtained by employing multilayer nanocrystal titania made of a bottom densely packed layer and a top open structure of varying thickness. The cells were assembled under ambient conditions using silver paste as counter electrode. All-solid-state dye-sensitized solar cells with alkyloxy-imidazolium iodide ionic polymer/SiO$_2$ nanocomposite electrolyte and triphenylamine-based metal-free organic dyes (TC15) were constructed [243]. By optimizing the content of I$_2$, 1,2-dimethyl-3-propylimidazolium iodide, and SiO$_2$ nanoparticles in the electrolyte, considerable ionic conductivity of 0.151 mS/cm was achieved due to the formation of high-efficiency electron exchange tunnels. The SEM analysis revealed a favorable interfacial contact between the electrolyte and the TiO2. This cell attained high energy conversion efficiency of 2.70 and 4.12% under the illumination intensities of 100 and 10 mW/cm^2, respectively. Three scenarios were tested while studying monolithic dye-sensitized solar cells: (a) no spacer layer, just direct contact between the titania and the graphite; (b) a large particle anatase titania scattering/spacer layer; and (c) an insulating zirconia layer [244].

The quasi-solid-state DSSCs using the gel electrolytes can alleviate problems of the organic solvent-based liquid-state electrolytes usually used, which result in practical limitations of sealing and long-term operation at higher temperatures [245]. However, this approach still retains a significant volume of volatile liquid encapsulated in the gel pores, resulting in a large increase in vapor pressure as the temperature is raised. Several attempts have been made to develop all-solid-state electrolytes such as p-type inorganic semiconductors (CuI, CuSCN, etc.) [246], organic hole transport materials (such as triarylamine and polythiophene-based derivatives) [247], and polymeric materials incorporating triiodide/iodide and/or inorganic nanoparticles [246, 248]. Among them, the nanocomposite polymer electrolytes exhibited considerable photovoltaic performance due to the favorable ionic conductivity (σ), mechanical stability, and interfacial contact between the electrolyte and the TiO_2 film [249].

Mesoporous ZnO electrodes from sol–gel-processed nanoparticles were sensitized with conventional ruthenium complexes and infiltrated with the solid-state hole transporter medium 2.2′,7,7′-tetrakis-(N,N-di-p-methoxyphenylamine)-9,9′-spirobifluorene (spiro-OMeTAD) [250]. The nature of the polymeric additive used in the initial ZnO formulation, as well as the ZnO electrode sintering treatment, was varied. It was shown that using ethyl cellulose in the initial ZnO formulation is responsible for an improved dye loading on the ZnO porous electrode. Using only 800 nm thick porous ZnO electrodes sensitized by N719, the best performing device exhibits a short-circuit current density of 2.43 mA/cm^2 under simulated solar emission of (100 mW/cm^2), associated with an overall power conversion efficiency of 0.50%. Miettunen et al. [251] studied different kinds of metals such as the *photoelectrode substrate*. Stainless steels (StSs), Inconel, and titanium substrates were tested to find stable substrate options. According to the authors' data, the oxide layer of Inconel substrates increased resistive losses, which caused a lower fill factor and photovoltaic efficiency compared to the Ti-based cells.

Data on optical fiber and tandem DSSCs, quantum dot solar cells, polymers, and fullerenes in solar cells, and fabrication of solar cell components will be considered in Chapter 6.

References

1 Goncalves, L.M., de Zea Bermudez, V., Ribeiroa, H.A., and Mendes, A.M. (2008) *Energy Environ. Sci.*, **1** (2), 655–667.

2 Shavaleevsky, O. (2008) *Pure App. Chem.*, **80** (10), 2079–2089.

3 Hoffmann, W. (2006) *Sol. Energy Mater. Sol. Cells*, **90**, 3285–3311.

4 Bisquert, J., Cahen, D., Hodes, G., Rühle, S., and Zaban, A. (2004) *J. Phys. Chem. B*, **108** (24), 8106–8118.

5 O'Regan, B. and Grätzel, M. (1991) *Nature*, **353** (6346), 737–740.

6 Lewcenko, N.A., Matthew J. Byrnes, M.J., Torben Daeneke, T., Mingkui Wang, M., Zakeeruddin, S.M., Grätzel, M., Spiccia, L. (2010) *J. Mater Chem*, **20**, 3694–3702.

7 Lewcenko, N.A., Byrnes, M.J., Daeneke, N., Wang, M., Zakeeruddin, S.M., Grätzel, M., and Leone Spiccia, L. (2010) *J. Mater. Chem.*, **20**, 3694–3702.

8 Weintraub, B., Wei, Y., and Wang, Z.L. (2009) *Angew. Chem. Int. Ed.*, **48** (47), 8981–8985.

9 Singh, P.K., Bhattacharya, B., and Nagarale, R.K. (2010) *J. Appl. Polym. Sci.*, **118** (5), 2976–2980.

10 Casanova, D., Rotzinger, F.P., and Grätzel, M. (2010) *J. Chem. Theor. Comput.*, **6** (4), 1219–1227.

11 McEvoy, A.J., and Grätzel, M. (2009) *Ceramic Materials in Energy Systems for Sustainable Development, Monographs in Materials and Society*, vol. 8, Academy of Ceramics, pp. 337–355.

12 Grätzel, M., and Durrant, J.R. (2008) *Series on Photoconversion of Solar Energy*

(eds M.D. Archer and A.J. Nozik), vol. 3, World Scientific Press, pp. 503–538.

13 Ryan, M. (2009) *Platinum Metals Rev.*, **53** (4), 216–218.

14 Lin, H., Wang, W.-L., Liu, Y.-Z., Li, X., and Li, J.-B. (2009) *Frontiers Mater. Sci. China*, **3** (4), 345–352.

15 Li, B., Wang, L., Kang, B., Wang, P., and Qiu, Y. (2006) *Sol. Energy Mater. Sol. Cells*, **90** (5), 549–573.

16 Asghar, M.I., Miettunen, K., Halme, J., Vahermaa, P., Toivola, M., Aitola, K., and Lund, P. (2010) *Energy Environ. Sci.*, **3** (4), 418–426.

17 Toivola, M., Halme, J., Miettunen, K., Aitola, K., and Lund, P.D. (2009) *Espoo Int. J. Energ. Res.*, **33** (13), 1145–1160.

18 Caramori, S., Cristino, V., Boaretto, R., Argazzi, R., Bignozzi, C.A., and Di Carlo, A. (2010) *Int. J. Photoenergy*, **2010**, 16.

19 Alibabaei, L., Wang, M., Giovannetti, R., Teuscher, J., di Censo, D., Moser, J.-E., Comte, P., Pucciarelli, F., Zakeeruddin, S.M., and Gratzel, M. (2010) *Energy Environ. Sci.*, **3** (7), 956–961.

20 Yanagida, S. (2006) *C. R. Chim.*, **9** (5–6), 597–604.

21 Calandra, P., Calogero, G., Sinopoli, A., and Hindawi, P.G. (2010) *Int. J. Photoenergy*, **15** (1), 1–15.

22 Chang, J.A., Rhee, J.H., Im, S.H., Lee, Y.H., Kim, H.-J., Seok, S.I., Nazeeruddin, M.K., and Gratzel, M. (2010) *Nano Lett.*, **10** (7), 2609–2612.

23 Nazeeruddin, M.K., Bessho, T., Le Ceveya, T., Itoa, S., Klein, C., De Angelis, F., Fantacci, S., Comtea, P., Liska, P., Imai, H., and Grätzel, M. (2007) *J. Photoch. Photobio. A*, **185** (2–3), 331–337.

24 Ikeda, N., Teshima, K., and Miyasaka, T. (2006) *Chem. Commun.*, 1733–1735.

25 Islam, A., Sugihara, H., Hara, K., Singh, L.P., Katoh, R., Yanagida, M., Takahashi, Y., Murata, S., and Arakawa, H. (2001) *J. Photochem. Photobiol. A*, **145** (1–2), 135–141.

26 Islam, A., Sugihara, H., Yanagida, S., Hara, R., Fujihashi, G., Tachibana, Y., Katoh, R., Murata, S., and Arakawa, H. (2002) *New J. Chem.*, **26** (8), 966–968.

27 Yamaguchi, T., Yanagida, M., Katoh, R., Sugihara, H., and Arakawa, H. (2004) *Chem. Lett.*, **33** (8), 986–988.

28 (a) You, Y. and Park, S.Y. (2005) *J. Am. Chem. Soc.*, **127** (36), 12438–12439; (b) Nazeeruddin, M.K., Péchy, P., Renouard, T., Zakeeruddin, S.M., Humphry-Baker, R., Comte, P., Liska, P., Cevey, L., Costa, E., Shklover, V., Spiccia, L., Deacon, G.B., Bignozzi, C.A., and Grätzel, M. (2001) *J. Am. Chem. Soc.*, **123**, 1613.

29 Kong, F.-T., Dai, S.-Y., and Wang, K.-J. (2007) *Chin. J. Chem.*, **25** (2), 168–171.

30 Gao, F., Wang, Y., Zhang, J., Shi, D., Wang, M., Humphry-Baker, R., Wang, P., Zakeeruddin, S.M., and Grätzel, M. (2008) *Chem. Commun.*, 2635–2637.

31 Funaki, T., Yanagida, M., Onozawa-Komatsuzaki, N., Kasuga, K., Kawanishi, Y., and Sugihara, H. (2009) *Inorg. Chim. Acta*, **362** (7), 2519–2522.

32 Nazeeruddin, M.K., Bessho, T., Ceveya, L., Itoa, S., Klein, C., De Angelis, F., Fantacci, S., Comtea, P., Liska, P., Imai, H., and Grätzel, M. (2007) *J. Photoch. Photobio. A* **185**, 331–337.

33 Abrahamsson, M., Heuer, W.B., and Meyer, G.J. (2009) Abstracts of Papers. 238th ACS National Meeting, Washington, DC, United States, PHYS-612.

34 Ardo, S., Sun, Y., Staniszewski, A., Castellano, F.N., and Meyer, G.J. (2010) *J. Am. Chem. Soc.*, **132** (19), 6696–6709.

35 Nazeeruddin, M.K., Humphry-Baker, R., Officer, D.L., Campbell, W.M., Burrell, A.K., and Grätzel, M. (2004) *Langmuir*, **20** (15), 6514–6517.

36 Wang, Q., Campbell, W.M., Bonfantani, E.E., Jolley, K.W., Officer, D.L., Walsh, P.J., Gordon, K., Humphry-Baker, R., Nazeeruddin, M.K., and Graetzel, M. (2005) *J. Phys. Chem. B*, **109** (32), 15397–15409.

37 Campbell, W.M., Jolley, K.W., Wagner, P., Wagner, K., Walsh, P.J., Gordon, K.C., Schmidt-Mende, L., Nazeeruddin, M.K., Wang, Q., Grätzel, M., and Officer, D.L. (2007) *J. Phys. Chem. C*, **111** (32), 11760–11762.

38 Campbell, W.M., Burrell, A.K., Officer, D.L., and Jolley, K.W. (2004) *Coordin. Chem. Rev.* (13–14), 1363–1379.

39 Brumbach, M.K., Boal, A.K., and Wheeler, D.R. (2009) *Langmuir*, **25** (18), 10685–10690.

40 Lind, S.J., Gordon, K.C., Gambhir, S., and Officer, D.L. (2009) *Phys. Chem. Chem. Phys.*, **11** (27), 5598–5607.

41 Lee, C.W., Lu, H.P., Lan, C.M., Huang, Y.L., Liang, Y.R., Yen, W.N., Liu., Y.C., Lin, Y.S., Diau, E.W., and Yeh, C.Y. (2009) *Chemistry*, **15** (6), 1403–1412.

42 Lin, H. and Imahori, H. (2011) *Key Eng. Mater.*, **451** (1), 29–40.

43 Ito, S., Zakeeruddin, I.S., Humphry-Baker, R., Liska, R., Charvet, P., Comte, M., Nazeeruddin, M.K., Péchy, P., Takata, M., Miura, H., Uchida, S., and Grätzel, M. (2006) *Adv. Mater.*, **18** (9), 1202–1205.

44 Teng, C., Yang, X., Yang, C., Li, S., Cheng, M., Hagfeldt, A., and Sun, L. (2010) *J. Phys. Chem. C* **114**, 9101–9110.

45 Wang, Z., Cui, Y.D.-H., Kasada, C., Shinpo, A., and Hara, K. (2007) *J. Phys. Chem. C*, **111** (19), 7224–7230.

46 Sayama, K., Tsukagoshi, S., Hara, K., Ohga, Y., Shinpou, A., Abe, Y., Suga, S., and Arakawa, H. (2002) *J. Phys. Chem. B*, **106** (6), 1363–1371.

47 Ma, X., Hua, J., Jin, Y., Meng, F., Zhan, W., and Tian, H. (2008) *Tetrahedron*, **64** (2), 345–350.

48 Chen, Y., Li, C., Zeng, Z., Wang, W., Wang, X., and Zhang, B. (2005) *J. Mater. Chem.*, **15**, 1654–1661.

49 Ferrere, S. and Gregg, B.A. (2002) *New J. Chem.*, **26** (16), 1155–1161.

50 Kitamura, T., Ikeda, M., Shigaki, K., Inoue, T., Anderson, N.A., Ai, X., Lian, T., and Yanagida, S. (2004) *Chem. Mater.*, **16** (9), 1806–1812.

51 Hara, K., Horiguchi, T., Kinoshita, T., Sayama, K., Sugihara, H., and Arakawa, H. (2000) *Chem. Lett.*, **29** (2), 316–317.

52 Xu, W., Peng, B., Chen, J., Liang, M., and Cai, F. (2008) *J. Phys. Chem. C*, **112** (3), 874–880.

53 Choi, H., Baik, C., Kang, S.O., Ko, J., Kang, M.S., Nazeeruddin, M.K., and Grätzel, M. (2008) *Angew. Chem., Int. Ed.*, **47** (2), 327–330.

54 Tian, H.N., Yang, X.C., Chen, R.K., Pan, Y.Z., Li, L., Hagfeldt, A., and Sun, L., *Chem. Commun.*, 3741–3743.

55 Chen, R., Yang, X., Tian, H., and Sun, L. (2007) *J. Photochem. Photobiol. A*, **189** (2), 295–300.

56 Alex, S., Santhosh, U., and Das, S. (2005) *Chem.*, **172** (1), 63–71.

57 Burke, A., Schmidt-Mende, L., Ito, S., and Grätzel, M. (2007) *Chem. Commun.*, 234–236.

58 Guo, M., Daio, P., Ren, Y.-J., Meng, F., Tian, H., and Cai, S.-M. (2005) *Sol. Enegy Mater. Sol. Cells.*, **88** (1), 23–35.

59 Wang, S., Li, F.-Y., and Huang, C.-H. (2000) *Chem. Commun.*, 2063–2064.

60 Hara, K., Dan-oh, Y., Kasada, C., Ohga, Y., Shinpo, A., Suga, S., Sayama, K., and Arakawa, H. (2004) *Langmuir*, **20** (10), 4205–4210.

61 Kim, S., Lee, J.K., Kang, S.O., Ko, J., Yum, J.-H., Fantacci, S., De Angelis, F., Di Censo, D., Nazeeruddin, M.K., and Grätzel, M. (2006) *J. Am. Chem. Soc*, **128** (51), 16701–16707.

62 Hwang, S., Lee, J.H., Lee, H., Kim, C., Park, C., Lee, M.H., Lee, W., Park, J., Kim, K., Park, N.G., and Kim, C. (2007) *Chem. Commun.*, 4887.

63 Moon, S.-J., Yum, J.-H., Humphry-Baker, R., Martin, K., Karlsson, K., Hagberg, D.P., Marinado, T., Hagfeldt, A., Sun, L., Grätzel, M., and Nazeeruddin, M.K. (2009) *J. Phys. Chem. C* **113** (38), 16816–16820.

64 Zhang, X.-H., Wang, Z.-S., Cui, Y., Koumura, N., Furube, A., and Hara, K. (2009) *J. Phys. Chem. C*, **113** (30), 13409–13415.

65 Abbotto, A., Manfredi, N., Marinzi, C., Angelis, F., Mosconi, E., Yum, J.H., Xianxi, Z., Nazeeruddin, M. K., Grätzel, M. (2009) *Energy & Environmental Science*, **2** (10), 1094–1101.

66 Gajjela, S.R., Ananthanarayanan, K., Yap, C., Grätzel, M., and Balaya, P. (2010) *Energy Environ. Sci* **3** (6), 838–845.

67 Kamat, P.V., Das, S., Thomas, K.G., and George, M.V. (1991) *Chem. Phys. Lett.*, **178** (1), 75–79.

68 Ajayaghosh, A. (2005) *Acc. Chem. Res.*, **38** (6), 449–459.

69 Choi, H., Kim, J.-J., Song, K., Ko, J., Nazeeruddin, M.K., and Grätzel, M. (2010) *J. Mater. Chem.*, **20** (16), 3280–3286.

70 Zhang, X.-H., Wang, Z.-S., Cui, Y., Koumura, N., Furube, A., and Hara, K. (2009) *J. Phys. Chem. C*, **113** (30), 13409–13415.

71 Chen, D.Y., Hsu, Y.Y., Hsu, H.C., Chen, B.S., Lee, Y.T., Fu, H., Chung, M.W., Liu, S.H., Chen, H.C., Ch, Y. and Chou, P.T. (2010) *J. Sol. Gel. Sci. Technol.*, **53** (3), 599–604;(2010) *Chem. Commun.*, 5256–5258.

72 Chen, D.Y., Hsu, Y.Y., Hsu, H.C., Chen, B.S., Lee, Y.T., Fu, H., Chung, M.W., Liu, S.H., Chen, H.C., Ch, Y., and Chou, P.T. (2010) *J. Sol. Gel. Sci. Technol.*, **53** (3), 599–604.

73 Feldt, S.M., Gibson, E.A., Gabrielsson, E., Sun, L., Boschloo, G., and Hagfeldt, A. (2010) *J. Am. Chem. Soc.*, **132** (46), 16714–16724.

74 Teng, C., Yang, X., Yang, C., Li, S., Cheng, M., Hagfeldt, A., and Sun, L. (2010) *J. Phys. Chem. C*, **114** (19), 9101–9110.

75 Marinado, T., Hahlin, M., Jiang, X., Quintana, M., Johansson, E.M.J., Gabrielsson, E., Plogmaker, S., Hagberg, D.P., Boschloo, G., Zakeeruddin, S.M., Grätzel, M., Siegbahn, H., Sun, L., Hagfeldt, A., and Rensmo, H. (2010) *J. Phys. Chem. C*, **114**, 11903–11910.

76 Lin, L.-Y., Tsai, C.-H., Wong, K.-T., Huang, T.-W., Hsieh, L., Liu, S.-H., Lin, H.-W., Wu, C.-C., Chou, S.-H., Chen, S.-H., and Tsai, A.-I. (2011) *J. Mater. Chem.*, **21**, 5950–5958.

77 Lo, C.-F., Hsu, S.-J., Wang, C.-L., Cheng, Y.-H., Lu, H.-P., Diau, Eric, W.-G., and Lin, C.-Y. (2010) *J. Phys. Chem. C*, **14**, 12018–12023.

78 Ono, T., Yamaguchi, T., and Arakawa, H. (2009) *Sol. Energy Mater. Sol. Cells*, **93** (67), 831–835.

79 Yamaguchi, T. and Arakawa, H. (2009) *Sol. Energy Mater. Sol. Cells*, **93** (67), 831–838.

80 Cappel, U.B., Feldt, S.M., Schoneboom, J., Hagfeldt, A., and Boschloo, G. (2010) *J. Am. Chem. Soc.*, **132** (26), 9096–9101.

81 Soo-Jin, M., Yafit, I., Yum, J.-H., Zakeeruddin, S.M., Hodes, G., and Grätzel, M. (2010) *J. Phys. Chem. Lett.*, **1** (10), 1524–1527.

82 Hodes, G. (2008) *J. Phys. Chem. C*, **112** (46), 17778–17787.

83 Robel, I., Subramanian, V., Kuno, M., and Kamat, P.V. (2006) *J. Am. Chem. Soc.*, **128** (7), 2385–2393.

84 Mora-Sero, I., Bisquert, J., Dittrich, Th., Belaidi, A., Susha, A.S., and Rogach, A.L. (2007) *J. Phys. Chem. C*, **111** (40), 14889–14892.

85 Leschkies, K.S., Divakar, R., Basu, J., Enache-Pommer, E., Boercker, J.E., Carter, C.B., Kortshagen, U.R., Norris, D.J., and Aydil, E.S. (2007) *Nano Lett.*, **7** (6), 1793–1798.

86 Lopez-Luke, T., Wolcott, A., Xu, L.-P., Chen, S., Wen, Z., Li, J., De La Rosa, E., and Zhang, J.Z. (2008) *J. Phys. Chem. C*, **112** (6), 1282–1288.

87 Lee, H.J., Yum, J.-H., Leventis, H.C., Zakeeruddin, S.M., Haque, S.A., Chen, P., Seok, S.I., Grätzel, M., and Nazeeruddin, M.K. (2008) *J. Phys. Chem. C*, **112** (30), 11600–11608.

88 Grätzel, M. (2009) *Acc. Chem. Res.*, **42** (11), 1788–1798.

89 Oja, I., Belaidi, A., Dloczik, L., Lux-Steiner, M.C., and Dittrich, T. (2006) *Semicond. Sci. Technol.*, **21** (4), 520–526.

90 Larramona, G., Chone, C., Jacob, A., Sakakura, D., Delatouche, B., Pere, D., Cieren, X., Nagino, M., and Bayon, R. (2006) *Chem. Mater.*, **18** (6), 1688–1696.

91 Diguna, L.J., Shen, Q., Kobayashi, J., and Toyoda, T. (2007) *Appl. Phys. Lett.*, **91**, 023116.

92 Ernst, K., Engelhardt, R., Ellmer, K., Kelch, C., Muffler, H.J., Lux-Steiner, M.C., and Konenkamp, R. (2001) *Thin Solid Films*, **387** (1), 26–28.

93 Page, M., Niitsoo, O., Itzhaik, Y., Cahen, D., and Hodes, G. (2009) *Solar Cells Energy Environ. Sci.*, **2** (2), 220–223.

94 Kaiser, I., Ernst, K., Fischer, C.H., Konenkamp, R., Rost, C., Sieber, I., and Lux-Steiner, M.C. (2001) *Sol. Energy Mater. Sol. Cells*, **67** (1), 89–96.

95 Lee, H.J., Yum, J.H., Leventis, H.C., Zakeeruddin, S.M., Haque, S.A., Chen, P., Seok, S.I., Gr∈azel, M., and Nazeeruddin, M.K. (2008) *J. Phys. Chem. C*, **112** (30), 11600–11608.

96 Koo, H., Park, J., Yoo, B., Yoo, K., Kim, K., and Park, N. (2008) *Inorg. Chem.*, **361** (3), 677–683.

97 Armel, V., Pringle, J.M., Forsyth, M., MacFarlane, D.R., Officer, D.L., and Wagner, P. (2010) *Chem. Commun.*, 3146–3148.

98 Zhu, K., Neale, N., Miedaner, A., and Frank, A. (2006) *Nano Lett.* (2006) (1), 69–74.

99 Wei, M., Konishi, Y., Zhou, H., Sugihara, H., and Arakawa, H. (2006) *J. Electrochem. Soc.* (2006) **153** (6), A1232–A1236.

100 Park, J., Lee, T., and Kang, M. (2008) *Chem. Commun.*, 2867–2869.

101 Adachi, M., Murata, Y., Takao, J., Jiu, J., Sakamoto, M., and Wang, F. (2004) *J. Am. Chem. Soc.*, **126** (45), 14943–14949.

102 Jiu, J., Isoda, S., Wang, F., and Adachi, M. (2006) *J. Phys. Chem. B*, **110** (5), 2087–2092.

103 Greene, L., Yuhas, B., Law, M., Zitoun, D., and Yang, P. (2006) *Inorg. Chem.*, **45** (19), 7535–7543.

104 Huisman, C., Schoonman, J., and Goossens, A. (2005) *Sol. Energy Mater. Sol. Cells*, **85** (1), 115–124.

105 Iijima, S. (1991) *Nature*, **354**, 56–58.

106 Huang, M., Mao, S., Feick, H., Yan, H., Wu, Y., Kind, H., Weber, E., Russo, R., and Yang, P. (2001) *Science*, **292** (5523), 1897–1899.

107 Lan, Z. and Wu, J. (2010) Faming Zhuanli Shenqing Gongkai Shuomingshu. CN 101857191 A 20101013.

108 Fan, J., Boettcher, S., and Stucky, G. (2006) *Chem. Mater.*, **18** (26), 6391–6396.

109 Greene, L., Johnson, J., Saykally, R., and Yang, P. (2005) *Nat. Mater.*, **4**, 455–459.

110 Hodes, G. (2008) *J. Phys. Chem. C*, **112** (46), 17778–17787.

111 Sayama, K., Sugihara, H., and Arakawa, H. (1998) *Chem. Mater.*, **10** (12), 3825–3832.

112 Wang, Z., Kawauchi, H., Kashima, T., and Arakawa, H. (2004) *Coord. Chem. Rev.*, **248** (13–14), 1381–1389.

113 Ito, S., Murakami, T., Comte, P., Liska, P., Grätzel, C., Nazeeruddin, M., and Grätzel, M. (2008) *Thin Solid Films*, **516** (14), 4613–4619.

114 Li, H., Jiang, B., Schaller, R., Wu, J., and Jiao, J., (2010) *J. Phys. Chem. C*, **114**, 11375–11386.

115 Yu, H., Zhang, S., Zhao, H., and Zhang, H. (2010) *Phys. Chem. Chem. Phys.*, **12** (25), 6625–6631.

116 Chiba, Y., Islam, A., Watanabe, Y., Komiya, R., Koide, N., and Han, L. (2006) *Jpn. J. Appl. Phys.*, **45** (25), L638–L640.

117 Sauvage, F., Di Fonzo, F., Li Bassi, A., Casari, C.S., Russo, V., Divitini, G., Ducati, C., Bottani, C.E., Comte, P., and Grätzel, M. (2010) *Nano Lett.*, **10** (7), 2562–2567.

118 Ghadiri, E., Taghavinia, N., Zakeeruddin, S.M., Grätzel, M., and Moser, J.-E. (2010) *Nano Lett.*, **10** (5), 1632–1638.

119 Han, T.H., Moon, H.-S., Hwang, J.O., Seok, S.I., Im, S.H., and Kim, S.O. (2010) *Nanotechnology*, **21** (18), 185601/1–185601/6.

120 Liu, W., Kou, D., Cai, M., Hu, L., Sheng, J., Tian, H., Jiang, N., and Dai, S. (2010) *J. Phys. Chem. C*, **114** (21), 9965–9969.

121 Wang, Z.-S., Yamaguchi, T., Sugihara, H., and Arakawa, H. (2005) *Langmuir*, **21** (10), 4272–4276.

122 Xu, C., Shin, P.H., Cao, L., Wu, J., and Gao, D. (2010) *Chem. Mater.*, **22** (1), 143–148.

123 Pandey, S.S., Sakaguchi, S., Yoshihiro, Y., and Hayase, S. (2010) *Org. Electron.*, **11** (3), 419–426.

124 Kim, J.-Y., Sekino, T., and Tanaka, S.-I. (2011) *International Journal of Applied Ceramic Technology*, **8**, 1353–1362.

125 Lai, Y.-H., Lin, C.-Y., Chen, H.-W., Chen, J.-G., Kung, C.-W., Vittal, R., and Ho, K.-C. (2010) *J. Mater. Chem.*, **20** (42), 9379–9385.

126 Shan, G.-B. and Demopoulos, G.P. (2010) *Adv. Mater.*, **22** (39), 4373–4377; Wang, G.Q., Gu, J.F., and Zhuo, S.P. (2010) *Chinese Chem. Lett.*, **21** (12), 1513–1516.

127 Fan, J., Boettcher, S., and Stucky, G. (2006) *Chem. Mater.*, **18** (26), 6391–6396.

128 Law, M., Greene, L., Johnson, J., Saykally, R., and Yang, P. (2005) *Nat. Mater.*, **4**, 455–459.

129 Wenger, B., Grätzel, M., and Moser, J.E. (2005) *J. Am. Chem. Soc.*, **127** (35), 12150–12151.

130 Tachibana, Y., Moser, J.E., Grätzel, M., Klug, D.R., and Durrant, J.R. (1996) *J. Phys. Chem.*, **100** (51), 20056–20062.

131 Ramakrishna, G., Verma, S., Jose, D.A., Kumar, D.K., Das, A., Palit, D.K., and Ghosh, H.N. (2006) *J. Phys. Chem. B*, **110**, 9012–9021.

132 O'Regan, B.C. and Durrant, J.R. (2009) *Acc. Chem. Res.*, **42** (11), 1799–1808.

133 Hagberg, D.P., Yum, J.H., Lee, H., De Angelis, F., Marinado, T., Karlsson, K.M., Humphry-Baker, R., Sun, L., Hagfeldt, A., Grätzel, M., and Nazeeruddin, M.K. (2008) *J. Am. Chem. Soc.*, **130** (19), 6259–6266.

134 Galoppini, E. (2004) *Coord. Chem. Rev.*, **248** (13–14), 1283–1297.

135 Kilsa, K., Mayo, E.I., Kuciauskas, D., Villahermosa, R., Lewis, N.S., Winkler, J.R., and Gray, H.B. (2003) *J. Phys. Chem. A*, **107** (18), 3379–3383.

136 Beek, W.J.E. and Janssen, R.A.J. (2004) *J. Mater. Chem.*, **14** (18), 2795–2800.

137 Chang, C.-W., Luo, L., Chou, C.-K., Lo, C.-F., Lin, C.-Y., Hung, C.-S., Lee, Y.-P., and Diau, E.W.-G. (2009) *J. Phys. Chem. C*, **113** (27), 11524–11530.

138 Beek, W.J.E. and Janssen, R.A.J. (2004) *J. Mater. Chem.*, **14** (18), 2795–2800.

139 Chang, C.-W., Luo, L., Chou, C.-K., Lo, C.-F., Lin, C.-Y., Hung, C.-S., Lee, Y.-P., and Diau, E.W.G. (2009) *J. Phys. Chem. C*, **113** (27), 11524–11531.

140 An, B.-K., Hu, W., Burn, P.L., and Meredith, P. (2010) *J. Phys. Chem. C*, **114** (41), 17964–17974.

141 Villanueva-Cab, J., Wang, H., Oskam, G., and Peter, L.M. (2010) *J. Phys. Chem. Lett.*, **1** (4), 748–751.

142 Bisquert, J. and Mora-Sero, I. (2010) *J. Phys. Chem. Lett.*, **1** (1), 450–455.

143 Wang, H. and Peter, L.M. (2009) *J. Phys. Chem. C*, **113** (42), 18125–18133.

144 Gonzalez-Vazquez, J.P., Anta, J.A., and Bisquert, J. (2010) *J. Phys. Chem. C*, **114** (14), 8552–8558.

145 Myllyperkio, P., Manzoni, C., Polli, D., Cerullo, G., and Korppi-Tommola, J. (2009) *J. Phys. Chem. C*, **113** (31), 13985–13992.

146 Zhu, K., Neale, N.R., Miedaner, A., and Frank, A.J. (2007) *Nano Lett.*, **7** (1), 69–74.

147 Wong, D., K.-P., Ku, C.-H., Chen, Y.-R., Chen, G.-R., Wu, J.-J. (2009) *ChemPhysChem*, **10** (15), 2698–2702.

148 Pivrikas, A., Sariciftci, N.S., Juska, G., and Osterbacka, R. (2007) *Prog. Photovolt. Res. Appl.*, **15** (8), 677–696.

149 Arndt, C., Zhokhavets, U., Mohr, M., Gobsch, G., Al-Ibrahim, M., and Sensfuss, S. (2004) *Synthetic Met.*, **147** (1–3), 257–260.

150 Shuttle, C.G., O'Regan, B., Ballantyne, A.M., Nelson, J., and Bradley, D.D.C. (2008) *Phys. Rev. B*, **78** (11), 113201–113205.

151 Dennler, G., Mozer, A.J., Juska, G., Pivrikas, A., Osterbacka, R., Fuchsbauer, A., and Sariciftfi, N.S. (2006) *Org. Electron.*, **7** (4), 229–234.

152 Juska, G., Sliauzys, G., Genevicius, K., Arlauskas, K., Pivrikas, A., Scharber, M., Dennler, G., Sariciftci, N.S., and Osterbacka, R. (2006) *Phys. Rev. B*, **74** (11), 115314–115318.

153 Pivrikas, A., Juska, G., Mozer, A.J., Scharber, M., Arlauskas, K., Sariciftci, N.S., Stubb, H., and Osterbacka, R. (2005) *Phys. Rev. Lett.*, **94** (17), 176806–174710.

154 Emil, E.-P., Bin, L., and Aydil, E.S. (2009) *Phys. Chem. Chem. Phys.*, **11**, 9648–9652.

155 Abrahamsson, M., Johansson, P.G., Ardo, S., Kopecky, A., Galoppini, E., and Meyer, G.J. (2010) *J. Phys. Chem. Lett.*, **1** (11), 1725–1728.

156 Wang, M., Chen, P., Humphry-Baker, R., Zakeeruddin, S.M., and Grätzel, M. (2009) *ChemPhysChem*, **10** (1), 290–299.

157 Privalov, T., Boschloo, G., Hagfeldt, A., Svensson, P.H., and Kloo, L. (2009) *J. Phys. Chem. C*, **113** (3), 783–790.

158 Wiberg, J., Marinado, T., Hagberg, D.P., Sun, L., Hagfeldt, A., and Albinsson, B. (2010) *J. Phys. Chem. B*, **114** (45), 14358–14363.

159 Marinado, T., Hagberg, D.P., Hedlund, M., Edvinsson, T., Johansson, E.M.J., Boschloo, G., Rensmo, H., Brinck, T., Sun, L.C., and Hagfeldt, A. (2009) *Phys. Chem. Chem. Phys.*, **11** (1), 133–141.

160 Zhu, K., Neale, N.R., Halverson, A.F., Kim, J.Y., and Frank, A.J. (2010) *J. Phys. Chem. C*, **114** (32), 13433–13441.

161 O'Regan, B.C., Lopez-Duarte, I., Martinez-Diaz, M.V., Forneli, A.,

Albero, J., Morandeira, A., Palomares, E., and Torres, T. (2008) *J. Am. Chem. Soc.*, **130** (10), 2906–2907.

162 Rowley, J. and Meyer, G.J. (2009) *J. Phys. Chem. C*, **113** (43), 18444–18447; Mora-Sero, I., Garcia-Belmonte, G., and Gimenez, S. (2009) *J. Phys. Chem.*, **113**, 7278–17290.

163 Garcia-Belmonte, G., Boix, P.P., Bisquert, J., Sessolo, M., and Bolink, H.J. (2010) *Sol. Energy Mater. Sol. Cells*, **94** (2), 366–375.

164 Uehara, S., Sumikura, S., Suzuki, E., and Mori, S. (2010) *Energy Environ. Sci.*, **3** (5), 641–644.

165 Staniszewski, A., Ardo, S., Sun, Y., Castellano, F.N., and Meyer, G.J. (2008) *J. Am. Chem. Soc.*, **130** (35), 11586–11587.

166 Bauer, C., Boschloo, G., Mukhtar, E., and Hagfeldt, A. (2002) *J. Phys. Chem. B*, **106** (49), 12693–12704.

167 Gardner, J.M., Giaimuccio, J.M., and Meyer, G.J. (2008) *J. Am. Chem. Soc.*, **130** (51), 17252–17253.

168 Wang, Q., Ito, S., Grätzel, M., Fabregat-Santiago, F., Mora-Sero, I., Bisquert, J., Bessho, T., and Imai, H. (2006) *J. Phys. Chem. B*, **110**, 19406–19411.

169 Wang, Q., Moser, J.E., and Grätzel, M.J. (2005) *Phys. Chem. B*, **109** (31), 14945–14953.

170 Wei, T.-H., Wan, C., and Wang, Y. (2006) *Appl. Phys. Lett*, **88** (10), 103122–103123.

171 Papageorgiou, N., Maier, W.M., and Grätzel, M. (1997) *J. Electrochem. Soc.*, **144** (3), 876–884.

172 Kitamura, T., Maitani, M., Matsuda, M., Wada, Y., and Yanagida, S. (2001) *Chem. Lett.*, **30** (10), 1054–1055.

173 Murakamim, T.N. and Grätzel, M. (2008) *Inorg. Chim. Acta*, **361** (3), 572–580.

174 Imoto, K., Takahashi, T., Yamaguchi, T., Komura, T., Nakamura, J., and Murata, K. (2003) *Sol. Energy Mater. Sol. Cells*, **79** (4), 459–469.

175 Saito, Y., Ogawa, A., Uchida, S., Kubo, T., and Segawa, H. (2010) *Chem. Lett.*, **39** (5), 488–489.

176 Nusbaumer, H., Zakeeruddin, S.M., Moser, J.-E., and Grätzel, M. (2003) *Chem. Eur. J.*, **9** (16), 3756–3763.

177 Sapp, S.A., Elliott, C.M., Contado, C., Caramori, S., and Bignozzi, C.A. (2002) *J. Am. Chem. Soc.*, **124** (37), 11215–112225.

178 Caramori, S., Cazzanti, S., Marchini, L., Argazzi, R., Bignozzi, C.A., Martineau, D., Gros, P.C., and Beley, M. (2008) *Inorg. Chim. Acta*, **361** (3), 627–634.

179 Fukuri, N., Masaki, N., Kitamura, T., Wada, Y., and Yanagida, S. (2006) *J. Phys. Chem. B*, **110** (50), 25251–25258.

180 Gorlov, M. and Kloo, L. (2008) *Dalton Trans.* (20), 2655–2659.

181 Wang, Z., Sayama, K., and Sugihara, H. (2005) *J. Phys. Chem. B*, **109** (47), 22449–22455.

182 Sapp, S., Elliott, C., Contado, C., Caramori, S., and Bignozzi, C. (2002) *J. Am. Chem. Soc.*, **124** (37), 11215–11222.

183 Cameron, P., Peter, L., Zakeeruddin, S., and Grätzel, M. (2004) *Coord. Chem. Rev.*, **248** (13–14), 1447–1453.

184 Campbell, W., Burrell, A., Officer, D., and Jolley, K. (2004) *Coord. Chem. Rev.*, **248** (13–14), 1363–1379.

185 Ito, S., Nazeeruddin, M., Liska, P., Comte, P., Charvet, R., Péchy, P., Jirousek, M., Kay, A., Zakeeruddin, M., and Grätzel, M. (2006) *Prog. Photovolt. Res. Appl.*, **14** (7), 589–601.

186 Gratzel, M. (2005) *MRS Bull.*, **30** (1), 23–27.

187 Oja, I., Belaidi, A., Dloczik, L., Lux-Steiner, M.C., and Dittrich, T. (2006) *Semicond. Sci. Technol.*, **21** (4), 520–526.

188 Larramona, G., Chone, C., Jacob, A., Sakakura, D., Delatouche, B., Pere, D., Cieren, X., Nagino, M., and Bayon, R. (2006) *Chem. Mater.*, **18** (6), 1688–1696.

189 Diguna, L.J., Shen, Q., Kobayashi, J., and Toyoda, T. (2007) *Appl. Phys. Lett.*, **91**, 023116.

190 Ernst, K., Engelhardt, R., Ellmer, K., Kelch, C., Muffler, H.J., Lux-Steiner, M.C., and Konenkamp, R. (2001) *Thin Solid Films*, **387** (1), 26–28.

191 Page, M., Niitsoo, O., Itzhaik, Y., Cahen, D., and Hodes, G. (2009) *Solar Cells. Energy Environ. Sci.*, **2** (2), 220–223.

192 Lee, H.J., Yum, J.H., Leventis, H.C., Zakeeruddin, S.M., Haque, S.A., Chen, P., Seok, S.I., Grätzel, M., and

Nazeeruddin, M.K. (2008) *J. Phys. Chem. C*, **112** (30), 11600–11608.

193 Kubo, W., Kitamura, T., Hanabusa, K., Wada, Y., and Yanagida, S. (2002) *Chem. Commun.*, 374–375.

194 Wang, P., Zakeeruddin, S., Moser, J., and Grätzel, M. (2003) *J. Phys. Chem. B*, **107** (48), 13280–13285.

195 Bonhote, P., Dias, A.P., Papageorgiou, N., Kalyanasundaram, K., and Gratzel, M. (1996) *Inorg. Chem.*, **35** (5), 1168–1178.

196 Santa-Nokki, H., Busi, S., Kallioinen, J., Lahtinen, M., and Korppi-Tommola, J. (2007) *J. Photochem. Photobiol. A*, **186** (1), 29–33.

197 Bai, Y., Cao, Y., Zhang, J., Wang, M., Li, R., Wang, P., Zakeeruddin, S.M., and Grätzel, M. (2008) *Nat. Mater.*, **7** (7), 626–630.

198 Xue, B., Wang, H., Hu, Y., Li, H., Wang, Z., Meng, Q., Huang, X., Sato, O., Chen, L., and Fujishima, A. (2004) *Photochem. Photobiol. Sci.*, **3**, 918–919.

199 Wang, P., Zakeeruddin, S.M., Moser, J.-E., Humphry-Baker, R., and Grätzel, M. (2004) *J. Am. Chem. Soc.*, **126** (23), 7164–7165.

200 Yamaguchi, T., Tobe, N., Matsumoto, D., Nagai, T., and Arakawa, H. (2010) *Sol. Energy Mater. Sol. Cells*, **94** (5), 812–816.

201 Papageorgiou, N. (2004) *Coord. Chem. Rev.*, **248**, 1421–1446.

202 Wang, Y.-F., Lei, B.-X., Hou, Y.-F., Zhao, W.-X., Liang, C.-L., Su†, C.-Y., and Kuang, D.-B. (2010) *Inorg. Chem.*, **49**, 1679–1686.

203 Murakami, T.N., Ito, S., Wang, Q., Nazeeruddin, M.K., Bessho, T., Cesar, I., Liska, P., Humphry-Baker, R., Comte, P., Péchy, P., and Grätzel, M. (2006) *J. Electrochem. Soc.*, **153** (12), A2255–A2261.

204 Huang, Z., Liu, X.H., Li, K.X., Li, D.M., Luo, Y.H., Li, H., Song, W.B., Chen, L.Q., and Meng, Q.B. (2007) *Electrochem. Commun.*, **9**, 596–598.

205 Lee, W.J., Ramasamy, E., Lee, D.Y., and Song, J.S. (2009) *ACS Appl. Mater. Interfaces*, **1** (60), 1145–1149.

206 Wu, J.H., Li, Q.H., Fan, L.Q., Lan, Z., Li, P.J., Lin, J.M., and Hao, S.C. (2008) *J. Power Sources*, **181** (1), 172–176.

207 Li, Q.H., Wu, J.H., Tang, Q.W., Lan, Z., Li, P.J., Lin, J.M., and Fan, L.Q. (2008) *Electrochem. Commun.*, **10** (9), 1299–1302.

208 Ma, T.L., Fang, X.M., Akiyama, M., Inoue, K., Noma, H., and Abe, E. (2004) *J. Electroanal. Chem.*, **574** (1), 77–83.

209 Kavan, L., Yum, J.H., and Gratzel, M. (2011) *Nano*, **5**, 165–72.

210 Li, G.-R., Wang, F., Jiang, Q.-W., Gao, X.-P., and Shen, P.-W. (2010) *Angew. Chem. Int. Ed.*, **49** (21), 3653–3656.

211 Sun, H., Luo, Y., Zhang, Y., Li, D., Yu, Z., Li, K., and Meng, Q. (2010) *J. Phys. Chem. C*, **114**, 11673–11679.

212 Wang, P., Zakeeruddin, S.M., Comte, P., Exnar, I., and Grätzel, M. (2003) *J. Am. Chem. Soc.*, **125** (5), 1166–1167.

213 Sirimanne, P.M., Winther-Jensen, B., Weerasinghe, H.C., and Yi-Bing Cheng, Y.B. (2010) *Thin Solid Films*, **518** (10), 2871–2875.

214 Lee, K.M., Chen, P.Y., Hsu, C.Y., Huang, H.H., Ho, W.H., Chen, H.C., and Ho, K.C. (2009) *J. Power Sources*, **188** (1), 313–318.

215 Caramori, S., Vito Cristino, V., Boaretto, R., Argazzi, R., Bignozzi, K.A., and Di Carlo, A. (2010) *Int. J. Photoenergy* **2010**, 16.

216 Ahmad, S., Yum, J.M., Xianxi, Z., Grätzel, M., Butt, H.J., and Nazeeruddin, M.K. (2010) *J. Mater. Chem.*, **20** (90), 1654–1658.

217 Lee, K.S., Lee, H.K., Wang, D.H., Park, N.-G., Lee, J.Y., Park, O.O., and Park, J.H. (2010) *Chem. Commun.*, 4505–4507.

218 Saito, Y., Ogawa, A., Uchida, S., Kubo, T., and Segawa, H. (2010) *Chem. Lett.*, **39** (5), 488–489.

219 Gregg, B.A., Pichot, F., Ferrere, S., and Fields, C.L. (2001) *J. Phys. Chem. B*, **105**, 1422–1429.

220 O'Regan, B. and Schwartz, D.T. (1995) *Chem. Mater.*, **7**, 1349–1355.

221 Hagen, J., Schaffrath, W., Otschik, P., Fink, R., Bacher, A., Schmidt, H., and Haarer, D. (1997) *Synth. Met.*, **89** (3), 215–220.

222 Bach, U., Lupo, D., Comte, P., Moser, J., Weissortel, F., Salbeck, J., Spreitzer, H., and Grätzel, M. (1998) *Nature*, **395**, 583–585.

223 Tennakone, K., Senadeera, G., Silva, D., and Kottegoda, I. (2000) *Appl. Phys. Lett.*, **77** (19), 2367–2369.

224 Kim, J.-Y., Sekino, T., and Tanaka, S.-I. (2011) *International Journal of Applied Ceramic Technology*, **8**, 1353–1362.

225 Kay, A. and Grätzel, M. (1996) *Sol. Energy Mater. Sol. Cells*, **44** (10), 99–117.

226 Imoto, K., Takahashi, K., Yamaguchi, T., Komura, T., Nakamura, J., and Murata, K. (2003) *Sol. Energy Mater. Sol. Cells*, **79** (4), 459–469.

227 Ikeda, N., Teshima, K., and Miyasaka, T. (2006) *Chem. Commun.*, 1733–173.

228 Wenger, S., Bouit, P.-A., Chen, Q., Teuscher, J., Di Censo, D., Humphry-Baker, R., Moser, J.E., Delgado, J.L., Martin, N., Zakeeruddin, S.M., and Grätzel, M. (2010) *J. Am. Chem. Soc.*, **132** (14), 5164–5169.

229 Goto, K., Kawashima, T., and Tanabe, N. (2006) *Sol. Energy Mater. Sol. Cells*, **90** (18), 3251–3260.

230 Mikroyannidis, J.A., Kabanakis, A., Balraju, P., and Sharma, G.D. (2010) *J. Phys. Chem. C*, **114** (28), 12355–12363.

231 Ito, S., Ha, N., Rothenberger, G., Liska, P., Comte, P., Zakeeruddin, S., Pechy, P., Nazeeruddin, M., and Grätzel, M. (2006) *Chem. Commun.*, 4004–4006.

232 Murakami, T., Kijitori, Y., Kawashima, N., and Miyasaka, T. (2004) *J. Photochem. Photobiol. A*, **164** (1–3), 187–191.

233 Jun, Y., Kim, J., and Kang, M. (2007) *Sol. Energy Mater. Sol. Cells*, **91**, 779–784.

234 Zhang, G.L., Bala, H., Cheng, Y.M., Shi, D., Iv, X.J., Yu, Q.J., and Wang, P. (2009) *Chem. Commun.*, 2198–2120.

235 Schmidt-Mende, L., Bach, U., Humphry-Baker, R., Horiuchi, T., Miura, H., Ito, S., Uchida, S., and Grätzel, M. (2005) *Adv. Mater.*, **1,7** (7), 813–818.

236 Snaith, H.J., Zakeeruddin, S.M., Schmidt-Mende, L., Klein, C., and Grätzel, M. (2005) *Angew. Chem., Int. Ed.*, **44** (36), 6413–6416.

237 O'Reagan, B. and Grätzel, M. (2005) *Nature*, **353** (6346), 737–740.

238 Wei, D., Unalan, H.E., Han, D., Zhang, Q., Niu, L., Amaratunga, G., and Ryhanen, T. (2008) *Nanotech.*, **19** (42), 424006.

239 Yang, Y., Tao, J., Jin, X., and Qin, Q. (2011) *Int. J. Photoenergy*, Vol. 2011, Article ID 405738, 9 pages

240 O'Regan, B., Lenzmann, F., Muis, R., and Wienke, J. (2002) *Chem. Mater.*, **14** (12), 5023–5029.

241 Moon, S.-J., Yafit Itzhaik, Y., Yum, J.-H., Zakeeruddin, S.M., Hodes, G., and Grätzel, M. (2010) *J. Phys. Chem. Lett.*, **1** (10), 1524–1527.

242 Balis, N., Dracopoulos, V., Antoniadou, M., and Lianos, P. (2010) *J. Photoch. Photobio. A*, **214** (1), 69–73.

243 Shi, J., Wang, L., Liang, Y., Peng, S., Cheng, F., and Chen, J. (2010) *J. Phys. Chem. C*, **114** (14), 6814–6821.

244 Thompson, S.J., Duffy, N.W., Bach, U., and Cheng, Y.-B. (2010) *J. Phys. Chem. C*, **114** (5), 2365–2369.

245 Huo, Z., Dai, S., Zhang, C., Kong, F., Fang, X., Guo, L., Liu, W., Hu, L., Pan, X., and Wang, K. (2008) *J. Phys. Chem. B*, **112** (12), 12927–12941.

246 Nogueira, A.F., Durrant, J.R., and De Paoli, M.A. (2001) *Adv. Mater.*, **13** (11), 826–829.

247 Fukuri, N., Masaki, N., Kitamura, T., Wada, Y., and Yanagida, S. (2006) *J. Phys. Chem. B*, **110** (50), 25251–25258.

248 Han, H., Liu, W., Zhang, J., and Zhao, X. (2005) *Adv. Funct. Mater.*, **15** (12), 1940–1944.

249 Liu, W., Zhang, J., and Zhao, X. (2005) *Adv. Funct. Mater.*, **15** (12), 1940–1944.

250 Boucharef, M., di Bin, C., Boumaza, M.S., Colas, M., Snaith, H.J., Ratier, B., and Boucle, J. (2010) *Nanotechnology*, **21** (20), 205203/1–205203/12.

251 Miettunen, K., Ruan, X., Saukkonen, T., Halme, J., Toivola, M., Guangsheng, H., and Lund, P. (2010) *J. Electrochem. Soc.*, **157** (6), B814–B819.

6
Dye-Sensitized Solar Cells II

6.1
Optical Fiber DSSC

The energy conversion efficiency of three-dimensional dye-sensitized solar cells (DSSC) in a hybrid structure that integrates optical fibers and nanowire (NW) arrays is greater than that of a two-dimensional device ([1–5] and references therein). The technical advantages of fiber devices include flexibility, light weight, and possibility to make large-area active surfaces that can be incorporated into wearable technology. In order to capture maximum amount of the incoming light, the semiconductor nanomaterials are used as scaffold to hold a large number of dye molecules in a 3D matrix, increasing the number of molecules for any given surface area of the cell. The advantage of a fiber optic solar cell system over a planar one is that light bounces around inside an optical fiber as it travels along its length, providing more opportunities to interact with the solar cell on its inner surface and producing more current. The total area of such a cell is higher, and increased surface area means improved light harvesting. Integrated self-monitoring optical transport fiber consists of a photonic bandgap optical fiber and temperature monitoring elements along the entire fiber length.

A tunable fiber photodetector has been made comprising an amorphous semiconductor core contacted by metallic NWs and surrounded by a cylindrical shell resonant optical cavity [4]. An optical fiber-based organic photovoltaic (PV) device using poly(3-hexylthiophene) and 1-(3-methoxycarbonyl)-propyl-1-phenyl-(6,6) C_{61} bulk heterojunction (BHJ) blends as the absorbing material has been fabricated onto a multimode optical fiber.

The optoelectronically active optical fiber, which was demonstrated in the work [2], consisted of dye-sensitized solar cell structure deposited on claddingless optical fiber (Figure 6.1). Both silica and plastic optical fibers were used as a substrate. DSC structure consisting of ZnO:Al transparent current collector layer, TiO_2 photoelectrode sensitized with ruthenium dye, gelatinized iodine electrolyte, and carbon-based counterelectrode was deposited layer by layer on top of the optical fiber. Maximum obtained short-circuit current, I_{sc}, was 26 nA/cm^2 and maximum open-circuit voltage, V_{oc}, was 0.44 V.

Solar Energy Conversion. Chemical Aspects, First Edition. Gertz Likhtenshtein.
© 2012 Wiley-VCH Verlag GmbH & Co. KGaA. Published 2012 by Wiley-VCH Verlag GmbH & Co. KGaA.

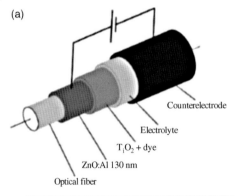

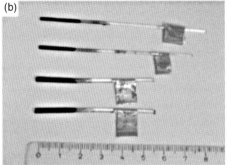

Figure 6.1 (a) Schematic drawing of the PV fiber. (b) Photograph of the PV fibers (without the CE contacts) of different diameters [2] Reproduced with permission from Elsevier.

Using poly(3-hexylthiophene) and 1-(3-methoxycarbonyl)-propyl-1-phenyl-(6,6) C_{61} bulk-heterojunction blends as the absorbing material, organic photovoltaic devices have been fabricated onto multimode optical fibers [3]. The behavior of the short-circuit current density, filling factor, and open-circuit voltage as the angle of the incident light onto the cleaved fiber face was varied. The authors suggested that the evanescent field at the interface between the fiber and the transparent contact may play a role in coupling light from the fiber into the device. An innovative hybrid structure that integrates optical fibers and nanowire arrays as three-dimensional DSSCs that have a significantly enhanced energy conversion efficiency was reported [1]. The DSSC hybrid structure integrates optical fibers and ZnO NWs grown using a chemical approach on the fiber surfaces. The design principle is shown in Figure 6.2. The main structure consists of a bundle of quartz fibers arranged such that the incident sunlight can enter the fibers from one end. The upper region of the fibers functions to effectively guide light for concealed and adaptable applications. The fiber surface was coated with a cladding layer of low refractive index for minimizing light loss. The DSSC was fabricated on the lower region of the fiber surface. This segment of the fiber was first coated with an indium tin oxide (ITO) layer that simultaneously acts as a conductive electrode and a high refractive index material that allows light to escape the fiber and enter the DSSC. The key principle of the

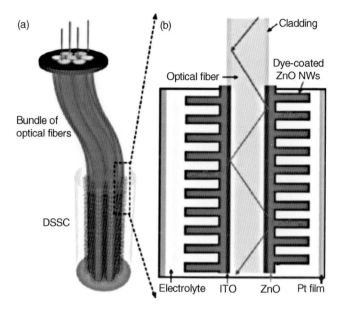

Figure 6.2 Design and principle of a three-dimensional DSSC. The cross-section of the fiber can be cylindrical or rectangular. (a) The 3D DSSC is composed of optical fibers and ZnO NWs are grown vertically on the fiber surface. The top segment of the bundled optical fibers utilizes conventional optical fibers and allows for remote transmission of light. The bottom segment consists of the 3D DSSC for solar power generation at a remote/concealed location. (b) Detailed structure of the 3D DSSC [1].

design was that the light entering from the axial direction inside the fiber experiences multiple internal reflections along the fiber. At each internal reflection at the fiber/ITO/ZnO NW interfaces, light will cross the interface to reach the dye molecules through the NWs. In comparison to the case of light illumination normal to the fiber axis from outside the device (2D case), the internal axial illumination enhances the energy conversion efficiency of a rectangular fiber-based hybrid structure by a factor of up to six for the same device. The absolute full-sun efficiency (AM 1.5 illumination, 100 mW/cm^2) was 120% higher than the highest value reported for ZnO NWs grown on a flat substrate surface and 47% higher than that of ZnO NWs coated with a TiO$_2$ film capability.

A hybrid structure that integrates optical fibers and nanowire arrays as three-dimensional (3D) DSSCs with significantly enhanced energy conversion efficiency was reported [6]. It was shown that ZnO NWs grow normal to the optical fiber surface and enhance the surface area for the interaction of light with dye molecules. The light illuminates the fiber from one end along the axial direction, and its internal reflection within the fiber creates multiple opportunities for energy conversion at the interfaces. The author's approach relied on aligned ZnO nanowire arrays grown on surfaces of a flat substrate, and a dye-sensitized solar cell was built on its top surface to convert solar energy and a piezoelectric nanogenerator was built on its bottom surface for harvesting ultrasonic wave energy from the surroundings. The two energy-harvesting approaches can work simultaneously or individually, and they can be integrated in parallel or series.

6.2
Tandem DSSC

The tandem approach, that is, stacking many solar cells together has been successfully used in conventional photovoltaic devices to maximize energy generation [7–13]. The maximum photoconversion efficiency of the photovoltaic device depends on the number of components that compose the cell. A tandem cell contains two or more different sensitizers (one on each photoelectrode); therefore, a larger portion of the solar spectrum could be harvested since it is easier to cover a broad absorption window with two different dyes having complementary absorption spectra than with a single one. In a tandem DSSC, the photovoltage is independent of the redox potential of the mediator and depends only on the difference between the potentials of the valence band of the p-semiconductor and the conduction band of the n-semiconductor.

In the pioneering work [8], nanostructured NiO film was prepared by depositing nickel hydroxide slurry on conducting glass and sintering at 500 °C to a thickness of about 1 μm The highest incident photon-to-current conversion efficiencies of tetrakis (4-carboxyphenyl)porphyrin (TPPC) and erythrosin B-coated NiO films were 0.24 and 3.44%, respectively. In sandwich solar cells with a platinized conducting glass as counterelectrode exposed to light from a sun simulator (light intensity: 68 mW/cm^2), a short-circuit cathodic photocurrent density (I_{SC}) of 0.079 mA/cm^2 and an open-circuit voltage (V_{OC}) of 98.5 mV for TPPC-coated NiO electrode were achieved. Similarly, $I_{SC} = 0.232$ mA/cm^2 and $V_{OC} = 82.8$ mV were registered when the NiO electrode was coated with erythrosin B. The first tandem DSSC composed of a TiO$_2$ photoanode sensitized with cis-di(thiocyanato)-N,N-bis(2,2'-bipyridyl-4,4'-dicarboxylic acid)-ruthenium(II) (N719) and a NiO photocathode sensitized with erythrosine B was reported by Lindquist and coworkers in 2000 [9]. The photoconversion efficiency was found to be low ($\eta = 0.39\%$ under AM 1.5). The V_{OC} of the tandem cell reached 732 mV, which was 82 mV higher than that of the n-DSCC with TiO$_2$ ($V_{OC} = 650$ mV) and than that of the p-DSSC with NiO ($V_{OC} = 83$ mV), proving that tandem DSSCs could deliver a larger photopotential than do single semiconductor DSSCs.

Various tandem and hybrid cells have been reported [10–15]. Tandem DSSCs connected in parallel were prepared in works [14, 15]. In the parallel connection, the current density was the sum of the current densities of the top and bottom cells. In order to obtain higher efficiency, series-connected tandem DSSCs were introduced [16]. In this system, the photovoltage become the sum of photovoltages of the top and bottom cells. The current density of both the top and the bottom cells was tuned to be identical, the top cell was required to have a high voltage and the bottom cell should have a longer wavelength absorption area compared to the top cell. The top cell was made up of a transparent cell and the bottom cell utilizes only the light

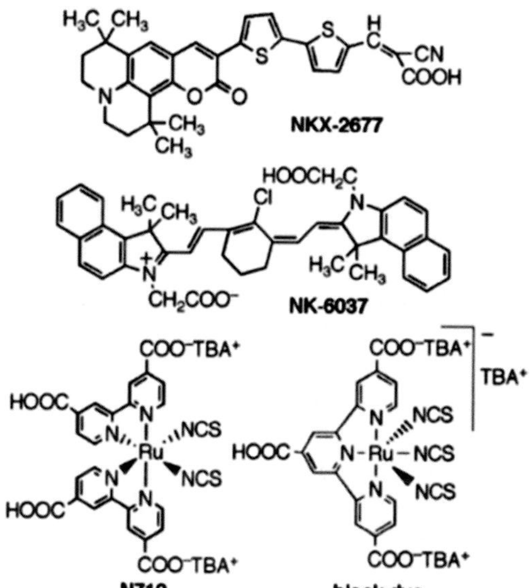

Figure 6.3 Structure-sensitizing dyes used in the work [16] Reproduced with permission from Elsevier.

passing through the top cell. Dye combinations of NKX-2677, N719, NK-6037, and black dye (Figure 6.3) were investigated. The best efficiency obtained in *our* this study is 10.4% ($J_{sc} = 10.8$ mA/cm^2, $V_{oc} = 1.45$ V, and FF $= 0.67$) for a series-connected tandem DSSC consisting of an N719 (Figure 6.4) top cell and a black dye bottom cell.

Dye-sensitized solar cells (containing dye-bilayer structure of black dye and NK3705 (3-carboxymethyl-5-[3-(4-sulfobutyl)-2(3*H*)-bezothiazolylidene]-2-thioxo-4-thiazolidinone, sodium salt) in one TiO$_2$ layer (2-TiO-BD-NK) were reported) [12]. The 2-TiO-BD-NK structure was fabricated by staining one TiO$_2$ layer with these two dyes, step by step, under a pressurized CO$_2$ condition. The dye bilayer structure was observed by using a confocal laser scanning microscope.

Tandem dye-sensitized solar cells (TDSC) consisting of two electrodes in one cell were reported [17]. A tandem cell (Cell TAN GF or Cell TAN St) has a floating electrode (bottom cell) and a TiO$_2$ electrode prepared on a F-doped SnO$_2$ glass substrate (top cell). The floating electrode was a flexible and self-standing composite film consisting

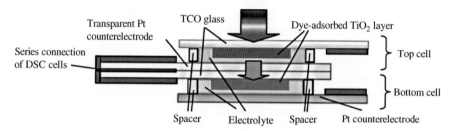

Figure 6.4 Structure of a series-connected DSSC [16] Reproduced with permission from Elsevier.

of a porous titania/dye layer supported by a glass mesh sheet or a stainless-steel mesh sheet. The incident photon-to-current conversion efficiency (IPCE) curves for Cell TAN GF and Cell TAN had two peaks corresponding to the visible absorption of the two dyes.

Nickel oxide (NiO) thin films were prepared onto ITO/glass substrates by spin coating, dipping, and electrochemically [18]. Studies of the morphological and structural properties of the films were done by atomic force microscopy (AFM). Photoelectrochemical and optical experiments were carried out in order to characterize the semiconductor properties of the nanostructured NiO thin films. The experiments were also done for Eosin B- and Erythrosin J-sensitized nanostructured NiO films. The photoelectrochemical results for all the bare NiO, NiO–Eosin B, and NiO–Erythrosin J/electrolyte (I_2/I^-) systems showed a p-type behavior. For the NiO/Erythrosin J system, the enhancement of the current under illumination in comparison to the dark current was about 200%.

Highly efficient photocathodes for dye-sensitized tandem solar cells were constructed [19]. The authors showed that p-DSSCs can convert absorbed photons to electrons with yields of up to 96%, resulting in a sevenfold increase in energy conversion efficiency compared to the previously reported photocathodes. The donor–acceptor dyes, studied as photocathodic sensitizers, comprised a variable-length oligothiophene bridge, which provides control over the spatial separation of the photogenerated charge carriers. As a result, charge recombination was decelerated by several orders of magnitude and tandem pn-DSSCs can be constructed that exceed the efficiency of their individual components. A tandem structure for improving the short-circuit performance of a dye-sensitized solar cell was presented [20]. Two dye-sensitized nanocrystal TiO_2 films were placed face-to-face as electrodes. As a counter-electrode, a Pt mesh with good transmittance was inserted between the electrodes. Two TiO_2 anodes were connected in parallel and in series with the Pt-mesh cathode. The front electrode thickness was increased according to the light model for the front and back electrodes described the output of each electrode well, and the efficiency of the tandem cell increased by 4.7% for a thickness of 7.8 μm.

In order to increase the range of absorption in a broad solar spectrum, a tandem structure with two different sensitizer dyes, ruthenium-complexed bipyridine carboxylates, especially cis-bis isothiocyanato-bis 2,2- bipyridyl-4,4- dicarboxylic acid ruthenium(II) (red dye) and triisothiocyanato 2,2,6,2-terpyridyl-4,4,4-tricarboxylic acid ruthenium(II) (black dye), in two different compartments of the cell, respectively, was realized [21]. Overall power conversion efficiencies as high as $\eta = 10.5\%$ and short-circuit current densities of J_{SC} mA/cm^2 = 21.1 were achieved under air mass 1.5 illumination with red dye and black dye in the upper and lower compartments of the cell, respectively. An efficient organic tandem solar cell combining a solid-state dye-sensitized with a $ZnPc/C_{60}$-based vacuum deposited bulk heterojunction solar cell was described [22]. Owing to an effective serial connection of both subcells and to the complementary absorption of the dyes used, a power conversion efficiency (PCE) of $\eta_p = (6.0 \pm 0.1)\%$ was achieved under simulated 100 mW/cm^2 AM 1.5 illumination. The device parameters were $V_{oc} = (1360 \pm 10)$ mV, $I_{sc} = (82 \pm 0.1)$ mA/cm^2, and FF $= (54 \pm 1)\%$.

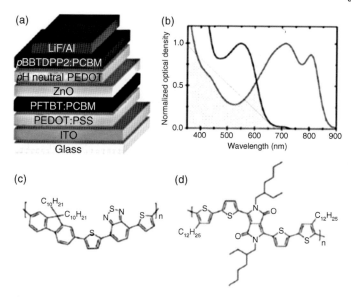

Figure 6.5 (a) Device structure of the polymer tandem solar cell. (b) Normalized absorption spectra of both individual active layers (PFTBT:PCBM and pBBTDPP2:PCBM). (c) Molecular structure of PFTBT and (d) pBBTDPP2 [24].

A tandem dye-sensitized solar cell was fabricated on a glass rod without a transparent conductive oxide layer (TCO-less GR-DSSC) [23]. Two model dyes having $\lambda_{max} = 429$ nm (Dye II) and $\lambda_{max} = 646$ nm (Dye I), respectively, were used to examine the potential of the cell. The incident photon-to-current conversion efficiency curve of the TCO-less GR-DSSC had two peaks at around 490 and 600–650 nm. Open-circuit voltage (V_{oc}) of the TCO-less GR-DSSC was 1.13 V, which was the sum of the V_{oc} (0.57 V) of each single cell. The results demonstrated that the cell has a tandem character. Tandem cells presented in the study [24] consisted of bulk heterojunction subcells comprising wide- and small-bandgap polymers as electron donors mixed with [6,6]-phenyl C_{61} butyric acid methyl ester (PCBM) as an electron acceptor (Figure 6.5). Poly[2,7-(9,9-didecylfluorene)-alt-5,5-(40,70-di-2-thienyl-20,10,30- benzothiadiazole)] (PFTBT) [9] was used as the wide-bandgap (1.95 eV) material and poly [3,6-bis(40-dodecyl-[2,20]bithiophenyl- 5-yl)-2,5-bis(2-ethylhexyl)-2,5-dihydropyrrolo [3,4-]pyrrole-1,4-dione] (pBBTDPP2) [10] for the small-bandgap (1.40 eV) cell. Mixed with PCBM, these polymers provided efficient single-junction cells, exhibiting high V_{oc} of about 1.0 and 0.6 V.

Efficient BHJ solar cells were obtained by mixing poly[2,7-(9,9-dihexylfluorene)-alt-bithiophene] (F6T2) and 6,6-phenyl C_{61} butyric acid methyl ester with variable weight ratios (Figure 6.6) [25]. Morphological images from atomic force microscopy and scanning electron microscopy (SEM) revealed that the phase separation in F6T2:PCBM blend films becomes pronounced with the increase in PCBM concentration, resulting in an increased fill factor up to 56.9%. At optimized F6T2:PCBM weight

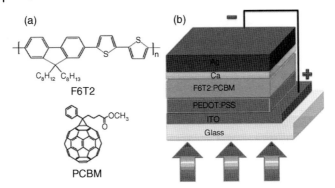

Figure 6.6 (a) Chemical structures of donor F6T2 and acceptor PCBM. (b) Structure of OSC [25]. Reproduced with permission from the American Chemical Society.

ratio (1 : 2), the single cell exhibits a highest power conversion efficiency of 2.46% and charge transport. The polymer–small molecule tandem cells constructed using F6T2: PCBM BHJ as the bottom cell and copper phthalocyanine (CuPc):fullerene (C_{60}) as the top cell showed an open-circuit voltage (V_{oc} 0.86 V, bottom cell, and 0.43 V, top cell).

In the study [26], a novel fabrication method for the preparation of mesoporous NiO films for tandem solar cell applications based on preformed NiO nanopowders with the dye coumarin 343 as sensitizer was presented. A series of different sensitizers and electrolytes were scrutinized for their application in dye-sensitized NiO photocathodes. Values for short-circuit current densities of 2.13 mA/cm^2 and overall energy conversion efficiencies of 0.033% under simulated sunlight (AM 1.5, 1000 W/m^2) were obtained. A tandem structure was introduced to improve the spectral response of dye-sensitized solar cells without losing their high external quantum yield [27]. Light-harvesting efficiency and IPCE were calculated using absorption spectra of dye-adsorbed TiO$_2$ electrode, electrolyte, and conducting glass support. The IPCE of single DSSCs fabricated using two typical ruthenium complexes was measured to present the improvement in the spectral response without losing their high external quantum yield by tandem structure. As a result, the tandem-structured cell exhibited higher photocurrent and conversion efficiency than each single DSSC mainly because of its extended spectral response.

FeS and FeS$_2$ nanosheet films as novel photocathodes for tandem dye-sensitized solar cells were synthesized on Fe substrates through 1-step hydrothermal treatment of a Fe foil and S powder in the presence or absence of hydrazine [28]. The resulting FeS$_x$ (x = 1, 2) nanosheet films were characterized and used as novel photocathodes in tandem solar cells with dye-sensitized TiO$_2$ nanorod films as the corresponding photoanode. In the case of the FeS nanosheet film photocathode, I_{sc} of 2.53 mA/cm^2, V_{oc} of 0.60 V, FF of 0.31, and conversion efficiency (η) of 1.32% were obtained under an illumination of 100 mW/cm^2.

An overview of the research carried out by a European consortium to develop and test new and improved ways to realize dye-sensitized solar cells with enhanced efficiencies and stabilities was presented [29]. Several new concepts and materials, fabrication protocols for TiO_2 and scatter layers, metal oxide blocking layers, strategies for cosensitization, and low-temperature processes of platinum deposition have been explored. The combined efforts have led to the attainment of maximum noncertified power conversion efficiencies under full sunlight of 11% for areas <0.2 cm² and 10.1% for a cell with an active area of 1.3 cm². Lifetime studies revealed negligible device degradation after 1000 h of accelerated tests under thermal stress at 80 °C in the dark and visible light soaking at 60 °C.

It was shown that implementation of a photocathode (sensitized p-SC) with a photoanode (sensitized TiO_2 photoanode) opens the possibility to fabricate tandem DSSCs [30]. The maximum photoconversion efficiency of the photovoltaic device in such a cell goes from 31% in a single-semiconductor photovoltaic cell to 42% in a tandem cell including two different semiconductors. In addition, in tandem DSSC, the photovoltage is independent of the redox potential of the mediator and depends only on the difference between the potentials of the valence band of the p-SC and the conduction band of the n-SC.

It was demonstrated that employing a molecular dyad (two peryleneimide-based dyes) and a cobalt-based electrolyte gave a threefold increase in open-circuit voltage for a p-type NiO device ($V_{OC} = 0.35$ V) and a fourfold better energy conversion efficiency [31]. Incorporating these improvements in a TiO_2/NiO tandem dye-sensitized solar cell resulted in a TDSC with a $V_{OC} = 0.91$ V. Three p-DSSCs were composed of a cobalt mediator and three different sensitizing dyes: coumarin 343 a typical sensitizer for p-type DSSCs and PI and PINDI, two peryleneimide-based dyes. The nanostructured NiO films were up to 5 μm thick. A schematic representation of a TDSC is shown in Figure 6.7. Absorption of light by the sensitizers, in separate

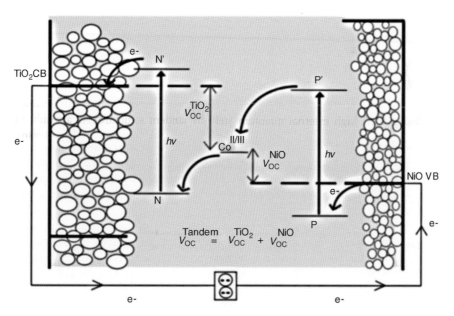

Figure 6.7 A schematic representation of a TDSC [31].

photoprocesses, causes an electron to be injected into the conduction band (CB) of the TiO_2 at the anode and a hole to be injected into the NiO at the cathode. These charges diffuse through the respective semiconductors to the SnO_2:F charge collector. To complete the circuit, the oxidized and reduced forms of the redox mediator regenerate the dye at the cathode and anode, respectively.

Two dye-sensitized solar cells were joined to form a tandem cell with two separate. absorption ranges for the two different absorber materials [32]. This enhanced the solar conversion efficiency and in particular the photovoltage of the DSSC. The DSSC tandem was realized as n–n junction. Dye molecules with elongated shapes were attached covalently by bridge group terminated by an anchor group to a desired chromophore. The morphology of 2%-efficient hybrid solar cells consisting of poly(3-hexylthiophene) as the donor and ZnO as the acceptor in the nanometer range by electron tomography was spatially resolved [33]. The morphology was statistically analyzed for spherical contact distance and percolation pathways. Together with solving the three-dimensional exciton diffusion equation, a consistent and quantitative correlation among the solar cell performance, the photophysical data, and the three-dimensional morphology has been obtained for devices with different layer thicknesses that enables differentiating between generation and transport as limiting factors to performance.

6.3
Quantum Dot Solar Cells

A quantum dot (QD) is a portion of matter (e.g., semiconductor) whose excitons are confined to all three spatial dimensions (http://en.wikipedia.org/wiki/Quantum_-dot) [34–37]. As a result, they have properties that are between those of bulk semiconductors and those of discrete molecules. The OD's theoretical thermodynamic efficiency might be as high as 44%, better than the original 31% calculated ceiling. These structures receive widespread attention for the development of promising third-generation photovoltaic devices due to the possibility of tailoring their optoelectronic properties by controlling size and composition. Owing to the size quantization property, the optical and electronic properties of the semiconductor materials can be tuned to the response of quantum dot solar cells. Creation of hot electrons and/or multiple exciton generation can further boost the performance of quantum dot-based devices through high-energy excitation. Electron transfer between CdSe and other molecular systems has demonstrated the ability to shuttle electrons across the semiconductor interface.

According to reviews on recent developments in the utilization of semiconductor quantum dots for light energy conversion [35–38], three major ways to utilize semiconductor dots in solar cell include (a) metal–semiconductor or Schottky junction photovoltaic cell, (b) polymer–semiconductor hybrid solar cell, and (c) quantum-dot-sensitized solar cell (QDSC). Modulation of band energies through size control opens new ways to control photoresponse and photoconversion efficiency of the solar cell. The reviews discussed various strategies to maximize

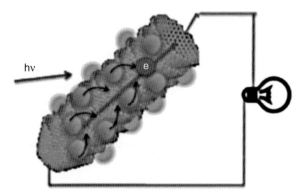

Figure 6.8 Illustration of stacked cup carbon nanotube-based solar cell. The "open-cup" structure provides a tubular carbon structure with many open functionalization sites [36]. Reproduced with permission from the American Chemical Society.

photoinduced charge separation and electron transfer processes for improving the overall efficiency of light energy conversion. The authors pointed out that capture and transport of charge carriers within the semiconductor nanocrystal network to achieve efficient charge separation at the electrode surface remains a major challenge.

By employing an electrophoretic deposition technique, Farrow and Kamat [36] cast stacked cup C nanotube (SCCNT)–CdSe composite films on optically transparent electrodes (OTEs). The quenching of CdSe emission in a QDSC, as well as transient absorption measurements, confirmed ultrafast electron transfer to SCCNTs. The rate constant for electron transfer increases from $9.51 \times 10^9\,s^{-1}$ to $7.04 \times 10^{10}\,s^{-1}$ as the size of CdSe nanoparticles decreases from 4.5 to 3 nm. Figure 6.8 illustrates a stacked cup carbon nanotube-based solar cell.

Different-sized CdSe quantum dots have been assembled on TiO_2 films composed of particle and nanotube morphologies using a bifunctional linker 3-mercaptopropionic acid [39]. Upon bandgap excitation, CdSe quantum dots inject electrons into TiO_2 nanoparticles and nanotubes. The maximum IPCE obtained with 3 nm-diameter CdSe nanoparticles was 35% for particulate TiO_2 and 45% for tubular TiO_2 morphology. The maximum IPCE observed at the excitonic band increased with decreasing particle size, whereas the shift in the conduction band to more negative potentials increased the driving force and favored fast electron injection. In the work [40], CdSe and CdTe nanocrystals were linked to nanostructured TiO_2 films using 3-mercaptopropionic acid as a linker molecule for establishing the interfacial charge transfer processes. These semiconductor nanocrystals exhibit markedly different external quantum efficiencies (70% for CdSe and 0.1% CdTe at 555 nm) when employed as sensitizers in quantum dot solar cells.

It was shown that incorporating colloidal CdSe quantum dots into CdSe nanowire-based photoelectrochem solar cells increases their incident photon-to-carrier conversion efficiencies from 13 to 25% at 500 nm [41]. The authors suggested that this effect stems from direct absorption and subsequent carrier generation by QDs and the overall IPCE increase occurs across the entire visible spectrum, even at wavelengths where the dots do not absorb light. This beneficial effect originates, in the authors' opinion, from an interplay between NWs and QDs where the latter fill voids between interconnected NWs, enabling better carrier transport to electrodes. Ligand exchange with 3-mercaptopropionic acid (MPA) was successfully used to tune the emission intensity of trioctylphosphineoxide/dodecylamine-capped CdSe quantum dots [42]. Addition of MPA to CdSe quantum dot suspension enhanced the deep trap emission with concurrent quenching of the band edge emission. Charge injection from excited CdSe quantum dots into nanostructured TiO_2 film was investigated [43]. The emission yield and the emission lifetime were shown to be modulated by increasing solution pH. Under such conditions, the conduction band of TiO_2 shifts 59 mV/pH unit to a more negative potential, thereby decreasing the driving force and thus decreasing the rate of nonradiative electron transfer from excited CdSe.

Quantum dot DSSCs have been manufactured using mesoporous TiO_2 electrodes coated with *in situ*-grown CdSe semiconductor nanocrystals by chemical bath deposition (CBD) [44]. Surface modification of the CdSe-sensitized electrodes by conformal ZnS coating and grafting of molecular dipoles (DT) was explored to both increase the injection from QDs into the TiO_2 matrix and reduce the recombination of the QD-sensitized electrodes. The most favorable sequence of the surface treatment (DT + ZnS) was shown to lead to a 600% increase in photovoltaic performance compared to the reference electrode (without modification): $V_{oc} = 0.488$ V, $J_{sc} = 9.74$ mA/cm^2, FF $= 0.34$, and efficiency $= 1.60\%$ under full 1 sun illumination.

A physical model based on recombination through a monoenergetic TiO_2 surface state with a ZnS coating and using polysulfide electrolyte as the redox couple that takes into account the effect of the surface coverage has been developed [45]. The three main methods of QD adsorption on TiO_2 and after ZnS coating are (i) *in situ* growth of QDs by CBD, (ii) deposition of presynthesized colloidal QDs by direct adsorption (DA), and (iii) deposition of presynthesized colloidal QDs by linker-assisted adsorption (LA). A systematic investigation by impedance spectroscopy of QDSCs prepared by these methods showed a decrease in the charge transfer resistance and increased electron lifetimes for CBD samples. Figure 6.9 shows a scheme of a TiO_2 nanostructured film deposited on transparent conducting oxide, sensitized with CdSe QDs and covered with a ZnS layer and energetic diagram of CdSe-sensitized TiO_2 in contact with electrolyte.

Colloidal quantum dot photovoltaic devices on transparent conductive oxides combining low-cost processability with quantum size-effect tunability to match absorption with the solar spectrum were fabricated [46]. The resultant depleted heterojunction solar cells provide a 5.1% air mass 1.5 power conversion efficiency. The devices employ IR bandgap size-effect-tuned PbS colloidal quantum dots, enabling broadband harvesting of the solar spectrum. The highest open-circuit voltages observed in solid-state colloidal quantum dot solar cells to date, as well as fill factors approaching 60%, through the combination of efficient hole blocking (heterojunction) were reported. TiO_2 nanotube arrays and particulate films were modified with CdS quantum dots with an aim to tune the response of the photoelectrochemical cell in the visible region [47]. These CdS nanocrystals, upon

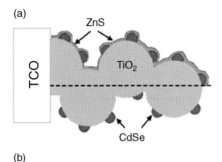

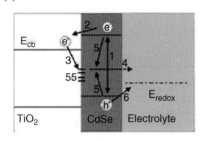

Figure 6.9 (a) Scheme of a TiO_2 nanostructured film deposited on transparent conducting oxide, sensitized with CdSe QDs (bottom) and covered with a ZnS layer (top). (b) Energetic diagram of CdSe-sensitized TiO_2 in contact with electrolyte, showing the main electronic processes at the interface in a QDSC: (1) photoexcitation, (2) electron injection, (3) trapping of a free electron at a surface state, (4) charge transfer of a trapped electron toward electrolyte acceptor, (5) recombination in the absorber semiconductor, and (6) hole extraction [45]. Reproduced with permission from the American Chemical Society.

excitation with visible light, inject electrons into the TiO_2 nanotubes and particles and thus enable their use as photosensitive electrodes. Maximum IPCE values of 55 and 26% were observed for CdS-sensitized TiO_2 nanotube and nanoparticulate architectures, respectively. The nearly doubling of IPCE observed with the TiO_2 nanotube architecture was attributed to the increased efficiency of charge separation and transport of electrons.

6.4
Polymers in Solar Cells

Polymer solar cells are a type of flexible solar cell [48–55] (http://en.wikipedia.org/wiki/Polymer_solar_cell). Polymer photovoltaic devices have drawn considerable attention as alternative inexpensive solar cells and conjugated polymers are promising candidates for use in low-cost electronics and photovoltaics that have reached power conversion efficiencies of 5% in recent studies. Compared to silicon-based devices, polymer solar cells are lightweight, are potentially disposable, are inexpensive to fabricate, are flexible, are customizable on the molecular level, and have lower potential for negative environmental impact. Deposition of organics by screen printing, doctor blading, inkjet printing, and spray deposition is possible because these materials can be made soluble. These techniques are required for the high-throughput roll-to-roll processing that will drive the cost of polymer-based PV down to a point where it can compete with current grid electricity. In addition, all these deposition techniques take place at low temperature, which allows devices to be fabricated on plastic substrates for flexible devices. The inherent economics of high-throughput manufacturing, light weight, and flexibility are qualities claimed to offer a reduction in the price of PV panels by reducing installation costs. Flexible PV also opens up niche markets like portable power generation and aesthetic PV in building design.

Present state-of-the-art techniques for making efficient polymer-based PV devices were reviewed [55]. The basic device operation, material requirements, and current technical challenges in making more efficient solar cells were discussed. The following polymer materials were used in polymer photovoltaics: 6,6-phenyl-C_{61}-butyric acid methyl ester; poly(2-methoxy-5-(3′,7′-dimethyloctyloxy)-1,4-phenylene-vinylene; regioregular poly(3-hexylthiophene); and poly[2,6-(4,4-bis-(2-ethylhexyl)-4H-cyclopenta[2,1-b;3,4-b]-dithiophene)-alt-4,7-(2,1,3-benzothiadiazole)]. Distinctive indium tin oxide nanorods were employed to serve as buried electrodes for polymer-based solar cells [56]. The embedded nanoelectrodes allowed three-dimensional conducting pathways for low-mobility holes, offering a highly scaffolded cell architecture in addition to bulk heterojunctions. As a result, the power conversion efficiency of a polymer cell with ITO nanoelectrodes was increased to 3.4 and 4.4% under 1-sun and 5-sun illumination conditions. Also, the corresponding device lifetime was prolonged twice as much to 110 min under 5-sun illumination.

In the review [57], local heterogeneity of electronic properties and performance in a wide range of systems, including model polymer–fullerene blends such as poly(3-hexylthiophene) (P3HT) and PCBM, newer polyfluorene copolymer–PCBM blends, and all polymer donor–acceptor blends, were discussed. Four types of cyclopenta-dithiophene (CDT)-based low-bandgap copolymers, poly[{4,4-bis(2-ethylhexyl)-4H-cyclopenta[2,1-b:3,4-b′]dithiophene-2,6-diyl}-alt-(2,2′-bithiazole-5,5′-diyl)] (PehCDT-BT), poly[(4,4-dioctyl-4H-cyclopenta[2,1-b:3,4-b′]dithiophene-2,6-diyl)-alt-(2,2′-bithiazole-5,5′-diyl)] (PocCDT-BT), poly[{4,4-bis(2-ethylhexyl)-4H-cyclopenta[2,1-b:3,4-b′]dithiophene-2,6-diyl}-alt-{2,5-di(thiophen-2-yl)thiazolo[5,4-d]thiazole-5,5′-diyl}] (PehCDT-TZ), and poly[(4,4-dioctyl-4H-cyclopenta[2,1-b:3,4-b′]dithiophene-2,6-diyl)-alt-{2,5-di(thiophen-2-yl)thiazolo[5,4-d]thiazole-5,5′-diyl}] (PocCDT-TZ), for use in photovoltaic applications were synthesized [58]. The intramolecular charge transfer interaction between the electron-sufficient CDT unit and the electron-deficient bithiazole (BT) or thiazolothiazole (TZ) units in the polymeric backbone induced a low bandgap (about 1.8 eV) and broad absorption that ranged from 300 to 700–800 nm. When the polymers were blended with PCBM, PehCDT-TZ exhibited the best performance with an open-circuit voltage of 0.69 V, short-circuit current of 7.14 mA/cm^2, and power conversion efficiency of 2.23% under air mass 1.5 global (1.5 G) illumination conditions (100 mW/cm^2).

The effect of metal oxide nanocomposites on the performance of bulk heterojunction polymer solar cells was investigated [59]. A photoactive layer composed of P3HT and PCBM was blended with a newly developed $ZrTiO_4/Bi_2O_3$ (BITZ) metal oxide nanocomposite. A short-circuit voltage of 9.90 mA/cm^2, an open-circuit voltage of 0.64 V, a fill factor of 0.60, and the overall power conversion efficiency of 3.72% were observed. The authors attributed the enhancement in performance to the improved absorption and fast transport of charge carriers by the suppressed recombination of photogenerated electrons and holes. In the work [60], solution-processed organic thin-film solar cells with triple-layered structures were fabricated by combining a hole transporting layer made of poly(3,4-ethylenedioxythiophene): poly(4-styrenesulfonate) (PEDOT:PSS), a light-harvesting layer assembled by layer-by-layer (LbL) deposition of poly(p-phenylenevinylene) (PPV) and PSS, and an electron-transporting layer of fullerene C_{60} dispersed in a polystyrene film. As a result of experiments, the exciton lifetime and diffusion constant were evaluated as 0.67 ± 0.02 ns and 8×10^{-4} cm^2 s^{-1}, respectively. Cationic and water-soluble CdSe nanorods (NRs) were synthesized and partnered with anionic polymers including poly(sodium 4-styrenesulfonate) (PSS) and two polythiophene-based photoactive polymers, sodium poly[2-(3-thienyl)-ethoxy-4-butylsulfonate (PTEBS) and poly[3-(potassium-6-hexanoate)thiophene-2,5-diyl] (P3KHT) [61]. A series of photovoltaic devices were fabricated on ITO electrodes using CdSe NRs in combination with PTEBS or P3KHT. In the work [62], polymer solar cells giving a high open-circuit voltage was fabricated using the thermocleavable polymer poly-(3-(2-methylhexylox-ycarbonyl)dithiophene) (P3MHOCT) as electron donors and unsubstituted poly-thiophene (PT) as reference.

Multilayer structures consisting of two blended layers of P3HT:PCBM and poly{N-[1-(2'-ethylhexyl)-3-ethylheptanyl]-dithieno[3,2-b:2',3'-d]pyrrole-2,6-diyl-alt-4,7-di(2-thienyl)-2,1,3-benzothiadiazole-5', 5'-diyl} (PDTPDTBT):PCBM were prepared by a thermal lamination technique (Figures 6.10 and 6.11) [63]. It was shown that the recombination rate at the interfaces was largely affected by the stacking order of the layers, resulting in the difference in the fill factors. The broadened light absorption range of the multilayer devices compared to that of single-layer P3HT:PCBM devices improved the short-circuit current from 8.83 to 9.41 mA/cm^2 because of the absorption of the PDTPDTBT:PCBM layer, resulting in a power conversion efficiency of 3.0% with an open-circuit voltage of 0.58 V and a fill factor of 54%.

Owing to the development of the bulk heterojunction structures, electron-donating and electron-accepting materials were blended together to form a single active layer. As a result, the power conversion efficiency of polymer solar cells has improved drastically and several groups have achieved a PCE of over 5% [64, 65]. Hierarchically macro-/mesoporous Ti-Si oxide photonic crystal (i-Ti-Si PC) with highly efficient photocatalytic activity has been synthesized by combining colloidal crystal template and amphiphilic triblock copolymer [66]. It was found that the thermal stability of mesoporous structures in the composite matrix were improved due to the introduction of silica acting as glue and linking anatase nanoparticles together, and the photocatalytic activity of the i-Ti-Si PCs was affected by the calcination conditions. When the energy of slow photon was optimized to the absorption region of TiO_2, a maximum enhanced factor of 15.6 was achieved in comparison to nanocrystal TiO_2 films (nc-TiO_2), which originated from the synergetic effect of SP enhancement and high surface area.

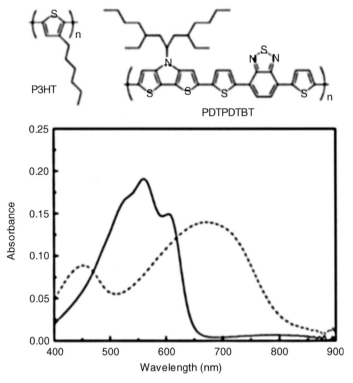

Figure 6.10 Molecular structures and absorption spectra of P3HT (solid line) and PDTPDTBT (dashed line) in thin films [63]. Reproduced with permission from the American Chemical Society.

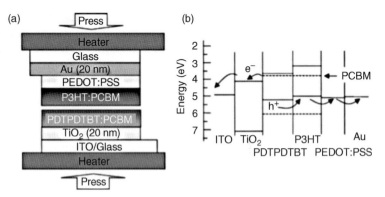

Figure 6.11 (a) Schematic representation of lamination of the device and (b) energy diagram of the materials [63]. Reproduced with permission from the American Chemical Society.

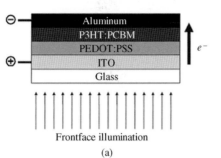

Figure 6.12 The Normal geometry employed for the model devices (a) comprise spin-coated PEDOT:PSS and P3HT-PCBM layers and evaporated aluminum. The inverted geometry employed for both model devices and fully R2R-coated modules (b). The model devices employed spin-coated ZnO, P3HT-PCBM, and PEDOT:PSS and evaporated silver, whereas the fully R2R-coated modules employed slot-die-coated ZnO, P3HT-PCBM, and PEDOT:PSS with the silver electrode being applied using screen printing (slot-die coating would also have been possible) [67]. Reproduced with permission from the Royal Chemical Society.

In the work [67], an inverted polymer solar cell geometry comprising five layers was optimized using laboratory-scale cells and the operational stability was studied under model atmospheres. The device geometry was substrate–ITO–ZnO–(active layer)–PEDOT:PSS–silver with P3HT-PCBM as the active layer (Figure 6.12). The inverted devices were compared with model devices with a normal geometry where the order of the layers was substrate–ITO–PEDOT:PSS–(active layer)–aluminum. Both device types were optimized to a PCE of 2.7% (1000 W/m^2, AM 1.5 G, $72 \pm 2\,^\circ$C) and were found to be stable in a nitrogen atmosphere during the test period of 200 h. The inverted model device was then used to develop a new process giving access to fully roll-to-roll (R2R)-processed polymer solar cells entirely by solution processing starting from a polyethyleneterephthalate (PET) substrate with a layer of indium tin oxide. The modules comprising eight serially connected cells gave power conversion efficiencies as high as 2.1% for the full module with 120 cm^2 active area (AM 1.5 G, 393 W/m^2) and up to 2.3% for modules with 4.8 cm^2 active area (AM 1.5 G, 1000 W/m^2).

To clarify the lifetime of the structure, two models that describe the lifetime dependence on bias voltage or carrier density were discussed [68]. The first was to treat the lifetime as a product of the chemical capacitance and recombination resistance. The second approach was based on a kinetic model that describes in detail the different processes governing the decay of the carrier population in a measurement of τ_n. The following main dynamic processes occurring in the blend layer with excess of holes and electrons were photogenerated into the P3HT HOMO and PCBM LUMO manifold, respectively. According to the models, charge carriers diffuse along the diode bulk and eventually recombine. The occupancy level of lowest unoccupied molecular orbital (LUMO) (HOMO) states was determined by competing photogeneration and recombination rates. As it was suggested, this in turn governs the achievable V_{oc} that depends on the splitting of the quasi-Fermi levels, $qV_{oc} = E_{Fn} - E_{Fp}$.

Chindaduang et al. [69] focused on the energy conversion improvement of dye-sensitized solar cells by using poly(ethylene oxide)-multiwalled carbon nanotube (PEO-MWCNT) electrolyte. Compared to the MWCNT-free solar cells, the addition of 0.05 wt% MWCNTs to the polymer electrolyte results in a dramatic increase in the short-circuit current (J_{sc}), consequently raising the device performance by approximately 9% under a direct light of the air mass 1.5 irradiation at 100 mW/cm². The role of the conductive carbon materials in the polymer electrolyte has been investigated by means of ionic conductometry, electrochemical impedance spectroscopy, and UV–visible spectroscopy. Synthesis and characterization of indeno[1,2-b]fluorene-based low bandgap copolymers for photovoltaic cells were reported [70]. Two types of indenofluorene-based low-bandgap conjugated polymers, poly(6,6',12,12'-tetraoctylindeno[1,2-b]fluorene-co-4,7-bis(2-thienyl)-2,1,3-benzothiadiazole) (PIF-DBT) and poly(6,6',12,12'-tetraoctylindeno[1,2-b]fluorene-co-5,7-dithien-2-yl-thieno[3,4-b]pyrazine) (PIF-DTP), were synthesized and characterized for use in plastic solar cells. The optical, electrochemical, and charge carrier mobility, and morphological and photovoltaic characteristics were studied. It was found that the polymers formed optical-quality films by spin casting. Photophysical studies revealed a low bandgap of 1.9 eV for PIF-DBT and 1.6 eV for PIF-DTP, which harvested the broad solar spectrum ranging from 300 to 650 nm (PIF-DBT) and from 300 to 800 nm (PIF-DTP) in the film. The field effect mobility measurements showed a hole mobility of apprximately 10^{-3} cm² V⁻¹ s⁻¹ for the copolymers. Among the polyindenofluorene copolymers, PIF-DBT50 (containing 50 mol% DBT) showed the best photovoltaic performance with an open-circuit voltage of 0.77 V, a short-circuit current of 5.50 mA/cm², and a PCE of 1.70% when the polymers were blended with PC71BM, under air mass 1.5 global (AM 1.5 G, 100 mW/cm²) illumination conditions.

In the study [71], dye-sensitized titanium dioxide surfaces were passivated by a trichloromethylsilane reaction in order to decrease the fast recombination rates when using the ferrocene redox couple. The formation and binding of poly(methylsiloxane) on the dye-sensitized TiO₂ surface was verified with IR spectroscopy and photoelectron spectroscopy. Photoelectrochemical characterization of the silanization method showed that the treatment decreased the recombination rate of photoinjected electrons with ferrocenium and improved the efficiency of the DSSC. Highly efficient plastic substrate DSSCs were developed using a press method [72]. The conversion efficiency was improved by optimizing the press conditions, the thickness of the TiO₂ layer, and the surface treatment of the plastic substrate. The best efficiency of a 0.25 cm² plastic substrate DSS Cs measured was $Z = 8.1\%$ under 100 mW/cm² (AM 1.5, 1 sun). A validated solar cell performance ($Z = 7.6\%$) for a 1.111 cm² plastic substrate DSSC was obtained from measurements at RCPV, AIST, Japan. Plastic substrate DSSCs fabricated by the proposed method are highly efficient and can be produced at low cost. FE-SEM micrographs of (a) a nonpressed TiO₂ photo-

Figure 6.13 FE-SEM micrographs of (a) a nonpressed TiO2 photoelectrode and (b) a TiO$_2$ photoelectrode pressed under 100 MPa [72]. Reproduced with permission from Elsevier.

electrode and (b) a TiO$_2$ photoelectrode pressed under 100 MPa are presented in Figure 6.13.

Tandem cells presented in the study [73], E67–E71, consisted of bulk heterojunction subcells comprising wide- and small-bandgap polymers as electron donors mixed with [6,6]-Ph C$_{61}$ butyric acid Me ester (PCBM) as an electron acceptor. Poly [2,7-(9,9-didecylfluorene)-*alt*-5,5-(4′,7′-di-2-thienyl-2′,1′,3′-benzothiadiazole)] (PFTBT) is used as the wide-bandgap (1.95 eV) material and poly[3,6-bis(4′-dodecyl-[2,2′]bithiophenyl-5-yl)-2,5-bis(2-ethylhexyl)-2,5-dihydropyrrolo(3,4-)pyrrole-1,4-dione] (pBBTDPP2) for the small-bandgap (1.40 eV) cell. When mixed with PCBM, these polymers provide efficient single-junction cells, exhibiting high V_{ocs} of about 1.0 and 0.6 V. A stack of solution-processed ZnO nanoparticles and pH-neutral poly (3,4-ethylenedioxythiophene):poly(styrene sulfonic acid) (PEDOT:PSS) was used as recombination layer. The combined tandem solar cell designs with the thicker nanolayers, 180 nm for the first cell and 125 nm for the second cell, achieved an open-circuit potential of 1.58 V, short-circuit current of 6.0 mA/cm^2, and an overall conversion efficiency of 4.9%, whereas the short-circuit current of the front cell was only 5.5 mA/cm^2, and the open-circuit potential for each of the single cells was less than 1 V.

Bulk heterojunctions formed by an interpenetrating blend of an optically active polymer and electron accepting molecules were reported to constitute a promising route toward cheap and versatile solar cells dyes [74]. In the paper [75], the origin of the improvement in short-circuit current of poly(3-hexylthiophene)/6,6-phenyl C$_{61}$-

butyric acid methyl ester (6,6)-phenyl-C_{61}-butyric acid methyl ester (PCBM)) solar cells with thermal annealing was examined. Transient absorption spectroscopy was employed to demonstrate that thermal annealing results in an approximate twofold increase in the yield of dissociated charges. It was found that the enhanced charge generation is correlated with a decrease in P3HT's ionization potential upon thermal annealing. These observations provided evidence that the LUMO level offset of annealed/P3HT/PCBM blends may be sufficient to drive efficient charge generation in polythiophene-based solar cells.

The effect of three-dimensional morphology on the efficiency of hybrid polymer solar cells was investigated [76]. The morphology of 2% efficient hybrid solar cells consisting of poly(3-hexylthiophene) as the donor and ZnO as the acceptor in the nanometer range by electron tomography was resolved. The morphology was statistically analyzed for spherical contact distance and percolation pathways. Together with solving the three-dimensional exciton diffusion equation, a consistent and quantitative correlation between solar cell performance, photophysical data, and the three-dimensional morphology has been obtained for devices with different layer thicknesses that enables differentiating between generation and transport as limiting factors to performance.

Hains et al. [77] described the design, synthesis, characterization, and organic photovoltaic (OPV) device implementation of a novel interfacial layer (IFL) for insertion between the anode and the active layer of P3HT:PCBM bulk heterojunction solar cells. Such devices exhibit an average PCE of 3.14%, a fill factor of 62.7%, an open-circuit voltage of 0.54 V, and a short-circuit current of 9.31 mA/cm^2, parameters rivaling those of optimized PEDOT:PSS-based device. The effect of processing on the performance in narrow-bandgap diketo-pyrrolo-pyrrole polymer solar cells was investigated [78]. It was shown that solar cells based on a new conjugated donor polymer with alternating quaterthiophene and diketo pyrrolo-pyrrole units in combination with C_{60} and C_{70} PCBM acceptors afforded high external quantum efficiencies over a broad spectral range into the near-IR. The cells provided power conversion efficiencies of up to 4.0% under simulated AM 1.5 G solar light conditions.

Photoconversion properties were demonstrated for a device based on a small dye molecule, absorbing light in the near-IR region, mixed with two organic charge transport materials and together forming a dye-sensitized organic bulk heterojunction [79]. The organic dye molecule, phthalocyanine (1,4,8,11,15,18,22,25-octabutoxy-29H,31H-phthalocyanine), mixed with a blend of poly(3-hexylthiophene) and 1-(3-methoxycarbonyl)-propyl-1-phenyl-(6,6)C_{61}, showed a photoconversion spectrum extended more than 150 nm toward longer wavelengths, compared to a device without such dye sensitization. In the dye-sensitized region of the photoconversion spectrum, the maximum internal quantum efficiency was estimated to be 40%. With higher dye concentrations, the internal quantum efficiency decreases. Transient laser spectroscopy measurements show that after excitation of the dye there is an electron transfer from the dye to PCBM and a subsequent hole transfer from the dye to P3HT, which results in a long-lived (P3HT$^+$/dye/PCBM$^-$) charge separated.

6.5
Fabrication of Solar Cell Components

A mesoporous nanocrystal graded TiO_2 film for DSSCs was prepared on either a transparent conductive oxide glass or a transparent conductive polymer (ITO/PEN, PEN = poly(ethylene naphthalene-2,6-dicarboxylate)) through a doctor blade method and then treated by chemical sintering at 150 °C [80]. High light-to-energy conversion efficiencies of 4.10 and 3.05% were obtained within the DSSCs fabricated by using the graded films on ITO/oxide glass and ITO/PEN, respectively. Mesoporous anatase TiO_2 nanocrystals with high crystallinity and large surface area were synthesized by a one-step method, combining hydrolysis and alcothermal processes, without the use of a templating agent [81]. The photovoltaic performance of photoanodes for dye-sensitized solar cell under 1 sun (AM 1.5 G) illumination gave a solar conversion efficiency as high as 6.06%, an enhancement of 0.20% with respect to that of a reference photoanode made of a TiO_2 (Degussa, P25) powder. Photoinduced hydrophilicity of polycrystalline $SrTiO_3$ thin films was reported [82]. It was shown that the films with smaller grains were highly hydrophilic, and there was a higher hydrophilicizing rate than those with larger grains.

A facile approach to preparation of a perpendicularly aligned and highly ordered TiO_2 nanorod/nanotube (NR/NT) adjacent film by directly anodizing a modified Ti foil was developed [83]. The Ti foil substrate was modified with a layer of crystal TiO_2 film via a hydrothermal process in 0.05 M $(NH_4)_2S_2O_8$. The resultant NR/NT film was used as a photoanode for photoactivity evaluation. In the work [84], a procedure was presented for fabricating multiwall anatase TiO_2 nanotubes by anodic alumina-templated atomic layer deposition synthesis in conjunction with alternating TiO_2/Al_2O_3 nanolaminate structures, followed by wet etching of the sacrificial alumina. The proposed structures multiplied the roughness factor of the anatase TiO_2 nanotubes by increasing the number of wall layers.

Hierarchical nanostructured spherical carbon with hollow macroporous core in combination with mesoporous shell has been explored to support Pt cathode catalyst with high metal loading in proton exchange membrane fuel cells [85]. The hollow core-mesoporous shell carbon (HCMSC)-supported Pt (60 wt%) cathode catalyst has demonstrated to enhance catalytic activity toward oxygen reduction and improve PEMFC polarization performance compared to carbon black Vulcan XC-72 (VC)-supported ones. In the work [86], the transparent titanium oxide (TiO_2) photoelectrodes for dye-sensitized solar cells were deposited on fluorine doped tin oxide substrate by a sol–gel dip coating technique with acetic acid as an acid catalyst. The surface morphology investigated by field emission scanning electron microscope and atomic force microscope measurements revealed that with the increase in annealing temperature, the surface of TiO_2 photoelectrode showed more porous structure with higher roughness. Measurement of optical properties and incident photon-to-current conversion efficiency indicated that under conditions of maximum dye incorporation in TiO_2 electrode, at 300 °C, solar cell efficiency and fill factor of this above-mentioned sample are 0.96 and 46.3%, respectively.

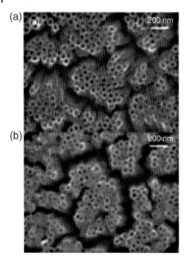

Figure 6.14 SEM images of the TiO$_2$ nanowire arrays (a) before In$_2$S$_3$ ALD coating and (b) after In$_2$S$_3$ ALD coating [87]. Reproduced with permission from the American Chemical Society.

In$_2$S$_3$ atomic layer deposition (In$_2$S$_3$ ALD) and its application as a sensitizer on TiO$_2$ nanotube arrays for solar energy conversion was reported [87]. The In$_2$S$_3$ ALD with indium acetylacetonate (In(acac)$_3$) and H$_2$S was studied with quartz crystal microbalance, X-ray reflectivity, and FTIR spectroscopy techniques. The photoelectrochemical properties of ALD-sensitized TiO$_2$ nanotube arrays with a Co^{2+}/Co^{3+} electrolyte were then characterized by measuring the photocurrent density versus voltage and the external quantum efficiency versus photon energy. The In$_2$S$_3$ ALD was displayed on initial Al$_2$O$_3$ ALD surfaces. The FTIR examinations revealed that the nucleation period on Al$_2$O$_3$ ALD surfaces may be related to the formation of Al(acac)$_3$ species that act to poison the initial Al$_2$O$_3$ ALD surface, while X-ray diffraction investigations indicated In$_2$S$_3$ ALD films that were consistent with X-ray photoelectron measurements. Scanning electron microscopy (Figure 6.14) and energy dispersive X-ray analysis imaging revealed In$_2$S$_3$ over the full length of the TiO$_2$ nanotube array after 175 cycles of In$_2$S$_3$ ALD at 150 °C under reactant exposure conditions that were self-limiting on flat substrates. In$_2$S$_3$ ALD was employed as a semiconductor sensitizer on TiO$_2$ nanotube arrays for solar conversion. The photoelectrochemical properties of these In$_2$S$_3$ ALD-sensitized TiO$_2$ nanotube arrays with a Co^{2+}/Co^{3+} electrolyte were characterized by measuring the photocurrent density versus voltage and the external quantum efficiency versus photon energy (Figure 6.15). A small quantum efficiency of <10% was observed that can be attributed to charge recombination losses and charge injection/collection processes.

It was shown [88] that micron-size electrodes made of multiwalled carbon nanotubes indicated Nernstian behavior and fast electron transfer kinetics for electrochemical reactions of Fe(CN)$_6^{3-/4-}$. This result was achieved suggesting the possibility of developing superior carbon electrodes based on carbon nanotubes for electrochemical applications. An inverted type organic bulk heterojunction solar cell

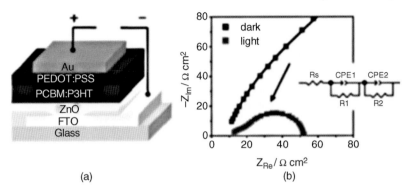

Figure 6.15 (a) An inverted type organic bulk heterojunction solar cell inserting zinc oxide as an electron collection electrode, fluorine-doped tin oxide (FTO)/ZnO/[6,6]-phenyl-C61-butyric acid methyl ester:regioregular poly(3-hexylthiophene) (PCBM:P3HT)/poly(3,4-ethylenedioxylenethiophene):poly(4-styrenesulfonic acid) (PEDOT:PSS)/Au; (b) results of the cell investigation by an alternating current impedance spectroscopy [89]. Reproduced with permission from the American Chemical Society.

inserting zinc oxide (ZnO) as an electron collection electrode, fluorine-doped tin oxide (FTO)/ZnO/[6,6]-phenyl-C_{61}-butyric acid methyl ester:regioregular poly(3-hexylthiophene) (PCBM:P_3HT)/poly(3,4-ethylenedioxylenethiophene):poly(4-styrenesulfonic acid) (PEDOT:PSS)/Au, was fabricated in air and characterized by an alternating current impedance spectroscopy (IS) (Figure 6.15) [89]. In the IS measurement, it was found that the depletion layer taking the photocurrent to the external circuit was formed in both the ZnO and the PCBM:P_3HT layers at the ZnO/PCBM:P_3HT interface.

Hagfeldt and coworkers introduced the press method involving the pressing of oxide layers for production of porous thin film following temperature preparation of nanostructured TiO_2 photoelectrodes [90]. Using conductive plastic substrate photoelectrodes prepared by this method, the authors achieved an efficiency of 5.5% under 10 mW/cm^2 (0.1 sun) irradiation. The liftoff process, in which the TiO_2 layer was first applied to a thin gold layer on a glass substrate, was developed [91]. Another concept for flexible DSSCs was the use of thin metal sheets as photoelectrode substrates, in which a conductive plastic film was used as the counterelectrode [92]. High-efficiency (7.2%) flexible dye-sensitized solar cells, with a Ti–metal substrate for nanocrystalline TiO_2 photoanode, and highly efficient (7.2%) flexible DSSCs with a Ti–metal substrate, less than 100 mW/cm^2 were reported. An efficiency of 8.6% using SiO_x and ITO-coated stainless substrates was reached in the study [93].

Constructing ordered sensitized heterojunctions using bottom-up electrochemical synthesis of p-type semiconductors in oriented n-TiO_2 nanotube arrays was reported [94]. The authors described a general synthesis strategy for spatially controlling the growth of p-type semiconductors in the nanopores of electrically conductive n-type materials. The strategy included the complete filling of the pore system of one semiconductor (host) material with nanoscale dimensions (<100 nm) with a different semiconductor (guest) material and the precluding electrochemical

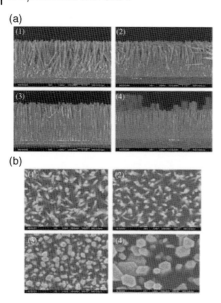

Figure 6.16 FESEM (a) cross-section and (b) top-view images of NW arrays with different thicknesses of seed layers: (1) 20, (2) 240, (3) 500, and (4) 1000 nm [95]. Reproduced with permission from Elsevier.

deposition process. As an illustration of this strategy, the authors reported on the facile electrochemical deposition of p-CuInSe$_2$ in nanoporous anatase n-TiO$_2$-oriented nanotube arrays and nanoparticle films. It was demonstrated that by controlling the ambipolar diffusion length, the p-type semiconductors can be deposited from the bottom-up, resulting in complete pore filling.

In the study [95], the vertical well-aligned ZnO NW arrays were prepared on ZnO/glass templates by an ACG method, where ZnO seed layers were grown on glass of different thicknesses of seed layers. The average diameter of nanowires was increased from 50 to 130 nm and the nanowire density was decreased from 110 to 60 mm^{-2}, while the seed layer thickness was varied from 20 to 1000 nm. Figure 6.16 shows top-view images of NW arrays with different thicknesses of seed layers. Hybrid polymer photovoltaic devices based on ZnO showed an n-type conductivity, and a semiconductor with a large exciton binding energy of 60 meV and E_g of 3.37 eV at room temperature was obtained.

CdSe nanocrystals have attracted much interest owing to their versatility and potential applications in nanotechnology [96, 97] and as a sensitizer for TiO$_2$ solar cells ([98, 99] references therein). For example, CdSe–TiO$_2$ nanocomposites were synthesized via aminolysis of Ti–oleate complexes in the presence of CdSe nanocrystals, and their application as sensitizers for TiO$_2$ solar cells was investigated [100]. The formation of CdSe–TiO$_2$ nanocomposites was confirmed using transmission electron microscopy and Raman spectroscopy. The emission spectrum of CdSe–TiO$_2$ nanocomposites revealed photoinduced charge separation at the CdSe–TiO$_2$ interface of the composite. An examination of the I–V properties of annealed

CdSe–TiO$_2$ nanocomposite-sensitized TiO2 films suggested that the TiO$_2$ phase enhances the electron conduction and protects CdSe against degradation during film annealing.

Influence of the nature of surface dipoles on observed photovoltage in dye-sensitized solar cells as probed by surface potential measurement was investigated [101]. A contact-free method of surface potential measurement using scanning Kelvin probe microscopy (SKPM) was conducted to probe the nature of nanocrystal TiO$_2$/dye interface. In combination with electric measurements, a correlation between the nature of sensitizing dye and the observed surface potential with open-circuit voltage (V_{oc}) was established. The effect of chenodeoxycholic acid (DCA) as coadsorbent with dyes on the enhancement of DSSC efficiency has been probed by SKPM, which revealed a positive shift of surface potential upon treatment of thin films of TiO$_2$ with DCA. Nakata *et al.* [102] described a new fabrication process for superhydrophilic–superhydrophobic patterns on a TiO$_2$ surface using a combination of an inkjet technique and the site-selective decomposition of a self-assembled monolayer by a photocatalytic reaction under UV irradiation. To induce high surface wettability, the authors carried out simple calcination of a Ti substrate. The substrate was oxidized to titanium oxide and had a vortex-like rough morphology, which was suitable for the formation of wettability patterns. This process was based on a TiO$_2$ surface and should offer a renewable, resource-saving, and environment-friendly methodology for the formation of wettability patterns.

6.6
Fullerene-Based Solar Cells

The development of polymer–fullerene plastic solar cells has made significant progress in recent years. Starting from the 1990s [103], these devices excelled by an efficient charge generation process as a consequence of a photoinduced charge transfer between the photoexcited conjugated polymer donor and the acceptor-type fullerene molecules [104, 105]. Due to the paramagnetic nature of the radical species, the photoinduced charge transfer can be analyzed by the help of light-induced electron spin resonance spectroscopy. Some novel applications of fullerene compounds were reviewed [106, 107]. Conjugated polymers such as polyfullerenes are intrinsic, quasi-one-dimensional semiconductors with bandgap energies covering the whole visible range of photon wavelengths [108].

Among the different organic photovoltaic devices, the conjugated polymer/fullerene bulk heterojunction approach is one of the most focused research interest today [109]. These devices are highly dependent on the solid-state nanoscale morphology of the two components (donor/acceptor) in the photoactive layer. The need for finely phase-separated polymer–fullerene blends is expressed by the limited exciton diffusion length present in organic semiconductors. Typical distances photoexcitations can travel within a pristine material are around 10–20 nm. Once the excitons reach the donor/acceptor interface, the photoinduced charge transfer results in the charge separation. After the charges have been separated, they require percolated pathways to the respective charge extracting electrodes in order to supply an external direct current.

Some novel applications of fullerene compounds, fullerene-based catalysts leading to photooxygenation and photodegradation under visible light irradiation and organic counter electrodes for a dye-sensitized solar cell, were reported [110]. The morphological features of the blends, probed by transmission electron microscopy, was correlated with the charge transfer exciton (CTE) emission intensity. The author observed that the CTE intensity (proportional to recombination) is independent of the morphology, whereas it is controlled by the intrachain conformation of the conjugated polymer. It was suggested that there are conjugated polymers with different degrees of rigidity and found that polythiophenes are particularly interesting for reducing CTE emission and thus loss of channels in solar cells. Several studies have highlighted the advantages of conjugated polymer/fullerene heterojunctions as effective material systems to harvest the visible part of the solar spectrum and convert energy into electrons and holes after exciton dissociation [111, 112]. In these systems, the working principle is based on the light absorption by conjugated polymer chromophores, charge separation at the interface with the fullerene, and charge carrier collection in the presence of the open-circuit voltage of the cell. For this type of solar cells, the limited exciton diffusion length (10 nm) enables a larger probability for charge transfer instead of exciton recombination in the polymer. Such morphology provides preferential percolation paths for electrons and holes in the fullerene and polymer domains, respectively.

Charge transfer and transport in polymer–fullerene solar cells donor–acceptor composite consisting of the polymer MDMO-PPV and the fullerene derivative PCBM were investigated [113]. In order to study the bulk charge transport properties, the authors carried out admittance spectroscopy on the polymer–fullerene solar cell device including a transparent semiconductor oxide front contact (ITO/PEDOT:PSS) and a metal back contact (Al). The temperature- and frequency-dependent device capacitance clearly uncovered two different defect states, the first, having an activation energy of 9 meV, indicated a shallow trap due to a bulk impurity, the second, having an activation energy of 177 meV. Photoconductive atomic force microscopy was used to map local photocurrents with 20 nm resolution in donor/acceptor blend solar cells of the conjugated polymer poly[2-methoxy-5-(3′,7′-dimethyloctyl-oxy)-1,4-phenylene vinylene] (MDMO-PPV) with the fullerene (6,6)-phenyl-C_{61}-butyric acid methyl ester (PCBM) spin-coated with various solvents [114]. Photocurrent maps under short-circuit conditions (zero applied bias) as well as under various applied voltages were presented. A significant variation in the short-circuit current between regions that appear identical in AFM topography was found. The authors suggested that these variations occur from one domain to another as well as on larger length scales incorporating multiple domains.

A new low-bandgap polymer fullerene solar cell (pBEHTB) with an absorption onset at 800 nm was reported [115]. When combined with a soluble fullerene derivative (PCBM), an efficient electron transfer occurred after excitation of the polymer. Bulk heterojunction solar cells have been prepared with a response up to 800 nm and an established PCE of 0.9%. The fabrication and measurement of solar cells with 6% power conversion efficiency using the alternating copolymer poly[N-9''-hepta-decanyl-2,7-carbazole-*alt*-5,5-(4',7'-di-2-thienyl-2',1',3'-benzothiadiazole)] (PCDTBT) in bulk heterojunction composites with the fullerene derivatives [6,6]-Ph C$_{70}$-butyric acid Me ester (PC70BM) were reported [116]. The PCDTBT/PC70BM solar cells exhibited $J_{SC} = 10.6$ mA/cm^2, $V_{OC} = 0.88$ V, FF $= 0.66$, and $\eta_e = 6.1\%$ under air mass 1.5 global (AM 1.5 G) irradiation of 100 mW/cm^2. The internal quantum efficiency was found to be close to 100.

Enhancing the thermal stability of polythiophene:fullerene solar cells by decreasing effective polymer regioregularity was reported [117]. Regioregularity has been shown to be an important aspect for the properties of poly(3-hexylthiophene) both in the pristine material and in bulk heterojunction photovoltaics with the fullerene derivative (6,6)-phenyl-C$_{61}$-butyric acid methyl ester (PCBM). A straightforward method was applied to adjust the effective regioregularity of poly(3-hexylthiophene) by introducing 3,4-dihexylthiophene into the polymer chain. The authors observed that polythiophene with 91% regioregularity maintains equivalent electronic properties and forms a more thermally stable interpenetrating network with PCBM than polythiophene with regioregularity >96%.

Koeppe *et al.* [118] presented a visible transparent semiconductor combinations, namely, zinc phthalocyanine or zinc naphthalocyanine together with soluble fullerenes in conjunction with a method for obtaining highly transparent thin metal films by tuning the interference patterns in the multilayer organic solar cell structures. In an optimal combination, solar cells with an efficiency of about 0.5% and a peak transparency of more than 60% in the visible part of the spectrum were fabricated. In the study [119], the origin of the improvement in short-circuit current of poly(3-hexylthiophene)/6,6-Ph C$_{61}$-butyric acid Me ester (P3HT/PCBM) solar cells with thermal annealing was examined. Transient absorption spectroscopy was employed to demonstrate that thermal annealing results in an approximately twofold increase in the yield of dissociated charges. It was shown that the enhanced charge generation is correlated with a decrease in ionization potential of P3HT upon thermal annealing. A phosphorus-containing fullerene derivative was successfully applied as an electron acceptor material in organic solar cells with power conversion efficiencies comparable to the values obtained using the state-of-the-art P3HT/PCBM composite system [120].

In the open-circuit voltage of polymer–fullerene solar cells, charge transfer absorption and emission were shown to be related to each other and V_{oc} in accordance with the assumptions of the detailed balance and quasi-equilibrium theory [121]. Devices based on poly[2-methoxy-5-(30,70-dimethyloctyloxy)-1,4-phenylene vinylene] (MDMO-PPV) and poly[2,7-(9-di-octylfluorene)-*alt*-5,5-(40;70-di-2-thienyl-20;10;30 benzothiadiazole)] (APFO3) were prepared using different polymer/fullerene stoichiometries. Optimal devices were obtained using a 1 : 4 polymer/fullerene weight ratio, resulting in a power conversion efficiency of 2% and 2.5%, respectively. The importance of the weak ground-state interaction between the polymer and the fullerene for the V_{oc} generation was underlined. In the study [122], in order to improve device performance and to better understand the relation between morphology and device operation, the optimization of solar cells based on poly(3-hexylthiophene) and [6,6]-Ph C_{61} butyric acid Me ester was performed by varying a specific cell parameter, namely, the concentration of the active layer components in the liquid phase before blend film deposition [122]. The study showed a significant increase in the short-circuit current, open-circuit voltage, and cell efficiency by properly choosing the formulation of the initial blend before film deposition. It was found that the optimized P3HT:PCBM device exhibited both slow recombination and high photocurrent generation associated with an overall power conversion efficiency of 4.25% under 100 mW/cm^2 illumination (AM 1.5 G).

Charge carrier concentration and temperature-dependent recombination in polymer–fullerene solar cells were reported [123]. Temperature-dependent transient photovoltage and photocurrent measurements on poly(3-hexylthiophene):[6,6]-phenyl-C_{61} butyric acid methyl ester bulk heterojunction solar cells were performed. A strong charge carrier concentration- and temperature-dependent Langevin recombination prefactor was found. The observed recombination mechanism was discussed in terms of bimolecular recombination and the experimental results were compared with charge carrier extraction by linearly increasing voltage measurements done on the same blend system. The authors explained the charge carrier dynamics, following an apparent order larger than two, by dynamic trapping of charges in the tail states of the Gaussian density of states.

Kim and Carroll in paper [124] reported that adding small amounts of Ag or Au nanoparticles (∼5–6 nm in diameter with a standard deviation of ∼20%) with a ligand shell of dodecylamine to poly(3-octylthiophene) (P3OT)/C_{60} blends is a suitable way to improve the device performance and attributed this effect to improved electron transport in the system. Recently, it was shown that the addition of small amounts of dodecylamine-capped Au nanoparticles to the active layer of organic bulk heterojunction solar cells consisting of P3OT and C_{60} has a positive impact on device performance due to improved electron transport [125]. Different strategies to incorporate colloidally prepared Au nanoparticles with a narrow size distribution into organic solar cells with the more common donor/acceptor system consisting of P3HT and PCBM were pursued. Au nanoparticles were prepared with either P3HT or dodecylamine as ligands. For all types of nanoparticles studied, the performance of the P3HT/PCBM solar cells was found to decrease with the Au particles as an additive to the active layer. A theory of the electronic structure and photophysics of 1 : 1 blends of derivatives of π-conjugated polyparaphenylenevinylene and fullerenes was developed [126]. Within Coulomb-correlated Hamiltonian applied to interacting chain of single-component π-conjugated polymers, an exciplex state of interfacial charge transfer complex was found that occurs below the polymer's optical exciton.

The role of morphology and polymer chain conformation in the charge transfer excitons in polymer/fullerene blends combining near-IR photoluminescence spectroscopy and transmission electron microscopy was revealed [127]. It was shown how carrier recombination through charge transfer excitons between conjugated polymers and fullerene molecules was mainly controlled by the intrachain conformation of the polymer and to a limited extent by the mesoscopic morphology of the blend. The photoluminescence intensity of the charge transfer exciton was correlated with the degree of intrachain order of the polymer. A schematic energy-level diagram for

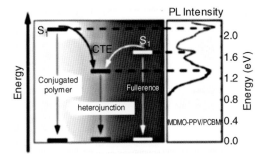

Figure 6.17 Schematic energy-level diagram for electronic excitations in conjugated polymer/fullerene blends and PL spectrum [127] Reproduced with permission from WILEY and son.

electronic excitations in conjugated polymer/fullerene blends and PL spectrum is presented in Figure 6.17.

Polymer:fullerene bulk heterojunction organic solar cells have achieved power conversion efficiencies up to 6.8% and are attracting a great deal of attention as a potential low-cost alternative to traditional inorganic photovoltaics [128–132]. It was shown that intercalation plays a key role in determining the optimal polymer: fullerene ratio since fullerenes must fill all available space between the polymer side chains prior to the formation of a pure electron transporting fullerene phase in blends with intercalation. Tuning the properties of BHJ by adjusting fullerene size to control intercalation was reported [132]. The authors compared intercalated and nonintercalated blends that use the same polymer, poly(2,5-bis(3-hexadecylthiophen-2-yl)thieno[3,2-b]thiophene (pBTTT7) with C_{16} side chains, and similar fullerene derivatives, phenyl-c71-butyric acid methyl ester (PC71BM) (NanoC) and its bisadduct, bisPC71BM (Figure 6.18), so that differences due to factors other than intercalation were minimized. It was shown that intercalation of fullerene derivatives between the side chains of conjugated polymers can be controlled by adjusting the fullerene size. It was found that the intercalated blends, which exhibit optimal solar cell performance at 1 : 4 polymer:fullerene by weight, had better photoluminescence quenching and lower absorption than the nonintercalated blends, which optimize at 1 : 1.

Honda et al. [133] reported multicolored dye-sensitized polymer/fullerene solar cells with two different near-IR dyes, silicon phthalocyanine bis(trihexylsilyl oxide) (SiPc) and silicon naphthalocyanine bis(trihexylsilyl oxide) (SiNc), enhanced power conversion efficiency up to 4.3%, compared to that of the individual ternary blend solar cells with a single dye under AM 1.5 G illumination. In the work [134], X-ray diffraction was used to demonstrate the formation of stable, well-ordered bimolecular crystals of fullerene intercalated between the side chains of the semiconducting polymer poly(2,5-bis(3-tetradecylthiophen-2-yl)thieno[3,2-b]thiophene. It is shown that fullerene intercalation occurred in blends with both amorphous and semicrystalline polymers when there is enough free volume between the side chains.

The influence of 1-(3-hexoxycarbonyl)propyl-1-phenyl-[6,6]-Lu3N@C81, Lu3N@-C80-PCBH on active layer morphology and the performance of OPV devices using

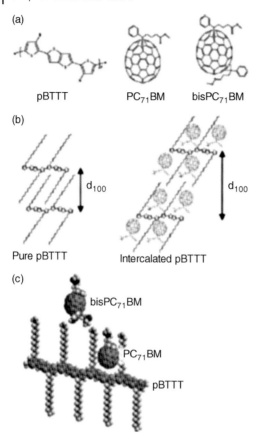

Figure 6.18 Molecular structures of pBTTT, PC71BM, and bisPC71BM (a), schematics showing possible structures for pure and intercalated pBTTT (b), and a space-filling ChemDraw model of pBTTT, PC71BM, and bisPC71BM to show their relative sizes (c). The second side group on bisPC71BM can attach to the fullerene at a number of different locations [132].

this material was reported [135]. Polymer/fullerene blend films with poly(3-hexylthiophene), donor material, and Lu3N@C80-PCBH, acceptor material, were studied using absorption spectroscopy, grazing incident X-ray diffraction, and photocurrent spectra of photovoltaic devices. The photovoltaic performance of these metalloendohedral fullerene blend films was found to be highly impacted by fullerene loading. Through properly matching the film processing and the donor/acceptor ratio, devices with power conversion efficiency greater than 4% were demonstrated. The open-circuit voltage of polymer:fullerene bulk heterojunction solar cells was investigated as a function of light intensity for different temperatures [136]. Devices consisted of a blend of a poly(p-phenylene vinylene) derivative as the hole conductor and 6,6-phenyl C_{61}-butyric acid methyl ester as the electron conductor. The observed photogenerated current and V_{oc} are at variance with classical p–n junction-based models. The influence of light intensity and recombination strength on V_{oc} is consistently explained by a model based on the notion that the quasi-Fermi levels are constant throughout the device, including both drift and diffusion of charge carriers.

Chapters 5 and 6 are key chapters of the book. They described principles and implementation of design and function of various dye-sensitized collar cells and its parts. Though the chapters demonstrate remarkable advantages in this area, a strategic goal to achieve the DSSC efficiency sufficient for commercial application does not yet fulfill. This would be a challenging problem for the next decades.

References

1 Weintraub, B., Wei, Y., and Wang, Z.L. (2009) *Angew. Chem. Int. Ed.*, **48** (47), 8981–8985, S8981/1–S8981/9.
2 Toivola, M., Ferenets, M., Lund, P., and Harlin, A. (2009) *Thin Solid Films*, **517** (8), 2799–2802.
3 Liu, J., Namboothiry, M.A.G., and Carroll, D.L. (2007) *Appl. Phys. Lett.*, **90** (6), 063501.
4 Bayindir, M., Sorin, F., Abouraddy, A.F., Viens, J., Hart, S.D., Joannopoulos, J.D., and Fink, Y. (2004) *Nature*, **431**, 826–829.
5 Bourzac, K. (2009) Wrapping solar cells around an optical fiber, Institute of Energy Efficiency, MIT, Boston.
6 Wang, Z.L. (2010) *Prepr. Symp. Am. Chem. Soc. Div. Fuel Chem.*, **55** (1), 323.
7 Uzaki, K., Pandey, S.S., Ogimi, Y., and Hayase, S. (2010) *Jpn. J. Appl. Phys.*, **49**, 082301 (5 pages).
8 He, J., Lindström, H., Hagfeldt, A., and Lindquist, S.E. (1999) *J. Phys. Chem.*, **103** (42), 8940–8943.
9 He, J., Lindström, H., Hagfeldt, A., and Lindquist, S.E. (2000) *Sol. Energy Mater. Sol. Cell*, **62**, 265–273.
10 Liska, P., Tampi, K.R., Graetzel, M., Bremaud, D., Rudmann, D., Upadhyaya, H.M., and Tiwari, A.N. (2006) *Appl. Phys. Lett.*, **88** (20), 203103.
11 Masatoshi, Y., Nobuko, O.-K., Mitsuhiko, K., Kazuhiro, S., and Hideki, S. (2010) *Sol. Energy Mater. Sol. Cells*, **94** (2), 297–302.
12 Inakazu, F., Noma, Y., Ogomi, Y., and Hayase, S. (2008) *Appl. Phys. Lett.*, **93** (9), 093304 (3).
13 Prasittichai, C. and Hupp, J.T. (2010) *J. Phys. Chem. Lett.*, **1**, 1611–1615.
14 Kubo, W., Sakamoto, A., Kitamura, T., Wada, Y., and Yanagida, S. (2004) *Photochem. Photobiol. A*, **164** (1), 33–39.
15 Uzaki, K., Pandey, S.S., Ogomi, Y., and Hayase, S., *Adv. Sci. Technol.*, **74**, 157–163.
16 Yamaguchi, T., Uchida, Y., Agatsuma, S., and Arakawa, H. (2009) *Sol. Energy Mater. Sol. Cells*, **93** (67), 733–736.
17 Uzaki, K., Pandey, S.S., Ogimi, Y., and Hayase, S. (2010) *Jpn. J. Appl. Phys.*, **49**, (8 Pt. 1), 082301/ 1–5.
18 Vera, F., Schrebler, R., Munoz, E., Suarez, C., Cury, P., Gomez, H., Cordova, R., Marotti, R.E., and Dalchiele, E.A. (2005) *Thin Solid Films*, **490** (2), 182–188.
19 Nattestad, A., Mozer, A.J., Fischer, M.K.R., Cheng, Y.B., Mishra, A., Baüerle, P., and Bach, U. (2010) *Nat. Mater.*, **9** (1), 31–35.
20 Murayama, M. and Mori, T. (2007) *J. Phys. D Appl. Phys.*, **40** (6), 1664–1668.
21 Dürr, M., Bamedi, A., Yasuda, A., and Nelles, G. (2004) *App. Phys. Lett.*, **84** (17), 3397–3399.
22 Bruder, I., Karlsson, M., Eickemeyer, F., Hwang, J., Erk, P., Hagfeldt, A., Weis, J., and Pschirer, N. (2009) *Sol. Energy Mater. Sol. Cells*, **93** (10), 1896–1899.
23 Usagawa, J., Pandey, S.S., Hayase, S., Kono, M., and Yamaguchi, Y. (2009) *Appl. Phys. Exp.*, **2** (6), 062203/ 1–062203/3.
24 Gilot, J., Wienk, M.M., and Janssen, A.J. (2010) *Adv. Mater.*, **22** (1), E67–E71.

25 Zhao, D., Tang, W., Ke, L., Tan, S.T., and Sun, X.W. (2010) *ACS Appl. Mater. Interfaces*, **2** (3), 829–837.

26 Nattestad, A., Ferguson, M., Kerr, R., Cheng, Y.-B., and Bach, U. (2008) *Nanotechnology*, **19** (29), 295304/1–295304/9.

27 Kubo, W., Sakamoto, A., Kitamura, T., Wada, Y., and Yanagida, S. (2004) *Appl. Phys. Lett.*, **4** (1), 9–10.

28 Hu, Y., Zheng, Z., Jia, H., Tang, Y., and Zhang, L. (2008) *J. Phys. Chem. C*, **112** (33), 13037–13042.

29 Kroon, J.M., Bakker, N.J., Smit, H.J.P., Liska, P., Thampi, K.R., Wang, P., Zakeeruddin, S.M., Grätzel, M., Hinsch, A., Hore, S., Würfel, U., Sastrawan, R., Durrant, J.R., Palomares, E., Pettersson, H., Gruszecki, T., Walter, J., Skupien, K., and Tulloch, G.E. (2007) *Prog. Photovolt. Res. Appl.*, **15** (1), 1–18.

30 Odobel, F., Le Pleux, L., Pellegrin, Y., and Blart, E. (2010) *Acc. Chem. Res.*, **43** (8), 1063–1071.

31 Gibson, E.A., Smeigh, A.L., Le Pleux, L., Fortage, J., Boschloo, G., Blart, E., Pellegrin, Y., Odobel, F., Hagfeldt, A., and Hammarstrom, L. (2009) *Angew. Chem. Int. Ed.*, **48** (24), 4402–4405.

32 Nattestad, A., Mozer, A.J., Fischer, M.K.R., Cheng, Y.B., Mishra, A., Bäuerle, P., and Bach, U. (2010) *Nat. Mater.*, **9** (1), 31–35.

33 Oosterhout, S.D., Wienk, M.M., Bavel, S.S., van Thiedmann, R., Koster, L.J.A., Gilot, J., Loos, J., Schmidt, V., and Janssen, R.A.J. (2009) *Nat. Mater.*, **8** (10), 818–824.

34 Murray, C.B., Kagan, C.R., and Bawendi, M.G. (2000) *Annu. Rev. Mater. Res.*, **30** (1), 545–610.

35 Kamat, P.V. (2008) *J. Phys. Chem. C*, **112** (46), 18737–18753.

36 Farrow, B. and Kamat, P.V. (2009) *J. Am. Chem. Soc*, **131** (31), 11124–11131.

37 Harris, C.T. and Kamat, P.V. (2009) *ACS Nano*, **3** (3), 682–690.

38 Barea, E.M., Shalom, M., Gimenez, S., Hod, I., Mora-Sero, I., Zaban, A., and Bisquert, J. (2010) *J. Am. Chem. Soc.*, **132** (19), 6834–6839.

39 Kongkanand, A., Tvrdy, K., Takechi, K., Kuno, M., and Kamat, P.V. (2008) *J. Am. Chem. Soc.*, **130** (12), 4007–4015.

40 Kamat, P.V. and Bang, J.H. (2009) Abstracts of Papers 238th ACS National Meeting, Washington, DC, United States, August 16–20, 2009 COLL-132.

41 Yu, Y., Kamat, P.V., and Kuno, M. (2010) *Adv. Funct. Mater.*, **20** (9), 1464–1472.

42 Baker, D.R. and Kamat, P.V. (2010) *Langmuir*, **26** (13), 11272–11276.

43 Chakrapani, V., Tvrdy, K., and Kamat, P.V. (2010) *J. Am. Chem. Soc.*, **132** (4), 1228–1229.

44 Barea, E.M., Shalom, M., Gimenez, S., Hod, I., Mora-Sero, I., Zaban, A., and Bisquert, J. (2010) *J. Am. Chem. Soc.*, **132** (19), 6834–6839.

45 Mora-Sero, I., Gimenez, S., Fabregat-Santiago, F., Gomez, R., Shen, Q., Toyoda, T., and Bisquert, J. (2009) *Acc. Chem. Res.*, **42** (11), 1848–1857.

46 Pattantyus-Abraham, A.G., Kramer, I.J., Barkhouse, A.R., Wang, X., Konstantatos, G., Debnath, R., Levina, L., Raabe, I., Nazeeruddin, M.K., Gratzel, M., and Sargent, E.H. (2010) *ACS Nano*, **4** (6), 3374–3380.

47 Baker, D.R. and Kamat, P.V. (2009) *Adv. Funct. Mater.*, **19** (5), 805–811.

48 Sariciftci, N.S. and Heeger, A.J. (1997) Photophysics, charge separation and device applications of conjugated polymer/fullerene composites, in *Handbook of Organic Conductive Molecules and Polymers*, vol. 1 (ed. H.S. Nalwa), John Wiley & Sons, Ltd., Chichester, Chapter 8, pp. 413–455.

49 Brabec, C.J., Sariciftci, N.S., and Hummelen, J.K. (2001) *Adv. Funct. Mater.*, **11** (1), 5–26.

50 Hoppe, H. and Sariciftci, N.S. (2008) Polymer solar cells, photoresponsive polymers, in *Advances in Polymer Science* (eds S.R. Marder and K.S. Lee), Springer, Berlin, pp. 1–86.

51 Luther, J., Nast, M., Fisch, M.N., Christoffers, D., Pfisterer, F., Meissner, D., and Nitsch, J. (2008) *Solar Technology*, Wiley-VCH Verlag GmbH, Weinheim.

52 Krebs, F.C. (2009) *Sol. Energy Mater. Sol. Cells*, **93** (4), 394–412.

53 Cai, W., Gong, X., and Cao, Y. (2010) *Sol. Energy Mater. Sol. Cells*, **94** (2), 114–127.

54 Nogueira, A.F., Durrant, J.R., and De Paoli, M.A. (2001) *Adv. Mater.*, **13** (11), 826–830.

55 Mayer, A., Scully, S., Hardin, B., Rowell, M., and McGehee, M. (2007) *Mater. Today*, **10** (1), 28–33.

56 Peichen, Y., Chang, C.-H., Su, M.-S., Hsu, M.H., and Wei, K.H. (2010) *Appl. Phys. Lett.*, **96** (15), 153307/1–153307/3.

57 Groves, C., Reid, O.G., and Ginger, D.S. (2010) *Acc. Chem. Res.*, **43** (5), 612–620.

58 Jung, I.H., Yu, J., Jeong, E., Kim, J., Kwon, S., Kong, H., Lee, K., Woo, H.Y., and Shim, H.-K. (2010) *Chem. Eur. J.*, **16** (12), 3743–3752.

59 Mohammed Hussain, A., Neppolian, B., Shim, H.-S., Kim, S.H., Kim, S.-K., Choi, H.-C., Kim, W.B., Lee, K., and Park, S.-J. (2010) *Jpn. J. Appl. Phys.*, **49** (4 Pt. 1), 042301/1–042301/4.

60 Masuda, K., Ikeda, Y., Ogawa, M., Benten, H., Ohkita, H., and Ito, S. (2010) *ACS Appl. Mater. Interfaces*, **2** (3), 829–837.

61 McClure, S.A., Worfolk, B.J., Rider, D.A., Tucker, R.T., Fordyce, J.A.M., Fleischauer, M.D., Harris, K.D., Brett, M.J., and Buriak, J.M. (2010) *ACS Appl. Mater. Interfaces*, **2** (1), 219–229.

62 Tromholt, T., Gevorgyan, S.A., Jørgensen, M., Krebs, F.C., and Sylvester-Hvid, K.O.S. (2010) *ACS Appl. Mater. Interfaces*, **1** (12), 2768–2777.

63 Nakamura, M., Yang, C., Zhou, E., Tajima, K., and Hashimoto, K. (2010) *ACS Appl. Mater. Interfaces*, **1** (12), 2703–2706.

64 Park, S.H., Roy, A., Beaupre, S., Cho, S., Coates, N., Moon, J.S., Moses, D., Leclerc, M., Lee, K., and Heeger, A.J. (2009) *Nat. Photonic.*, **3** (5), 297–302.

65 Gao, J., Hummelen, J.C., Wudl, F., and Heeger, A.J. (1995) *Science*, **270** (5243), 1789–1791.

66 Liu, J., Li, M., Wang, J., Song, Y., Jiang, L., Murakami, T., and Fujishima, A. (2009) *Environ. Sci. Technol.*, **43** (24), 9425–9431.

67 Krebs, F.C., Gevorgyan, S.A., and Alstrup, J. (2009) *J. Mater. Chem.*, **19** (30), 5442–5451.

68 Bisquert, J., Fabregat-Santiago, F., Mora-Sero, I., Garcia-Belmonte, G., and Gimenez, S. (2009) *J. Physic. Chem. C*, **113** (40), 7278–17290.

69 Chindaduang, A., Duangkaew, P., Pratontep, S., and Tumcharern, G. (2010) *Adv. Mater. Res.*, **93–94** (1), 31–34.

70 Kim, J., Kim, S.H., Jung, I.H., Jeong, E., Xia, Y., Cho, S., Hwang, I.-W., Lee, K., Suh, H., Shim, H.-K., and Woo, H.Y. (2010) *J. Mater. Chem.*, **20** (8), 1577–1586.

71 Feldt, S.M., Cappel, U.B., Johansson, E.M.J., Boschloo, G., and Hagfeldt, A. (2010) *J. Phys. Chem. C* **114** (23), 10551–10558.

72 Yamaguchi, T., Tobe, N., Matsumoto, D., Nagai, T., and Arakawa, H. (2010) *Sol. Energy Mater. Sol. Cells*, **94** (5), 812–816.

73 Gilot, J., Wienk, M.M., and Janssen, R.A.J. (2010) *Adv. Mater.*, **22** (8), E67–E71.

74 Park, S.H., Roy, A., Beaupre, S., Cho, S., Coates, N., Moon, J.S., Moses, D., Leclerc, M., Lee, K., and Heeger, A.J. (2009) *Nat. Photonics*, **3** (5), 297–302.

75 Clarke, T.M., Ballantyne, A.M., Nelson, J., Bradley, D.D.C., and Durrant, J.R. (2008) *Adv. Funct. Mater.*, **18** (24), 4029–4035.

76 Oosterhout, S.D., Wienk, M.M., van Bavel, S.S., Thiedmann, R., Koster, L.J.A., Gilot, J., Loos, J., Schmidt, V., and Janssen, R.A.J. (2009) *Nat. Mater.*, **8** (10), 818–824.

77 Hains, A.W., Ramanan, C., Irwin, M.D., Liu, J., Wasielewski, M.R., and Tobin, J.M. (2010) *ACS Appl. Mater. Interfaces*, **2** (1), 175–185.

78 Wienk, M.M., Turbiez, M., Gilot, J.J., and Rene, A.J. (2008) *Adv. Mater.*, **20** (13), 2556–2560.

79 Johansson, E.M.J., Yartsev, A., Rensmo, H., and Sundström, V. (2009) *J. Phys. Chem. C*, **113** (7), 3014–3020.

80 Li, X., Lin, H., Li, J., Wang, N., Lin, C., and Zhang, L. (2008) *J. Photoch. Photobio. A*, **195** (2–3), 247–253.

81 Parmar, K.P.S., Ramasamy, E., Lee, J.-W., and Lee, J.-S. (2009) *Scripta Mater.*, **62** (5), 223–226.

82. Katsumata, K., Shichi, T., and Fujishima, A. (2010) *J. Ceram. Soc. Jpn.*, **118** (1), 43–47.
83. Zhang, H., Liu, P., Liu, X., Zhang, S., Yao, X., An, T., Amal, R., and Zhao, H. (2010) *Langmuir*, **26** (13), 11226–11232.
84. Bae, C., Yoon, Y., Yoo, H., Han, D., Cho, J., Lee, B.H., Sung, M.M., Lee, M.G., Kim, J., and Shin, H. (2009) *Chem. Mater.*, **21** (13), 2574–2576.
85. Fang, B., Kim, J.H., Kim, M., Kim, M., and Yu, J.S. (2009) *Phys. Chem. Chem. Phys.*, **11** (9), 1380–13877.
86. Hossain, M.F., Biswas, S., Shahjahan, M., Majumder, A., and Takahashi, T. (2009) *J. Vac. Sci. Technol. A*, **27** (4), 1042–1046.
87. Sarkar, S.K., Kim, J.Y., Goldstein, D.N., Neale, N.R., Zhu, K., Elliott, C.M., Frank, A.J., and George, S.M. (2010) *J. Phys. Chem. C*, **114** (17), 8032–8039.
88. Lee, W.J., Ramasamy, E., Lee, D.Y., and Song, J.S. (2009) *ACS Appl. Mater. Interfaces*, **1** (6), 1145–1149.
89. Kuwabara, T., Kawahara, Y., Yamaguchi, T., and Takahashi, K. (2010) *ACS Appl. Mater. Interfaces*, **1** (10), 2107–2110.
90. Boschloo, G., Lindström, H., Magnusson, E., Holmberg, A., and Hagfeldt, A. (2002) *J. Photochem. Photobiol. A.*, **148** (1), 11–15.
91. Durr, M., Schmid, A., Obermaier, M., Rosselli, S., Yasuda, A., and Nelles, G. (2005) *Nat. Mater.*, **4** (8), 607–611.
92. Ito, S., Ha, N.-L.C., Rothenberger, G., Liska, P., Comte, P., Zakeeruddin, S.M., Pechy, P., Nazeeruddin, M.K., and Grätzel, M. (2006) *Chem. Commun.* (14), 4004–4006.
93. Park, J.H., Jun, Y., Yun, H.-G., Lee, S.-Y., and Kang, M.G. (2008) *J. Electrochem. Soc.*, **155** (7), F145–F149.
94. Wang, Q., Zhu, K., Neale, N.R., and Frank, A.J. (2009) *Nano Lett.*, **9** (2), 806–813.
95. Ji, L.W., Peng, S.M., Wu, J.S., Shih, W.S., Wu, C.Z., and Tang, I.-T. (2009) *J. Phys. Chem. Solids*, **70** (10), 1359–1362.
96. Huynh, W.U., Dittmer, J.J., and Alivisatos, A.P. (2002) *Science*, **295** (5564), 242.
97. Wang, P., Abrusci, A., Wong, H.M.P., Svensson, M., Andersson, M.R., and Greenham, N.C. (2006) *Nano Lett.*, **6** (8), 1789–1793.
98. Robel, I., Subramanian, V., Kuno, M., and Kamat, P.V. (2006) *J. Am. Chem. Soc.*, **128** (7), 2385–2393.
99. Lee, H.J., Yum, J.-H., Leventis, H.C., Zakeeruddin, S.M., Haque, S.A., Chen, P., Seok, S.I., Gratzel, M., and Nazeeruddin, M.d.K. (2008) *J. Phys. Chem. C*, **112** (30), 11600–11608.
100. Kim, J.Y., Choi, S.B., Noh, J.H., Sung Hun Yoon, S.H., Lee, S., Noh, T.H., Frank, A.J., and Hong, K.S. (2009) *Langmuir*, **25** (9), 5348–5351.
101. Pandey, S.S., Sakaguchi, S., Yamaguchi, Y., and Hayase, S. (2010) *Org. Electron.*, **11** (3), 419–426.
102. Nakata, K., Nishimoto, S., Yuda, Y., Ochiai, T., Murakami, T., and Fujishima, A. (2010) *Langmuir*, **26** (14), 11628–11630.
103. Gevaert, M. and Kamat, V. (1992) *J. Phys. Chem.*, **96** (38), 9883–9891.
104. Honda, S., Nogami, T., Ohkita, H., Benten, H., and Ito, S. (2009) *ACS Appl. Mater. Interfaces*, **1** (4), 804–810.
105. Cravino, A. and Sariciftci, N.S. (2009) Polyfullerenes for organic photovoltaics, in *Fullerene Polymers* (eds N. Martin and F. Giacalone), Wiley-VCH Verlag GmbH, Weinheim, pp. 171–187.
106. Hallermann, H., Kriegel, I., Da Como, E., Berger, J.M., von Hauff, E., and Feldmann, J. (2009) *Adv. Funct. Mater.*, **19** (22), 3662–3668.
107. Green, M.A., Emery, K., Hishikawa, Y., and Warta, W. (2009) *Prog. Photovoltaics*, **17** (1), 85–94.
108. Sariciftci, N.S. (1995) *Prog. Quant. Electron.*, **19** (1), 131–137.
109. Hoppe, H. and Sariciftci, N.S. (2006) *J. Mater. Chem.*, **16** (1), 45–61.
110. Tetsuo, H. (2007) Some novel application of fullerene compounds, in *Fullerene Research Advances* (ed. C.N. Kramer), Nova Science Publishers, pp. 33–53.
111. Yu, G., Gao, J., Hummelen, J.C., Wudl, F., and Heeger, A.J. (1995) *Science*, **270** (5243), 1789–1781.
112. Kim, Y., Cook, S., Tuladhar, S.M., Choulis, S.A., Nelson, J., Durrant, J.R.,

Bradley, D.D.C., Giles, M., McCulloch, I., Ha, C.S., and Ree, M. (2006) *Nat. Mater.*, **5** (3), 197–203.

113 Parisi, J., Dyakonov, V., Pientka, M., Riedel, I., Deibel, C., Brabeca, C.J., Sariciftcic, N.S., and Hummelenc, J.C. (2002) *Z. Naturforsch. A*, **57** (12), 995–1000.

114 Coffey, D.C., Reid, O.G., Rodovsky, D.B., Bartholomew, G.P., and Ginger, D.S. (2007) *Nano Lett.*, **7** (3), 738–744.

115 Wienk, M.M., Struijk, M.P., Ridolfi, G., Broeren, A.A.C., and Janssen, R.A.J. (2005) *Organic Photovoltaics VI (Proceedings of SPIE)*, SPIE, pp. 593810/1–593810/3.

116 Park, S.H., Roy, A., Beaupre, S., Cho, S., Coates, N., Moon, J.S., Moses, D., Leclerc, M., Lee, K., and Heeger, A.J. (2009) *Nat. Photonics*, **3** (5), 297–303.

117 Sivula, K., Luscombe, C.K., Thompson, B.C., and Fréchet, J.M.J. (2006) *J. Am. Chem. Soc.*, **128** (43), 13988–13989.

118 Koeppe, R., Hoeglinger, D., Troshin, P.A., Lyubovskaya, R.N., Razumov, V.F., and Sariciftci, N.S. (2009) *ChemSusChem*, **2** (4), 309–313.

119 Clarke, T.M., Ballantyne, A.M., Nelson, J., Bradley, D.D.C., and Durrant, J.R. (2008) *Adv. Funct. Mater.*, **18** (24), 4029–4035.

120 Troshin, P.A., Romanov, I.P., Susarova, D.K., Yusupov, G.G., Gubaidullin, A.T., Saifin, A.F., Zverev, V.V., Lyubovskaya, R.N., Razumov, V.F., and Sinyashin, O.G. (2010) *Mendeleev Commun.*, **20** (3), 137–139.

121 Vandewal, K., Tvingstedt, K., Gadisa, A., Olle Inganä, O., and Manca, J.V. (2009) *Nat. Mater.*, **8** (11), 904–909.

122 Radbeh, R., Parbaile, E., Boucle, J., Di Bin, C., Moliton, A., Coudert, V., Rossignol, F., and Ratier, B. (2010) *Nanotechnology*, **21** (3), 035201/1–035201/8.

123 Foertig, A., Baumann, A., Rauh, D., Dyakonov, V., and Deibel, C. (2009) *Appl. Phys. Lett.*, **95** (1–3), 052104.

124 Kim, K. and Carroll, D.L. (2005) *Appl. Phys. Lett.*, **87** (1–3), 203113.

125 Topp, K., Borchert, H., Johnen, F., Tunc, A.V., Knipper, M., von Hauff, E., Parisi, J., and Al-Shamery, K. (2010) *J. Phys. Chem. A*, **114** (11), 3981–3989.

126 Aryanpour, K., Psiachos, D., and Mazumdar, S. (2010) *Phys. Rev. B Condens. Matter Mater. Phys.*, **81** (8), 085407/1–085407/9.

127 Hallermann, M., Kriegel, I., Da Como, E., Berger, J.M., von Hauff, E., and Feldmann, J. (2009) *Adv. Funct. Mater.*, **19** (22), 3662–3668.

128 Park, S.H., Roy, A., Beaupre, S., Cho, S., Coates, N., Moon, J.S., Moses, D., Leclerc, M., Lee, K., and Heeger, A.J. (2009) *Nat. Photonics*, **3** (5), 297–302.

129 Thompson, B.C. and Frechet, J.M. (2008) *J. Angew. Chem. Int. Ed.*, **47** (1), 58–77.

130 Mayer, A.C., Toney, M.F., Scully, S.R., Rivnay, J., Brabec, C.J., Scharber, M., Koppe, M., Heeney, M., McCulloch, I., and McGehee, M.D. (2009) *Adv. Funct. Mater.*, **19** (8), 1173–1179.

131 Raffaelle, R.P., Anctil, A., DiLeo, R., Merrill, A., Petritchenko, O., and Landi, B.J. (2008) Conference Record of the IEEE Photovoltaic Specialists Conference 33rd, pp. 1106–1111.

132 Cates, N.C., Gysel, R., Beiley, Z., Miller, C.I., Toney, M.F., Heeney, M., McCulloch, I., and McGehee, M.D. (2009) *Nano Lett.*, **9** (12), 4553–4557.

133 Honda, S., Ohkita, H., Benten, H., and Ito, S. (2010) *Chem. Commun.*, **46** (35), 6596–6598.

134 Mayer, A.C., Toney, M.F., Scully, S.R., Rivnay, J., Brabec, C.J., Scharber, M., Koppe, M., Heeney, M., McCulloch, I., and McGehee, M.D. (2009) *Adv. Funct. Mater.*, **19** (8), 1173–1179.

135 Ross, R.B., Cardona, C.M., Swain, F.B., Guldi, D.M., Sankaranarayanan, S.G., Van Keuren, E., Holloway, B.C., and Drees, M. (2009) *Adv. Funct. Mater.*, **19** (14), 2332–2338.

136 Koster, L.J.A., Mihailetchi, V.D., Ramaker, R., and Blom, P.W.M. (2005) *Appl. Phys. Lett.*, **86** (1–3), 1235099.

7
Photocatalytic Reduction and Oxidation of Water

7.1
Introduction

Photochemical splitting of water into H_2 and O_2 using solar energy is a process of great economic and environmental interest. Hydrogen is the most promising fuel of the future owing to its carbon-free, high-energy content and potential to be efficiently converted into either electrical or thermal energy to replace finite fossil fuels with abundant, renewable, environment-friendly energy sources. Photocatalytic water splitting and H_2 evolution using abundant compounds as electron donors are expected to contribute to the development of a clean and simple system for solar hydrogen production and a solution to global energy and environmental issues in the future.

The construction of an efficient device for the photosplitting of water to hydrogen is one of the most important subjects from the viewpoint of solar light energy utilization and storage.

Fujishima and Honda in 1972 demonstrated a photoelectrochemical cell consisting of a TiO_2 photoanode and a Pt cathode to decompose water into hydrogen and oxygen under ultraviolet irradiation with an external bias [1]. Since this discovery, over 130 inorganic materials have been developed as catalysts for this reaction ([2–12] and references therein). The development of a photoelectrochemical device in which sunlight efficiently splits water into O_2, at the anode, and hydrogen, H_2, at the cathode to drive this energetically highly unfavorable reaction (ΔH^0) 572 kJ/mol) and breaking the four strong O–H bonds (enthalpy ΔH^0 494 kJ/mol each) remains one of the most challenging problems.

Biochemistry and early theoretical analysis [13–16] showed the direction to solve the problem of photocatalysis of the watter splitting. It was unequivocally proved that effective water splitting into O_2 and H_2 in the photosynthetic systems occurs in the manganase clusters of high redox potential (Section 7.2.5). The key of the process is the four-electronic oxidation of two water molecules in the cluster preceded by four one-electron photochemical oxidative stages (Section 7.2.5).

Solar Energy Conversion. Chemical Aspects, First Edition. Gertz Likhtenshtein.
© 2012 Wiley-VCH Verlag GmbH & Co. KGaA. Published 2012 by Wiley-VCH Verlag GmbH & Co. KGaA.

7.2
Photocatalytic Dihydrogen Production

In photogenerated catalysis, the photocatalytic activity depends on the ability of the catalyst to create electron–hole pairs. In artificial systems of water splitting, in a semiconductor the reaction occurs in three steps: (1) the photocatalyst absorbs photon energy and generates photoexcited electron–hole pairs in the bulk, (2) the photoexcited carriers separate and migrate to the surface, and (3) adsorbed species are reduced and oxidized by the photogenerated electrons and holes to produce H_2 and O_2, respectively.

7.2.1
Photocatalytic H_2 Evolution over TiO_2

Photocatalytic H_2 evolution over aqueous TiO_2 suspension, with methanol as hole scavenger, was systematically studied as a function of anatase and rutile phase compositions [17]. The highly crystalline, flame-synthesized TiO_2 nanoparticles (22–36 m^2/g) were designed to contain 4–95 mol% anatase, with the remaining being rutile. The amount of photocurrent generated under applied potential bias increased with increasing anatase content. The synergistic effects in terms of H_2 evolution were observed for a wide range of anatase contents, from 13 to 79 mol%. At the optimal 39 mol% anatase, the photocatalytic activity was enhanced by more than a factor of 2 with respect to the anatase- and rutile-rich phases. The authors thought that the synergistic effect in these mixed anatase–rutile phases originates from the efficient charge separation across phase junctions. A photoelectrochemical system with applied potential bias and a suspension-type photocatalytic system without applied potential bias are presented in Figure 7.1.

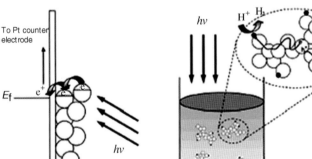

Figure 7.1 Differerence in the charge transport in (a) a photoelectrochemical system, in which photocurrent was collected under applied potential bias, and that of (b) a suspension-type photocatalytic system, in which photogenerated charges percolate to different sites within an aggregate [17]. Reproduced with permission from the American Chemical Society.

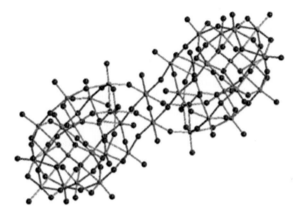

Figure 7.2 Molecular structure of compound I with 50% probability ellipsoids [19].

S-doped TiO_2 was studied as a potential catalyst for photoelectrochemical hydrogen generation [8]. Three preparation techniques were used: (1) ball milling S powder with Degussa P25 powder (P25), (2) ball milling thiourea with P25, and (3) a sol–gel technique involving Ti(IV) butoxide and thiourea. The resulting powders were heat treated and thin-film electrodes were prepared. Visible UV spectroscopy performed on bulk powders confirmed the extension of absorption into the visible region, but on thin-film electrodes (0.5 μm) the absorption coefficients were very small in the visible region. Enhancement of photocatalytic hydrogen production rate under UV/solar radiation using photosensitized TiO_2/RuO_2–$MV^{2+\bullet}$ was demonstrated [18]. The under-test photosensitizers (copper phthalocyanine, ruthenium bipyridyl, and Eosin Y) were added to slurry of TiO_2/RuO_2 semiconductor containing Me viologen (MV^{2+}) as an electron relay.

To improve the visible light absorption in the hydrogen evolution from water, Kato et al. [19] focused on the synthesis of Dawson-type dirhenium(V)-oxido-bridged POM, $[O\{ReV(OH)(\alpha2\text{-}P2W17O61)\}2]14-$ (**1**). The black purple compound I (Figure 7.2) was grafted onto a TiO_2 surface by electrostatic binding to give TiO_2 with a cationic quaternary ammonium moiety (I-grafted TiO_2) to form $\{TiO_2\}5500[\equiv Si(CH_2)_3N(CH_3)_3Cl]^7[\equiv Si(CH_2)_3N^+(CH_3)_3]7(K_7[O\{Re(OH)(\alpha_2\text{-}P_2W_{17}O_{61})\}_2]_7^-)$. With I-grafted TiO_2, hydrogen evolution from water vapor under irradiation with visible light (>400 nm and >420 nm) was achieved (Scheme 7.1).

7.2.2
Miscellaneous Semiconductor Photocatalysts for H_2 Evolution

Under visible light irradiation CdSe nanoribbons were found to photocatalyze H_2 evolution from aqueous sodium sulfite/sulfide solution with a quantum efficiency of 9.2% at 440 nm [20]. Photoelectrochemical measurements showed that the activity of nano-CdSe is caused by a raised flatband potential (−0.55 V, NHE), which follows from the increased bandgap (2.7 eV) of this quantum confined material. When the nanoribbons were chemically linked to MoS_2 nanoplates that were obtained by exfoliation and ultrasonication of bulk MoS_2, the activity increases almost four times, depending on the mass percentage of MoS_2. Photochemical properties of cadmium sulfide semiconductor nanoparticles (NPs) fixed on the outer or inner surfaces of the lipid vesicle membranes were studied [21]. The methods were found for efficient transmembrane electron transfer catalyzed by semiconductor nanoparticles in lipid vesicles. Experiments were carried out to create the electron transport chain based on CdS/Cu_xS nanoheterostructures. New electron carriers, menadione and $SiW_{12}O_4$ heteropoly anions, were used for electron transfer through lipid membranes. It is shown that heteropoly anions immobilized on the vesicle surface provide efficient conjugation of electron transport chains. This enabled formation of an active photocatalytic system for reduction of water into hydrogen.

It was shown that a bulk heterojunction photocatalyst of interfacing $CaFe_2O_4$ and $MgFe_2O_4$ nanoparticles is highly active for hydrogen production from water under visible light because the exciton easily reaches the interface and dissociates to minimize recombination [22]. Uniform hexagonal WO_3 nanowires with a diameter of 5–10 nm and lengths of up to several micrometers were synthesized and its electrocatalytic activity for hydrogen evolution reaction were investigated [23]. Photocatalytic hydrogen evolution by tungsten oxide dispersed in mesoporous MCM-48 (W-MCM-48) was studied [24]. The W-MCM-48 showed high activity for the hydrogen evolution. The local environment of incorporated W-oxide species and the mechanism of photocatalytic hydrogen evolution were explored. Microporous zeolitic inorganic membranes that are highly stable, chemically manipulated, and capable of maintaining charge balance were employed for the photochemical generation of long-lived separated charge pairs [25]. To generate H_2 from H_2O, procedures have been developed for synthesizing semiconductor nanoparticles of cadmium sulfide (CdS) and titanium dioxide (TiO_2) within the pore structure of zeolite Y membranes. Using nanoparticles bound inside zeolite Y crystallites, the dynamics of the electron/hole injection from CdS to TiO_2 were studied.

The efficiency of the charge injection process and the ability to move charge into the zeolite membrane and $(Ga_{1-x}Zn_x)(N_{1-x}O_x)$ with RuO_2, transition metal mixed oxides containing Cr, and noble metal/Cr_2O_3 (core/shell) nanoparticles as cocatalysts were defined by transient spectroscopy [26]. The open porous metal–organic frameworks (MOFs) that function as an activity site for the reduction of water into hydrogen molecules in the presence of $Ru(bpy)_3^{2+}$, MV^{2+}, and EDTA–2Na under visible light irradiation were described [27]. This activity was found to be highly efficient: the turnover number based on MOFs and the apparent quantum yield were 8.16 and 4.82%, respectively. The material enabled the adsorption of various gases into its pores; its hydrogen uptake capacity is 1.2 wt% at 77.4 K. According to Lei et al. [28], the dinuclear Ru^{II}–Pd^{II} complex (Figure 7.3) showed efficient H_2 production in the presence of triethylamine as a sacrificial electron and proton donor under visible light irradiation. XPS and TEM analyses revealed that photoreduction of Pd^{II} to Pd^0 causes dissociation of Pd from the complex to form colloids that are suggested to be the actual catalyst for H_2 production.

Zhang et al. [29] discussed charge transfer at semiconductor particles (quantum size effects, photonic efficiency, and quantum yield in photocatalytic systems), single

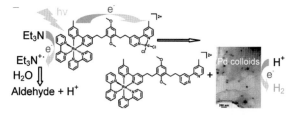

Figure 7.3 Dihydrogen evolution in the Ru^{II}–Pd^{II} complex Pd colloid system [27]. Reproduced with permission from the Royal Chemical Society.

semiconductor photocatalysts (TiO$_2$, ZnO, SnO$_2$, WO$_3$, Fe$_2$O$_3$, and CdS), and coupled semiconductor photocatalysts (TiO$_3$/WO$_3$, TiO$_2$/SnO$_2$, TiO$_2$/CdS, Ag/TiO$_2$, Au/TiO$_2$, and Pt/TiO$_2$). Hexaniobate nanoscrolls (NS-H$_4$Nb$_6$O^{17}) and acid-restacked calcium niobate nanosheets (R-HCa$_2$Nb$_3$O$_{10}$) (Figure 7.4) were compared as oxide semiconductors in photocatalytic assemblies for H$_2$ production using ethylenediaminetetraacetic acid (EDTA) as a sacrificial electron donor and platinum (Pt) nanoparticles as catalysts [30]. Ru(bpy)$_3^{2+}$ (Ru^{2+}) and Ru(bpy)$_2$(4,4'-(PO$_3$H$_2$)$_2$bpy)$^{2+}$ (bpy) 2,2'-bipyridine (RuP^{2+}) were employed as visible light sensitizers. RuP^{2+}, which is anchored by a covalent linkage to the NSH$_4$Nb$_6$O^{17} surface, functioned more efficiently than the electrostatically bound Ru^{2+} complex. More efficient electron injection from the excited sensitizer to RuP^{2+}-sensitized NS-H$_4$Nb$_6$O^{17} was found. R-HCa$_2$Nb$_3$O$_{10}$ produced H$_2$ photocatalytically using visible light ($\lambda > 420$ nm) with initial apparent quantum yields of 20–25%.

Potassium niobate nanoscrolls incorporating rhodium hydroxide nanoparticles were prepared for photocatalytic hydrogen evolution [31]. Well-dispersed rhodium trihydroxide nanoparticles (below 1 nm) were deposited into the interlayer galleries of scrolled K$_4$Nb$_6$O$_{17}$ nanosheets. The unmodified nanoscrolls were proved to be

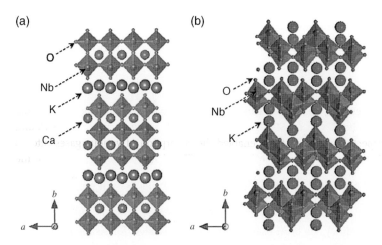

Figure 7.4 Schematic illustration of the structures of (a) KCa$_2$Nb$_3$O$_{10}$ and (b) K$_4$Nb$_6$O$_{17}$ [29]. Reproduced with permission from the American Chemical Society.

good catalysts for UV light-driven hydrogen evolution from aqueous methanol solutions, and their activity was significantly improved by anchoring a small amount of $Rh(OH)_3$ or Rh_2O_3 (0.1 wt% Rh) to the surface. The dinuclear RuII–PdII complex showed an efficient H_2 production in the presence of triethylamine as a sacrificial electron and proton donor under visible light irradiation. XPS and TEM analyses revealed that photoreduction of Pd^{II} to Pd^0 causes dissociation of Pd from the complex to form colloids that are suggested to be the actual catalyst for H_2 production.

A simplified method for synthesis of band structure-controlled $(CuIn)_xZn_2(1-x)S_2$ solid solution photocatalyst with high activity of photocatalytic H_2 evolution under visible light irradiation was developed [32]. $(CuIn)_xZn_2(1-x)S_2$ ($x = 0.01$–0.5) microspheres were prepared by a hydrothermal method. They were characterized by X-ray diffraction (XRD), SEM, transmission electron microscopy (TEM), UV–visible diffuse reflectance spectra (UV–vis), Raman scattering spectra, and Brunauer–Emmett–Teller (BET) surface area measurement. It was found that the $(CuIn)_xZn_2(1-x)S_2$ samples formed solid solution only in the presence of surfactant cetyltrimethylammonium bromide that led to an increase in the surface area for the $(CuIn)_xZn_2(1-x)S_2$ solid solution.

7.2.3
Photocatalytic H_2 Evolution from Water Based on Platinum and Palladium Complexes

A photoinduced hydrogen producing system containing an electron donor (d), photosensitizer (s), and a photocatalysis is presented in Figure 7.5 [33].

In recent efforts for light-driven H_2 formation attention was paid to employ the square planar platinum(II) complexes with terpyridyl acetylide chromophore as photosensitizer, including liberated H_2 from water [33–37]. Xu et al. [33] found cyclometalated platinum(II) complex, [ClPt(C^N^NPhMe)] (1) (HC^N^NPhMe = 4-(p-tolyl)-6-phenyl-2,20-bipyridine) to be an efficient photosensitizer for photocatalytic hydrogen evolution from water and obtained 98 turnovers of hydrogen after 10 h irradiation of solution containing MV^{2+} (4,40-dimethyl-2,20-bipyridinium).

A homogeneous system for the photogeneration of hydrogen from water based on a platinum(II) terpyridyl acetylide chromophore and a molecular cobalt catalyst was developed [35, 36]. The complex $[Co(dmgH)_2pyCl]^{2+}$ (1; dmgH = dimethylglyoximate, py = pridine) has been used as a molecular catalyst for visible light-driven hydrogen production in the presence of [Pt(tolylterpyridine)(phenylacetylide)]$^+$ (3) as a photosensitizer and triethanolamine (TEOA) as a sacrificial reducing agent.

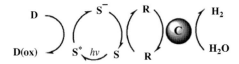

Figure 7.5 Photoinduced hydrogen producing system containing an electron donor (d) and photosenthitizer (s) and a photocatalysis [33]. Reproduced with permission from Elsevier.

Irradiation of the reaction solution initially containing 1.6×10^{-2} M TEOA, 1.1×10^{-5} M of **3**, and 1.99×10^{-4} M of cocatalyst **1** in MeCN/water (3:2 v/v) at pH = 8.5 for 10 h with $\lambda > 410$ nm yielded 400 turnovers of H_2. When TEOA is 0.27 M, ~1000 turnovers were obtained after 10 h of irradiation. Spectroscopic study of the photolysis solutions suggested that H_2 formation proceeds via Co(I) and protonation to form Co(III) hydride species.

7.3
Water Splitting into O_2 and H_2

7.3.1
Thermodynamics and Feasable Mechanism of the Water Splitting

Photocatalytic water splitting and H_2 evolution using abundant compounds as electron donors are expected to contribute to construction of a clean and simple system for solar hydrogen production, and a solution of global energy and environmental issues in the future. Different approaches to solar water splitting including semiconductor particles as photocatalysts and photoelectrodes, molecular donor–acceptor systems linked to catalysts for hydrogen and oxygen evolution, photovoltaic cells coupled directly or indirectly to electrocatalysts, and multinuclear complexes of paramagnetic metals have been proposed and implemented ([12, 37–48] and references therein).

In biological photosynthesis, water splitting with evolution of dioxygen occurs under the action of a relatively mild oxidant, the cation radical of chlorophyll P680 [sup +] (1.3–1.4 V) [49]. The potentials of oxidation of water by the one-, two-, and four electron mechanisms are equal to 2.7 (the hydroxyl radical), 1.36 (hydrogen peroxide), and 0.81 eV (dioxygen), respectively. Enclosed in parenthesis are the products evolved at the most endothermic step. It was evident [13–16] that the smoothest thermodynamic profile of the reaction is provided by the four-electron mechanism, which can be realized under mild condition, as a key step, in biological and artificial systems in most probably tetranuclear manganese clusters. The presence of such complexes in photosynthetic WOS and in model reactions has been proved experimentally (Section 7.2.5).

According to the studies [13–16], the photoevolution dioxygen from water may be described as a sequence of elementary steps: four one-electron steps of oxidation of the manganese cluster by the chlorophyll cation followed by one four-electron step of O_2 evolution. The following simplified Scheme 1 was visualized [14, 15]:

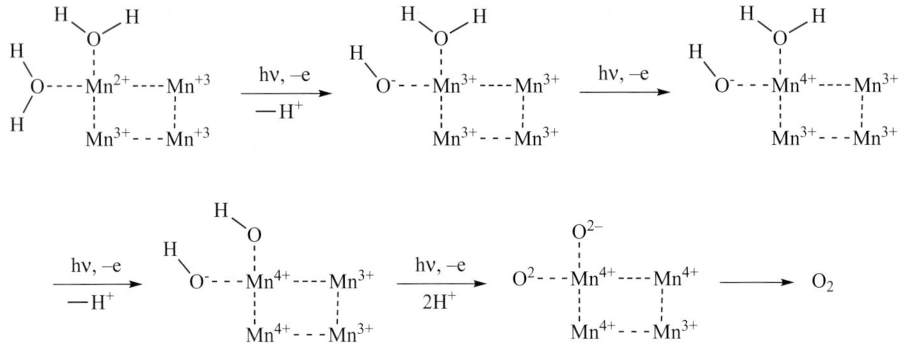

Scheme 7.1

Each one-electron step may be accompanied by the elimination of a proton, which contributes to the preservation of the total charge of the complex and considerably simplifies the last, key step of the process.

The multielectron nature of the process does not impose any additional theoretical restriction on its rate. The substrate–metal interactions in such complexes occur via multiorbital binding with high degree of orbital overlap and the electron transfer resonance integral (V) is high enough to provide four-electron adiabatic mechanism of the process and low energy activation. It is necessary to stress that in the clusters, electron transfer from a substrate to an adjusted metal atom is accompanied by simultaneous, compensative shift of the electron cloud. The strong delocalization of electrons in clusters reduces to a minimum the reorganization energy (λ) in the classical Marcus equation (1.6) and relatively high factor of nuclear synchronization. All these factors ensure high rate of multielectron processes.

7.3.2
Mn Clusters as Water Oxidizing Photocatalysts

7.3.2.1 Structure and Catalytic Activity of Cubane Manganese Clusters

To date, the most efficient system that uses solar energy to oxidize water is the photosystem II water oxidizing complex (WOC) or oxygen evolution complex (OEC), which is found within naturally occurring photosynthetic organisms (Section 7.2.5). The catalytic core of this enzyme is a $CaMn_4O_x$ cluster, which has been conserved since the emergence of this type of photosynthesis about 2.5 billion years ago. The key features that facilitate the catalytic success of the PSII-WOC offer important lessons for the design of abiological water oxidation catalyst.

Hydrogen generation by water splitting in manganese clusters has been studied extensively in order to convert the solar energy to chemical energy. There are now over 100 structures of tetra-Mn-oxo complexes in the CCDC database containing $M_4(OX)_4$ cores (cubans), unsymmetrical cube-like cores, or the planar butterfly core $[Mn_4O_2]^{n+\bullet}$ ([12, 37–48] and references therein). Many physical techniques have provided important insight into the OEC structure and function, including XRD and extended X-ray absorption fine structure (EXAFS) spectroscopy as well as time-resolved mass spectrometry (MS), ESR (EPR) spectroscopy, and Fourier transform IR spectroscopy applied in conjunction with mutagenesis studies.

The development of bioinspired Mn_4O_4-cubane water oxidation catalysts taking a leaf from photosynthesis, namely, the photosystem II water oxidizing complex (PSII-WOC), the most efficient system that uses solar energy to oxidize water, was reviewed [12]. The chemical principles of water oxidnation capabilities of structurally related synthetic manganese–oxo complexes, particularly those with a cubical Mn_4O_4 core ("cubanes") were examined. It was shown that the $[Mn_4O_4]^{6+}$ cubane core assembles spontaneously in solution from monomeric precursors or from $[Mn_2O_2]^{3+}$ core complexes in the presence of metrically appropriate bidentate chelates, for example, diarylphosphinates (ligands of $Ph_2PO_2^-$ and 4-phenyl-substituted derivatives), which bridge pairs of Mn atoms on each cube face ($Mn_4O_4L_6$). A prototypical molecular manganese–oxo cube $[Mn_4O_4]^{n+}$ in a family of "cubane" complexes $[Mn_4O_4L_6]$, where L is a diarylphosphinate ligand (p-R-$C_6H_4)_2PO_2$ (R = H, alkyl, OMe), was synthesized. Cubanes were found to be ferocious oxidizing agents, stronger than analogous complexes with the $[Mn_2O_2]^{3+}$ core. The cubane core topologically was found to be structurally suited to releasing O_2, and it does so in high yield upon removal of one phosphinate by photoexcitation in the gas phase or thermal excitation in the solid state.

A prototypical molecular manganese–oxo cube $[Mn_4O_4]^{n+}$ in a family of "cubane" complexes $[Mn_4O_4L_6]$, where L is a diarylphosphinate ligand (p-R-$C_6H_4)_2PO_2$ (R = H, alkyl, Oe), was synthesized (Figure 7.6) [50]. The diphenylphosphinate complex assembled spontaneously from manganese(II) and permanganate salts in high yield in nonaqueous solvents.

A photoelectrochemical cell was designed that catalyzes the photooxidation of H_2O using visible light as the only energy source [51]: A molecular catalyst, $[Mn_4O_4L_6]^+$, with L = bis(methoxyphenyl)phosphinate, was synthesized from earth-abundant elements. A photochemical charge separation system, $Ru^{II}(bipy)_2(bipy(COO))_2$,

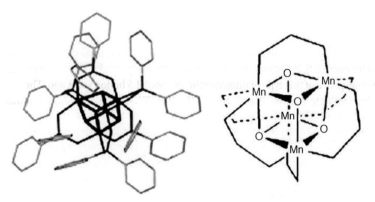

Figure 7.6 X-ray crystal structure of a complex synthesized by Brimblecombe et al. [50].

adhered to TiO$_2$-coated FTO conductive glass, and the molecular catalyst embedded in a p-conducting Nafion membrane were described. Electrochemical investigation of Mn$_4$O$_4$-cubane water oxidizing clusters, attached to an electrode surface and suspended within a Nafion film, was performed [52]. The authors described a detailed electrochemical study of two bioinspired Mn–oxo complexes, [Mn$_4$O$_4$L$_6$] (L = diphenylphosphinate (1) and bis(p-methoxyphenyl)phosphinate (2)). These complexes contain a cubic [Mn$_4$O$_4$]$^{6+}$ core stabilized by phosphinate ligands. A comparison of catalytic photocurrent generated by films deposited by electrode immobilization revealed that doping of the catalyst in Nafion resulted in higher photocurrent than was observed for a solid layer of cubane on an electrode surface. It was shown that both compounds undergo a two-electron, chemical irreversible reduction in CH$_2$Cl$_2$, with a mechanism that is dependent on scan rate and influenced by the presence of a proton donor.

The attachment of a bifunctional iodo-organo-phosphinate compound to gold (Au) surfaces via chemisorption of the iodine atom was described and used to chelate a redox-active metal cluster via the phosphinate group. XPS, AFM, and electrochemical measurements showed that (4-iodo-phenyl)phenyl phosphinic acid (IPPA) forms a tightly bound self-assembled monolayer (SAM) on Au surfaces [43]. The surface coverage of an IPPA monolayer on Au was found to be 0.40 ± 0.03 nmol/cm^2, corresponding to 0.4 monolayers. It is shown that the Au/IPPA SAM adsorbs Mn$_4$O$_4$(Ph$_2$PO$_2$)$_6$ from solution by a phosphinate exchange reaction to yield Au/IPPA/Mn$_4$O$_4$(Ph$_2$PO$_2$)$_5$ SAM. Electrochemistry confirmed that Mn$_4$O$_4$(Ph$_2$PO$_2$)$_6$ is anchored on the Au/IPPA surface and that redox chemistry can be mediated between the electrode and the surface-attached complex.

7.3.2.2 Catalytic Activity and Mechanism of WOS in Manganese Clusters

Cubanes [Mn$_4$O$_4$]$^{6+}$ were found to be ferocious oxidizing agents [12]. Catalytic evolution of O$_2$ and protons from water exceeding 1000 turnovers was achieved by suspending the oxidized cubane, [Mn$_4$O$_4$L$_6$]$^+$, into a proton conducting membrane (Nafion) preadsorbed on a conducting electrode and electroxidizing the photoreduced butterfly complexes by the application of an external bias. Catalytic water oxidation was obtained using sunlight as the only source of energy by replacing the external electric bias with redox coupling to a photoanode incorporating a Ru (bipyridyl) dye. The importance of proton-coupled electron transfer (PCET) in redox reactions with H atom donors was stressed [12]. A general scheme of the O$_2$ evolution during WOS on manganese cluster accompanied by PCET is presented in Figure 7.7.

The pinned butterfly species has Mn oxidation states of 2Mn^{II}$_2$Mn^{III}, determined by EPR spectroscopy. Its solution structure with Mn oxidation states of 2Mn^{II}$_2$Mn^{II}I was based on evidence obtained via multiple spectroscopic techniques (EPR, NMR, ESI-MS, and FTIR). Four sequentially formed hydrogenated intermediates were isolated and characterized. A suggested detailed scheme of the O$_2$ generation from the corner oxygen atoms yielding the butterfly by intramolecular bond formation, preceded by the release of a phosphinate anion, is shown in Figure 7.8.

7.3 Water Splitting into O$_2$ and H$_2$

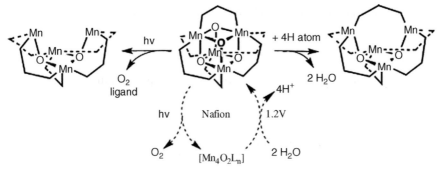

Figure 7.7 General scheme of the O$_2$ evolution during WOS on manganese cluster [12]. Reproduced with permission from the American Chemical Society.

It was shown [53] that the bioinspired Mn–oxo cubane complex, [Mn$_4$O$_4$L$_6$]$^+$ (1b + , L = (p-MeOPh)$_2$PO$_2$), is able to electrooxidize H$_2$O at 1.00 V (versus Ag/AgCl) under illumination by UV–visible light when suspended in a proton-conducting membrane (Nafion, a sulfonated tetrafluoroethylene-based fluoropolymer copolymer) coated onto a conducting electrode. A schematic showing the conceptual similarity of the photoanode **1b** + − Nafion/Ru(4)-TiO$_2$ to the PSII-WOC is given in Figure 7.9. Electrochemical measurements and UV–visible, NMR, and EPR spectroscopies were interpreted to indicate that 1b + is the dominant electroactive species in the Nafion. The observation of a possible intermediate and free phosphinate ligand within the Nafion suggested a catalytic mechanism involving photolytic disruption of a phosphinate ligand, followed by O$_2$ formation, and subsequent reassembly of the cubane structure. Catalytic turnover frequencies of 20–270

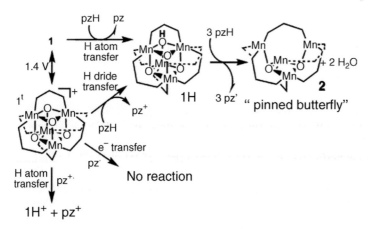

Figure 7.8 PCET reactions of the cubane model complexes Mn$_4$O4L$_6$, **1a** and **1a+** (L-) (C$_6$H$_5$)$_2$PO$_2$ [12]. Reproduced with permission from the American Chemical Society.

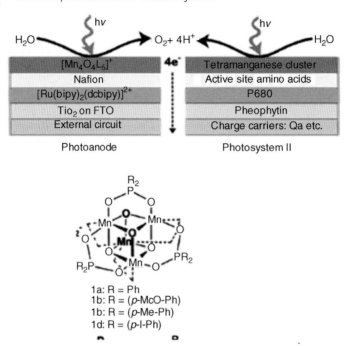

Figure 7.9 Schematic showing the conceptual similarity of the photoanode **1b** + –Nafion/Ru(4)–TiO$_2$, with the PSII-WOC [53]. Reproduced with permission from the American Chemical Society.

molecules of O$_2$ h^{-1} catalyst^{-1} at an overpotential of 0.38 V plus light (275–750 nm) and turnover numbers >1000 molecules of O$_2$ catalyst-1 were observed.

The hydrophobic cubane molecules that are shown in Figure 7.9 protected reactive intermediates while simultaneously allowing access to water molecules in restricted channels and diffusion of bulk water.

Sustained water oxidation photocatalysis by a bioinspired manganese cluster was demonstrated in Ref. [50]. Suggested reaction pathways, which accumulated numerous experimental data, and a possible photocatalytic cycle for the cluster are presented in Figure 7.10.

A series of synthesis of model complexes, spectroscopic characterization of these systems, and probe of the reactivity in the context of water oxidation were performed [54]. The authors described how models have made significant contributions ranging from understanding the structure of the water oxidation center (e.g., contributions to defining a tetrameric Mn$_3$Ca cluster with a dangler Mn) to the ability to discriminate between different mechanistic proposals. Photocatalytic overall water splitting promoted by two different cocatalysts for hydrogen and oxygen evolution under visible light was reported [55]. A proof-of-concept using GaN:ZnO loaded with Rh/Cr$_2$O$_3$ (core/shell) and Mn$_3$O$_4$ NPs as H$_2$ and O$_2$ evolution promoters under irradiation with light (λ > 420 nm) was formulated and implemented. Figure 7.11 shows time courses of H$_2$ and O$_2$ evolution using modified GaN:ZnO

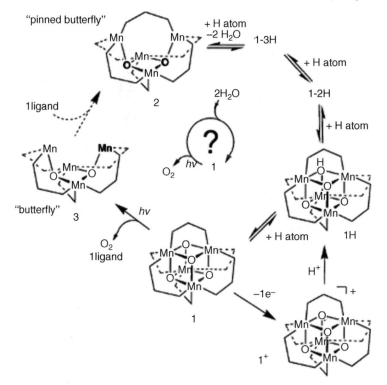

Figure 7.10 Reaction pathways and a possible photocatalytic cycle for 1. Observed reduction reactions of 1 and 1+ in solution and the gas-phase photodissociation to yield O_2. Reverse arrowsshow the proposed reoxidation steps that regenerate the cubanes [50].

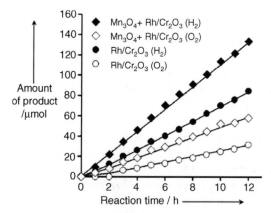

Figure 7.11 Time courses of H_2 and O_2 evolution using modified GaN:ZnO catalysts under visible light (l > 420 nm) Mn loading: 0.05 wt% [55].

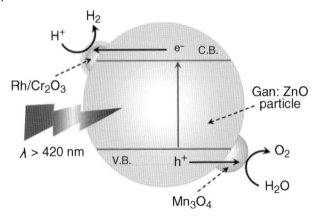

Figure 7.12 A proposed reaction mechanism for visible light-driven overall water splitting on GaN: ZnO modified with Mn_3O_4 and Rh/Cr_2O_3 (core/shell) nanoparticles [55].

catalysts and Figure 7.12 illustrates a suggested reaction mechanism for visible light-driven overall water splitting.

Although the quantum yield of this system was relatively low (1%), this work is the first demonstration that loading two different kinds of cocatalysts can effectively promote overall water splitting, along with direct evidence of the functionality of each cocatalyst in the reaction.

7.3.3
Heterogeneous Catalysts for WOS

7.3.3.1 General
Critical reviews of heterogeneous photocatalyst materials for water splitting [38, 56, 57, 59] showed the basis of photocatalytic water splitting and experimental points and provided surveys of heterogeneous photocatalyst materials for water splitting into H_2 and O_2 and H_2 or O_2 evolution from an aqueous solution containing a sacrificial reagent. The reviews described many oxides consisting of metal cations with d^0 and d^{10} configurations, for example, metal (oxy)sulfide and metal (oxy)nitride photocatalysts. A general scheme of heterogeneous water splitting is presented in Figure 7.13 [38].

As illustrated in Figure 7.14, the successful systems of WOS can be primarily divided into two approaches [58]. One approach is to split water into H_2 and O_2 using a single visible light-responsive photocatalyst with a sufficient potential to achieve overall water splitting. The other approach ("Z-scheme") applies a two-step excitation mechanism using two different photocatalysts. The advantages of a Z-scheme water splitting system are that a wider range of visible light is available because the energy required to drive each photocatalyst can be reduced and that the separation of evolved H_2 and O_2 is possible in principle.

An inorganic catalyst for photochemical splitting of water is iilustrated in Figure 7.15 [3].

7.3 Water Splitting into O_2 and H_2 | 249

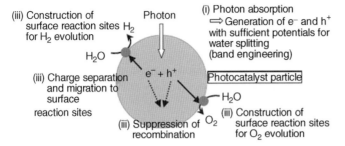

Figure 7.13 Heterogeneous process of photocatalytic water splitting [38]. Reproduced with permission from the Japan Institute of Energy.

According to Maeda et al. [55], the water splitting reaction on semiconductors occurs in three steps: (1) the photocatalyst absorbs photon energy greater than the bandgap energy of the material and generates photoexcited electron–hole pairs in the bulk, (2) the photoexcited carriers separate and migrate to the surface without recombination, and (3) adsorbed species are reduced and oxidized by the photogenerated electrons and holes to produce H_2 and O_2, respectively. The first two steps are strongly dependent on the structural and electronic properties of the photocatalyst, while the third step is promoted by an additional catalyst (called cocatalyst). Visible light water splitting using dye-syensitized oxide semiconductors was reported [10]. The authors described approach to two problems in solar water

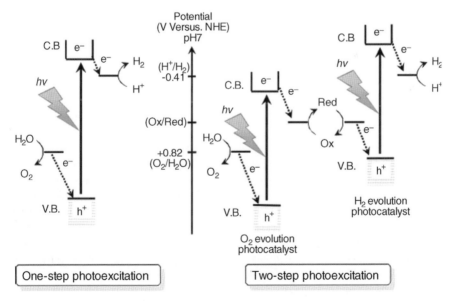

Figure 7.14 Schematic energy diagrams of photocatalytic water splitting for one-step and two-step photoexcitation systems [58]. Reproduced with permission from American Chemical Society.

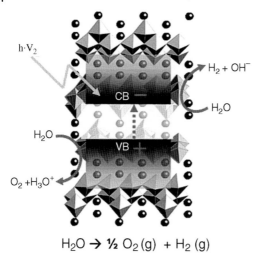

Figure 7.15 Illustration photocatalytic water splitting on a semiconductor [3]. Reproduced with permission from the American Chemical Society.

splitting: the organization of molecules into assemblies that promote long-lived charge separation and catalysis of the electrolysis reactions, in particular the four-electron oxidation of water. They underlined the quantum efficiency and degradation in terms of competing kinetic pathways for water oxidation, back electron transfer, and decomposition of the oxidized dye molecules.

Effects of doping of metal cations into wide bandgap semiconductor photocatalysts on morphology, visible light response, and photocatalytic performance were studied [59]. It was found that doping of lanthanide and alkaline earth ions improved activity of a $NaTaO_3$ photocatalyst for water splitting. Lanthanum was the most effective dopant. The $NaTaO_3$:La with a NiO cocatalyst gave 56% of a quantum yield at 270 nm.

Water oxidation photocatalysts on surface of semiconductors have been studied using a range of metal oxides including TiO_2 ([3, 60–64] and references therein).

7.3.3.2 Photocatalysts Based on Titanium Oxides

Zeolite-based materials that show photocatalytic properties for water splitting under the visible light have been synthesized by incorporating titanium dioxide (TiO_2), heteropolyacid (HPA), and transition metal cobalt (Co) [65]. Hydrogen generation to the tune of 2730 μmol/h/g TiO_2 has been obtained for the composite photocatalyst synthesized. This composite photocatalyst showed improvement in hydrogen evolution rate over other TiO_2-based visibly active photocatalyst previously reported.

A survey of solar hydrogen production systems such as the combination process of solar cell and water electrolysis and one-step photoelectrochemical water splitting using a p-GaAs-n-GaAs/p-GaInP$_2$ photoelectrode was introduced and photocatalytic and photoelectrochemical approaches to solar hydrogen production were described [38]. In photoelectrochem system using an oxide semiconductor film

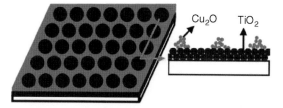

Figure 7.16 Scheme of TiO$_2$ film is covered by a Cu$_2$O microgrid [66]. Reproduced with permission from the American Chemical Society.

photoelectrode, solar energy conversion efficiency of H$_2$ and O$_2$ production of mesoporous TiO$_2$ (anatase) and WO$_3$ film were 0.32–0.44% at applied potential of 0.35 V versus NHE and 0.44% at 0.9 V, respectively. This efficiency of tandem cell system composed of a WO$_3$ film photoelectrode and a two series-connected DSSC ($V_{oc} = 1.4$ V) was 2.5–3.3%, and was suggested to be attractive for practical H$_2$ production process. Artificial inorganic leafs were developed by organizing light-harvesting, photoinduced charge separation, and catalysis modules (Pt/N-TiO$_2$) into leaf-shaped hierarchical structures using natural leaves as biotemplates [57]. The enhanced light-harvesting and photocatalytic water-splitting activities were caused by the reproduction of the leafs' complex structures and self-doping of nitrogen during synthesis.

A heavy-loading photocatalyst configuration (Figure 7.16) in which 51% of the surface of the TiO$_2$ film is covered by a Cu$_2$O microgrid was fabricated [66]. The coupled system showed higher photocatalytic activity under solar light irradiation than TiO$_2$ and Cu$_2$O films. This improved performance was due to the efficient charge transfer between the two phases, and the similar opportunity has to be exposed to irradiation and adsorbates.

An all-inorganic photocatalytic unit consisting of a binuclear TiOCr charge transfer chromophore coupled to an Ir oxide nanocluster has been assembled on the pore surface of mesoporous silica AlMCM-41 [67]. When exciting the Ti(IV)OCr(III) → Ti(III)OCr(IV) metal-to-metal charge transfer chromophore of an aqueous suspension of Ir$_x$O$_y$–TiCr–AlMCM-41 powder with visible light, oxygen evolution with a quantum efficiency of at least 13% was detected by Clark electrode measurements. *In situ* Fourier transform Raman and X-band electron paramagnetic resonance spectroscopy revealed the formation of superoxide species. The use of H$_2$^{18}O confirmed that the superoxide species originates from oxidation of water. The results indicated efficient photocatalytic oxidation of water at Ir oxide nanoclusters followed by trapping of the evolving O$_2$ by transient Ti(III) centers to yield superoxide.

7.3.3.3 Miscellaneous Semiconductors for the WOS Catalysis

Sustainable hydrogen production through photoelectrochemical water splitting using hematite (α-Fe$_2$O$_3$) was demonstrated [68]. The mesoporous hematite (α-Fe$_2$O$_3$) photoelectodes preparation by a solution-based colloidal method was reported. Sustainable hydrogen production through photoelectrochemical water splitting using hematite (α-Fe$_2$O$_3$) was demonstrated. The method yielded water-splitting photocurrents of 0.56 mA/cm^2 under standard conditions (AM 1.5 G 100 mW/cm^2, 1.23 V versus reversible hydrogen electrode, RHE) and over 1.0 mA/cm^2 before the dark current onset (1.55 V versus RHE). XPS and magnetic measurements using a SQUID magnetometer linked this effect to the diffusion and incorporation of dopant atoms from the transparent conducting substrate. Sustainable hydrogen production through photoelectrochemical water splitting using hematite (α-Fe$_2$O$_3$) was demonstrated.

Several polynuclear transition metal complexes, including our own dinuclear di-μ-oxo manganese compound [H$_2$O(terpy)Mn^{III}(μ-O)$_2$Mn^{IV}(terpy)H$_2$O](NO$_3$)$_3$ (**1**, terpy = 2,2':6',2"-terpyridine), have been reported to be homogeneous catalysts for water oxidation [69]. This study reported the covalent attachment of **1** onto nanoparticulate TiO$_2$ surfaces using a robust chromophoric linker **L**. **L**, a phenylterpy ligand attached to a 3-phenyl-acetylacetonate anchoring moiety via an amide bond, absorbs visible light and leads to photoinduced interfacial electron transfer into the TiO$_2$ conduction band. The authors characterized the electronic and structural properties of the **1**–**L**–TiO$_2$ assemblies by using a combination of methods, including computational modeling and UV–visible, IR, and EPR spectroscopies. The Mn(III,IV) state of **1** can be reversibly advanced to the Mn(IV,IV) state by visible light photoexcitation of **1**–**L**–TiO$_2$ NPs and recombines back to the Mn(III,IV) state in the dark in the absence of electron scavengers.

The effects of surface modification and reaction conditions on the photoelectrochemical properties of polycrystalic Cu(In,Ga)Se$_2$ (CIGS) thin films for water splitting were studied [70]. CIGS modified with platinum particles (Pt/CIGS) generated a cathodic photocurrent at potentials up to $+0.4$ V versus RHE at pH = 9.5. A polycrystalline CIGS thin film modified with platinum particles (Pt/CIGS) was found to generate a cathodic photocurrent at potentials up to $+0.4$ V versus RHE in Na$_2$SO$_4$ aqueous solution at pH = 9.5. This photocurrent persisted for 16 h at the negative potential of -0.24 V versus RHE, resulting in a turnover number of over 500. The CdS-inserted electrode (Pt/CdS/CIGS) generated a cathodic photocurrent with a 0.3 V higher onset potential than Pt/CIGS, and had an IPCE of 59%. Ir oxide nanoparticles stabilized by a heteroleptic Ru tris(bipyridyl) dye were used as sensitizers in photoelectrochemical cells consisting of a nanocrystal anatase anode and a Pt cathode [71]. The dye coordinated the IrO$_2 \cdot n$H$_2$O nanoparticles through a malonate group and the porous TiO$_2$ electrode through phosphonate groups. Under visible illumination ($\lambda > 410$ nm) in pH 5.75 aqueous buffer, O$_2$ was generated at anode potentials of -325 mV versus Ag/AgCl and H was generated at the cathode. The internal quantum yield for photocurrent generation was 0.9%.

A NiO (0.2 wt%)/NaTaO$_3$:La (2%) photocatalyst with a 4.1 eV bandgap showing high activity for water splitting into H$_2$ and O$_2$ with an apparent quantum yield of 56% at 270 nm was prepared and investigated [72]. Visible light-driven photocatalysts have also been developed through band engineering by doping of metal cations, forming new valence bands with Bi$_{6s}$, Sn$_{5s}$, and Ag$_{4d}$ orbitals, and by making solid solutions between ZnS with wide-bandgap and narrow-bandgap semiconductors. Overall water splitting under visible light irradiation has been achieved by construction of a Z-scheme photocatalysis system employing the visible light-driven photocatalysts for H_2 and O_2 evolution, and the Fe^{3+}/Fe^{2+} redox couple as an electron relay. In the work [73], it was shown that nanostructured Mn oxide clusters supported on mesoporous silica KIT-6 efficiently evolve O_2 in aqueous solution under mild conditions. For driving the catalyst with visible light, the established $Ru^{+2}(bpy)_3$–persulfate sensitizer system was used. Nanostructured Mn oxide clusters supported on mesoporous silica KIT-6 were characterized for generation of O_2 in aqueous solution under visible light using tris(2,2′-bipyridine)Ru^{2+} photosensitizer and $S_2O_8^{2+}$ electron acceptor. A turnover frequency (TOF) of $1140 s^{-1}$ per nanocluster catalyst (pH 5.8, RT, 350 mV overpotential) was achieved. Projected on a plane, this corresponds to a TOF of $1 s^{-1} nm^{-2}$ [7]. The authors concluded that a 100 such nanocluster catalysts stacked in a nanoporous scaffold, which can be readily achieved, will result in a TOF of $100 s^{-1} nm^{-2}$, the rate required for keeping up with the solar flux (1000 W/m^2, AM 1.5).

Light, inexpensive, effective nanostructured Co_3O_4 clusters in mesoporous silica as a nanometer-sized multielectron catalyst (Figure 7.17) made of a first-row transition metal oxide that efficiently evolves oxygen from water were reported [74]. The nanorod bundle structure of the catalyst results in a very large surface area, an important factor contributing to the high turnover frequency. Oxygen evolution

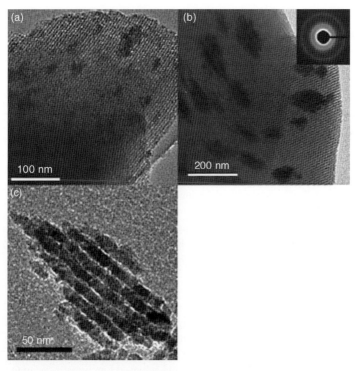

Figure 7.17 TEM images of (a) SBA-15/Co_3O_4 4% loading, (b) SBA-15/Co_3O_4 8% loading, (c) Co_3O_4 nanocluster (8% sample) after removal of the SBA-15 silica material using aqueous NaOH as etching reagent. The inset in (b) shows the SAED pattern [74].

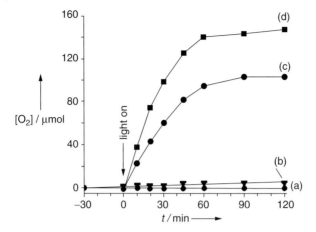

Figure 7.18 Oxygen evolution in aqueous suspensions (40 ml) of (a) SBA-15/NiO (8%), (b) micrometer-sized Co_3O_4 particles, (c) SBA-15/Co_3O_4 (8%), and (d) SBA-15/Co_3O_4 (4%). Measurements were conducted at pH 5.8 and 228 °C. Catalysis was initiated by Ar ion laser emission at 476 nm (240 mW). Experimental details of the oxygen detection method are described in Ref. [74].

photocatalyzed by Co_3O_4 clusters in aqueous suspensions is demonstrated in Figure 7.18.

A combinatorial approach was carried out to systematically study visible light responsiveness of Fe-Ti-M (M: various metal elements) oxides for photoelectrochemical water splitting [75]. Among the 25 elements tested, strontium was the most effective. A ternary metal oxide $Fe_{86}·1Ti_9·6Sr_4·3O_x$ was identified as a new lead structure for a visible light responsive, n-type semiconductor. The authors have conducted various kinds of characterization of the Fe-Ti-Sr oxide semiconductor and discussed the reason why Sr in the Fe-Ti oxide gave the highest photocurrent.

The nanorod bundle structure of the catalyst resulted in a very large surface area, an important factor contributing to the high efficiency. Effects of electrolyte addition on photocatalytic activity of $(Ga_{1-x}Zn_x)(N_{1-x}O_x)$ modified with either Rh_2-yCr_yO_3 or RuO_2 nanoparticles as cocatalysts for overall water splitting under visible light ($\lambda > 400$ nm) were investigated [76]. The cocatalyst Rh_2-yCr_yO_3 was confirmed to selectively promote the photoreduction of H^+, while RuO_2 functions as both H_2 evolution site and as efficient O_2 evolution site. The addition of an appropriate amount of NaCl or A_2SO_4 (A = Li, Na, or K) to the reactant solution was found to increase activity by up to 75% compared to the case without additives. Direct splitting of seawater to produce H_2 and O_2 was also demonstrated using Rh_2-yCr_yO_3-loaded $(Ga_{1-x}Zn_x)(N_{1-x}O_x)$ catalyst under visible light. Tabata et al. [77] reported the modification of tantalum nitride (Ta_3N_5), which has a bandgap of 2.1 eV, with nanoparticulate iridium (Ir) and rutile titania (R-TiO_2). It was shown that this system achieved functionality as an O_2 evolution photocatalyst in a two-step water-splitting system with an IO_3^-/I^- shuttle redox mediator under visible light ($\lambda > 420$ nm) in combination with a Pt/ZrO_2/

TaON H_2 evolution photocatalyst. It was suggested that the loaded Ir nanoparticles acted as active sites to reduce IO_3^- to I^-, while the R-TiO_2 modifier suppressed the adsorption of I^- on Ta_3N_5, allowing to evolve O_2 in the two-step water-splitting system.

A two-step photocatalytic water splitting (Z-scheme) system consisting of a modified ZrO_2/TaON species (H_2 evolution photocatalyst), an O_2 evolution photocatalyst, and a reversible donor/acceptor pair was investigated [58]. Among the O_2 evolution photocatalysts and redox mediators examined, Pt-loaded WO_3 (Pt/WO_3) and the IO_3^-/I^- pair were, respectively, found to be the most active components. Combining these two components with Pt-loaded ZrO_2/TaON achieved stoichiometric water splitting into H_2 and O_2 under visible light, achieving an apparent quantum yield of 6.3% under irradiation by 420.5 nm monochromatic light under optimal conditions, six times greater than the yield achieved using a TaON analogue. The authors suggested that the high activity of this system is due to the efficient reaction of electron donors (I^- ions) and acceptors (IO_3^- ions) on the Pt/ZrO_2/TaON and Pt/WO_3 photocatalysts, respectively, which suppresses undesirable reverse reactions involving the redox couple that would otherwise occur on the photocatalysts.

It was shown that artificial reaction centers, where electrons are injected from a dye molecule into the conduction band of nanoparticulate titanium dioxide on a transparent electrode, coupled to catalysts, such as platinum or hydrogenase enzymes, can produce hydrogen gas [78]. Oxidizing equivalent from such reaction centers can be coupled to iridium oxide nanoparticles, which can oxidize water. This system uses sunlight to split water into oxygen and hydrogen fuel, but efficiencies are low, and an external electric potential is required. Particles of noble metals (Pt, Rh, Au, Ag) or transitionmetal oxides (NiO$_x$, RuO_2, $Rh_{2-y}Cr_yO_3$) have been investigated as cocatalysts providing hydrogen evolution sites for photocatalytic overall water splitting. [79].

The use of a 2D carbon nanostructure, graphene, as a support material for the dispersion of Pt nanoparticles (Figure 7.19) [80], provided ways to develop advanced

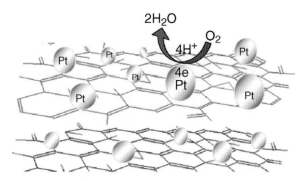

Figure 7.19 Dispersion of Pt nanoparticles on a two-carbon sheet (graphene) to facilitate an electrocatalytilitic process [80]. Reproduced with permission from the American Chemical Society.

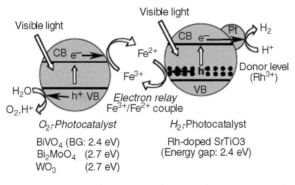

Figure 7.20 Water splitting system by a two-photon process with visible light response [38]. Reproduced with permission from the Chemical Society of Japan.

electrocatalyst materials for fuel cells. Platinum nanoparticles were deposited onto graphene sheets by means of borohydride reduction of H_2PtCl_6 in a graphene oxide (GO) suspension. The partially reduced GO-Pt catalyst was deposited as films onto glassy carbon and carbon. Nearly 80% enhancement in the electrochemically active surface area can be achieved by exposing partially reduced GO-Pt films with hydrazine followed by heat treatment (300 °C, 8 h).

The water splitting under visible light irradiation has been achieved by the Z-scheme system consisted of a Fe^{3+}/Fe^{2+} redox couple as an electron relay and two powdered heterogeneous photocatalysts have been developed [38] (Figure 7.20). The ($Pt/SrTiO_3$:Rh)–($BiVO_4$) system showed the highest activity with 0.3% of an apparent quantum yield at 440 nm. It can use visible light up to 520 nm.

Materials described in this chapter were proved to act as powerful photocatalysts for water reduction and splitting under visible light irradiation; however, the search for other materials as visible light response photocatalysts is still an interesting and promising objective in both basic and applied aspects.

References

1 Fujishima, A. and Honda, K. (1972) *Nature*, **238**, 37–38.
2 Nakabayshi, S., Fujishima, A., and Honda, K. (1983) *Chem. Phys. Lett.*, **102**, 464–467.
3 Osterloh, F.E. (2008) *Chem. Mater.*, **20** (1), 35–54.
4 Kudo, A. and Miseki, Y. (2009) *Chem. Soc. Rev.*, **38**, 253–278.
5 Maeda, K. and Domen, K. (2007) *J. Phys. Chem. C*, **111** (22), 7851–7861.
6 Turner, J. (2004) *Science*, **305** (5686), 972–974.
7 Gratzel, M. (2001) *Nature*, **414**, 338–344.
8 Randeniya, L.K., Murphy, A.B., and Plumb, I.C. (2008) *J. Mater. Sci.*, **43** (4), 1389–1399.
9 Youngblood, W.J., Hata, H., and Mallouk, T.E. (2009) *J. Phys. Chem. C*, **113** (18), 7962–7969.
10 Youngblood, W.J., Lee, S.-H.A., Kazuhiko, M., and Mallouk, T.E. (2009) *Acc. Chem. Res.*, **42** (12), 1966–1973.
11 Seger, B. and Kamat, P.V. (2009) *J. Phys. Chem. C*, **113** (19), 7990–7995.
12 Dismukes, G.C., Brimblecombe, R., Felton, G.A.N., Pryadun, R.S., Sheats, J.E.,

Spiccia, L., and Swiegers, G.F. (2009) *Acc. Chem. Res.*, **42** (12), 1935–1943.

13 Semenov, N.N., Shilov, A.E., and Likhtenshtein, G.I. (1975) *Dokl. Akad. Nauk SSSR*, **221** (6), 1374–1377.

14 Likhtenshtein, G.I. (1979) *Multinuclear Redox Metalloenzymes*, Nauka, Moscow.

15 Likhtenshtein, G.I. (1988) *Chemical Physics of Redox Metalloenzymes*, Springer, Heidelberg.

16 Likhtenshtein, G.I. (2003) *New Trends in Enzyme Catalysis and Mimicking Chemical Reactions*, Kluwer Academic/Plenum Publishers, NY.

17 Kho, Y.K., Iwase, A., Teoh, W.Y., Madler, L., Kudo, A., and Amal, R. (2010) *J. Phys. Chem. C*, **114** (6), 2821–2829.

18 Nada, A.A., Hamed, H.A., Barakat, M.H., Mohamed, N.R., and Veziroglu, T.N. (2008) *Int. J. Hydrogen Energy*, **33** (13), 3264–3269.

19 Kato, C.N., Hara, K., Hatano, A., Goto, K., Kuribayashi, T., Hayashi, K., Shinohara, A., Kataoka, Y., Mori, W., and Nomiya, K. (2008) *Eur. J. Inorg. Chem* (20), 3134–3141.

20 Frame, F.A. and Osterloh, F.E. (2010) *J. Phys. Chem. C*, **114** (23), 10628–10633.

21 Vasil'tsova, V. and Parmon, V.N. (1999) *Kinet. Catal.*, **40** (1), 62–70.

22 Kim, H.G., Borse, P.H., Jang, J.S., Jeong, E.D., Jung, O.-S., Suh, Y.J., and Lee, J.S. (2009) *Chem. Commun.*, 5889–5891.

23 Phuruangrat, A., Ham, D.J., Hong, S.J., Thongtem, S., and Lee, J.S. (2010) *J. Mater. Chem.*, **20** (9), 1683–1690.

24 Zhao, D., Rodriguez, A., and Koodali, R.T. (2010) Abstracts of papers. 239th ACS National Meeting, San Francisco, CA, United States, March 21–25, 2010, FUEL-312.

25 White, J.C. and Dutta, P.K. (2008) Abstracts. 40th Central Regional Meeting of the American Chemical Society, Columbus, OH, United States, June 10–14, CRM-423.

26 Sun, X., Maeda, K., Le Faucheur, M., Teramura, K., and Domen, K. (2009) *J. Phys. Chem. C*, **113** (18), 7962–7969.

27 Yusuke, K., Konomi, S., Yuhei, M., Kazuki, M., Hiroshi, T., Shuichi, N., and Wasuke, M. (2009) *Energy Environ. Sci.*, **2**, 397–400.

28 Lei, P., Hedlund, M., Lomoth, R., Rensmo, H., Johansson, O., and Hammarström, L. (2008) *J. Am. Chem. Soc.*, **130** (1), 26–27.

29 Zhang, H., Chen, G., and Bahnemann, D.W. (2009) *J. Mater. Chem.*, **19** (29), 5089–5095.

30 Maeda, K., Eguchi, M., Lee, S.H.A., Youngblood, W.J., Hata, H., and Mallouk, T.E. (2009) *J. Phys. Chem. C*, **113** (19), 7962–7969.

31 Ma, R., Kobayashi, Y., Youngblood, W.J., and Mallouk, T.E. (2008) *J. Mater. Chem.*, **18** (48), 5982–5985.

32 Zhang, X., Du, Y., Zhou, Z., and Guo, L. (2010) *Int. J. Hydrogen Energy*, **35** (8), 3313–3321.

33 Xu, Q., Fu, W.-F., Zhang, G., Bian, Z., Zhang, J., Xu, H., and Xu, W. (2008) *Catal. Commun.*, **10** (1), 49–52.

34 Han, X., Wu, L.-Z., Si, G., Pan, J., Yang, Q.-Z., Zhang, L.-P., and Tung, C.-H. (2007) *Chem. Eur. J.*, **13** (4), 1231–1239.

35 Du, P., Knowles, K., and Eisenberg, R. (2008) *J. Am. Chem. Soc.*, **130** (38), 12576–12577.

36 Du, P.W., Schneider, J., Jarosz, P., Zhang, J., Brennessel, W.W., and Eisenberg, R. (2007) *J. Phys. Chem. B*, **111** (24), 6887–6894.

37 Elvington, M., Brown, J., Arachchige, S.M., and Brewer, K.J. (2007) *J. Am. Chem. Soc.*, **129**, 10644–16645.

38 Kudo, A., Kato, H., and Issei Tsuji, I. (2004) *Chemistry Letters*, l.**33** (12) 1534–539

39 Zhou, H., Li, X., Fan, T., Osterloh, F.E., Ding, J., Sabio, E.M., Zhang, D., and Guo, Q. (2010) *Adv. Mater.*, **22** (9), 951–956.

40 Yu, Y., Dubey, M., Bernasek, S.L., and Dismukes, G.C. (2007) *Langmuir*, **23** (15), 8257–8263.

41 Du, P., Schneide, J., Li, F., Zhao, W., Patel, U., Castellano, F.N., and Eisenberg, R. (2008) *J. Am. Chem. Soc.*, **130** (15), 5056–5058.

42 Brimblecombe, R., Bond, A.M., Dismukes, G.C., Swiegers, G.F., and Spiccia, L. (2009) *Phys. Chem. Chem. Phys.*, **11** (30), 6441–6449.

43 Yu, Y., Dubey, M., Bernasek, S.L., and Dismukes, G.C. (2007) *Langmuir*, **23** (15), 8257–8263.

44 Arakawa, H. (2009) *J. Jpn. Inst. Energy* (2009) **88** (5), 405–412.
45 Swiegers, G.F., Huang, J., Brimblecombe, R., Chen, J., Dismukes, G.C., Mueller-Westerhoff, U.T., Spiccia, L., and Wallace, G.G. (2009) *Chem. Eur. J.*, **15** (19), 4746–4759.
46 Brimblecombe, R., Swiegers, G.F., Dismukes, G.C., and Spiccia, L. (2008) *Angew. Chem. Int. Ed.*, **47**, 7335–7338.
47 Dasgupta, J., Tyryshkin, A.M., Baranov, S.V., and Dismukes, G.C. (2009) *Appl. Magn. Reson.*, **37** (1–4), 137–150.
48 Yusuke, K., Konomi, S., Yuhei, M., Kazuki, M., Hiroshi, T., Shuichi, N., and Wasuke, M. (2009) *Energy Environ. Sci.*, **2**, 397–400.
49 Rappaport, F., Guergova-Kuras, M., Nixon, P.J., Diner, B.A., and Lavergne, J. (2002) *Biochemistry*, **41** (26), 8518–8527.
50 Brimblecombe, R., Swiegers, G.F., Dismukes, G.C., and Spiccia, L. (2008) *Angew. Chem. Int. Ed.*, **47**, 7335–7338.
51 Brimblecombe, R., Koo, A., Dismukes, G.C., Swiegers, G.F., and Spiccia, L. (2010) *J. Am. Chem. Soc.*, **132** (9), 2892–2894.
52 Brimblecombe, R., Bond, A.M., Dismukes, G.C., Swiegers, G.F., and Spiccia, L. (2009) *Phys. Chem. Chem. Phys.*, **11** (30), 6441–6449.
53 Brimblecombe, R., Kolling, D.R.J., Bond, A.M., Dismukes, G.C., Swiegers, G.F., and Spiccia, L. (2009) *Inorg. Chem.*, **48** (15), 7269–7279.
54 Meelich, K., Zaleski, C.M., and Pecoraro, V.L. (2008) *Philos. Trans. R. Soc. Lond. B Biol. Sci.*, **363** (1494), 1271–1281.
55 Maeda, K., Xiong, A., Yoshinaga, T., Ikeda, T., Sakamoto, N., Hisatomi, T., Takashima, M., Lu, D., Kanehara, M., Setoyama, T., Teranishi, T., and Domen, K. (2010) *Angew. Chem. Int. Ed.*, **49** (24), 4096–4099.
56 Akihiko, K. and Yugo, M. (2009) *Chem. Soc. Rev.*, **38**, 253–278.
57 Zhou, H., Li, X., Fan, T., Osterloh, F.E., Ding, J., Sabio, E.M., Zhang, D., and Guo, Q. (2010) *Adv. Mater.*, **22** (9), 951–956.
58 Maeda, K., Higashi, M., Lu, D., Abe, R., and Kazunari, D. (2010) *J. Am. Chem. Soc.*, **132** (16), 5858–5868.
59 Kudo, A., Niishiro, R., Iwase, A., and Kato, H. (2007) *Chem. Phys.*, **339** (1–3), 104–110.
60 Palmas, S., Polcaro, A.M., Ruiz, J.R., Da Pozzo, A., Mascia, M., and Vacca, A. (2010) *Int. J. Hydrogen Energy*, **35** (13), 6561–6570.
61 Gratzel, M. (2005) *Chem. Lett.*, **34** (1), 8–11.
62 Kanan, M. and Nocera, D. (2008) *Science*, **321**, 1072–1076.
63 Licht, S. (2001) *J. Phys. Chem. B*, **105** (27), 6281–6294.
64 Yagi, M., Syouji, A., Yamada, S., Komi, M., Yamazaki, H., and Tajima, S. (2009) *Photochem. Photobiol. Sci.*, **8** (1), 139–147.
65 Dubey, N., Rayalu, S.S., Labhsetwar, N.K., and Devotta, S. (2008) *Int. J. Hydrogen Energy*, **33** (21), 5958–5966.
66 Zhang, J., Zhu, H., Zheng, S., Pan, F., and Wang, T. (2009) *ACS Appl. Mater. Interfaces*, **1** (10), 2111–2114.
67 Han, H. and Frei, H. (2008) *J. Phys. Chem. C*, **112** (41), 16156–16159.
68 Sivula, K., Zboril, R., Le, F.F., Robert, R., Weidenkaff, A., Tucek, J., Frydrych, J., and Gratzel, M. (2010) *J. Am. Chem. Soc.*, **132** (21), 7436–7444.
69 Li, G., Sproviero, E.M., McNamara, W.R., Snoeberger, R.C., III, Crabtree, R.H., Brudvig, G.W., and Batista, V.S. (2010) *J. Phys. Chem. B*, **114** (45), 14214–14222.
70 Yokoyama, D., Minegishi, T., Maeda, K., Katayama, M., Kubota, J., Yamada, A., Konagai, M., and Domen, K. (2010) *Electrochem. Commun.*, **12** (6), 851–853.
71 Youngblood, W.J., Lee, S.-H.A., Kobayashi, Y., Hernandez-Pagan, E.A., Hoertz, P.G., Moore, T.A., Moore, A.L., Gust, D., and Mallouk, T.E. (2009) *J. Am. Chem. Soc.*, **131** (3), 926–927.
72 Kudo, A., Kato, H., and Tsuji, I. (2004) *Chem. Lett.*, **33** (12), 1534–1537.
73 Jiao, F. and Frei, H. (2010) *Chem. Commun.*, **46** (17), 2920–2922.
74 Jiao, F. and Frei, H. (2009) *Angew. Chem. Int. Ed.*, **48** (10), 1841–1844.
75 Kusama, H., Wang, N., Miseki, Y., and Sayama, K. (2010) *J. Comb. Chem.*, **12** (3), 356–362.
76 Maeda, K., Masuda, H., and Domen, K. (2009) *Catal. Today*, **147** (3–4), 173–178.

77 Tabata, M., Maeda, K., Higashi, M., Lu, D., Takata, T., Abe, R., and Domen, K. (2010) *Langmuir*, **26** (12), 9161–9165.
78 Gust, D., More, T.A., and More, A.L. (2009) *Acc. Chem. Res.*, **42** (12), 1890–1898.
79 Masaaki, Y., Kazuhiro, T., Kazuhiko, M., Akio, I., Jun, K., Yoshihisa, S., Yasunari, I., and Kazunari, D. (2009) *J. Phys. Chem. C*, **113** (23), 10151–10157.
80 Seger, B. and Kamat, P.V. (2009) *J. Phys. Chem. C*, **113** (19), 7990–7995.
81 Hideki, K., Mikihiro, H., Ryoko, K., Yoshiki, S., and Akihiko, K., *Chem. Lett.*, **33**, 1348–1349.

Conclusions

The main outcome of photosynthesis is the oxidation of water and the synthesis of glucose from carbon dioxide at the expense of sunlight energy. On the scale of the Earth, annually about 50 billion tons of carbon from carbon dioxide is bound into forms that provide energy and structural material for all living organisms. World energy consumption is about $4.7 \; 10^{20}$ J (450 quadrillion Btu) and is expected to grow about 2% each year for the next 25 years [1]. Nowadays, renewable sources comprise about 13% of all energy production. Photovoltaics or solar cells account for no more than 0.04% and most probably only in 2030 will this figure reach 1% [2].

Electron transfer, which is one of the most ubiquitous and fundamental phenomena in chemistry, physics, and biology is a key elementary act in the light energy conversion in natural and artificial systems. The Marcus classical work on electron transfer in polar media in the past decades has stimulated vigorous development of the process theory. The present status of the theory of electron transfer between donor acceptor in pairs and between a dye and a semiconductor was reviewed in brief. The contribution of driving force, electron coupling, nuclear reorganization, vibrational modes, electron–proton transfer coupling, spin effects on charge separation in electron transfer has been discussed. Special attention was devoted to the long-distance electron transfer and specificity of electrochemical electron transfer. The latter processes play a pivotal role in the system of light energy conversion.

Nature's solar energy storage systems found in photosynthetic organisms are remarkable among the existing photochemical structures. The following three aspects of fundamental importance were considered: (1) light-harvesting (antenna) complex, (2) the structure and action mechanism of the system of conversion of light energy into chemical energy in the primary charge photoseparation in bacterial and plant photosynthesis, and (3) the structure and the possible mechanisms of the participation of polynuclear manganese systems in the photooxidation of water.

Light energy conversion in the photosynthetic reaction centers is characterized by the high energetic efficiency and the quantum yield close to 100%. Such a result was achieved by fulfilling several principal conditions: (1) light harvesting is very effective, (2) the donor and acceptor groups are disposed at a certain optimum distance relative to each other, (3) the electron transfer driving force (ΔG_0) and redox potentials of primary donor D^* and primary acceptor A are optimal, (4) nanosecond molecular

Solar Energy Conversion. Chemical Aspects, First Edition. Gertz Likhtenshtein.
© 2012 Wiley-VCH Verlag GmbH & Co. KGaA. Published 2012 by Wiley-VCH Verlag GmbH & Co. KGaA.

dynamics of media, and (4) charge-separated pair D^+A^-, keeping strong chemical reactivity, are isolated from side reactions. These lessons from the Nature should be taken into consideration in designing any artificial photochemical system of the light energy conversion.

Main purposes of the investigation of photoelectron transfer in donor–acceptor pairs in solutions and on templates are (a) a deeper understanding of elementary processes in photosynthetic systems and (b) development of donor–acceptor systems that have long lifetime charge separation state without loss of energy for sequential electron transfer. In this direction, two types of these systems were synthesized and investigated. In dyads and triads of different structures of donor and acceptor segments and a bridge between them, ET kinetics was investigated in detail using nano, pico and fepto time-resolved fluorescence and optical techniques and electron spin resonance. In the second type of donor–acceptor pairs, chemical reactions followed after the charge separation. In such systems, the photooxidation of a stable *sacrificial* substrate resulted in the formation of reduced acceptor, which accumulates reducing energy. Special sections described recently discovered effects of spin catalysis on electron transfer and charge separation lifetime and the role of molecular dynamic factors in electron transfer in proteins.

Electron transfer between a dye and semiconductor is one of the key steps in dye-sensitized collar cells. The dynamics of interfacial electron transfer from a variety of dyes in sensitized TiO_2 and miscellaneous semiconductors of different microstructures, prehistory, Fermi level, and conditions were described. Redox processes on metal surfaces, which are important in DSSCs, were considered. Insights into electron transfer kinetics at the electrode–protein, electrode–ferrocene, and electrode–dendrimer interface were presented.

The foregoing information is a background for central chapters of the book devoted to structure and action mechanisms of the dye-sensitized solar cells. A brief description of main principles of solar cells preceded a section on classical Gertzel cell. Further progress in construction of major components of the Gertzel cells such as a photosensitive dye as electron donor in excited state, a substrate as a transparent electrode, a semiconductor as an anode, a counterelectrode as a cathode, connecting wires, a charge carrier bearing negative charge in a specific media (M), which separate cathode and anode, was reviewed. Several principal innovations including the use of semiconductor sensitizers, 3D matrix, ionic liquids, nanostructured semiconductors, porous nanocrystalline materials, nanoscale carbon/semiconductor composites, solid-state dye-sensitized solar cell occupy great deal of attention, recently. In the best dye-sensitized solar cells, the light conversion efficiency has reached about 12%.

Further development of photosensitive dye as electron donor in excited state may be achieved using dyads and triads that have long lifetime charge separation state (D^+A^-) without loss of energy for sequential electron transfer. In such systems, the higher efficiency of light conversion as a result of photoelectron transfer to cathode is expected compared to that for a monomolecular dye.

The energy conversion efficiency of three-dimensional dye-sensitized solar cells in a hybrid structure that integrates optical fibers and nanowire arrays was shown to be greater than that of a two-dimensional device. Other innovations – the tandem and hybrid approaches – stacking many solar cells together have been successfully used in conventional photovoltaic devices including DSSCs to maximize energy generation. Quantum dots are receiving widespread attention for the development of promising third-generation photovoltaic devices. Polymer photovoltaic devices have also drawn considerable attention as alternative inexpensive solar cells, and conjugated polymers are promising candidates for use in low-cost electronics and photovoltaics. The development of polymer–fullerene plastic solar cells has made significant progress in recent years.

The construction of an efficient device for photosplitting of water to hydrogen is one of the most important subjects from the viewpoint of solar light energy utilization and storage. Numerous studies in this direction were performed. As a strategy for an effective visible light harvesting, spectral sensitization of wide-bandgap semiconductors by dye molecules has been studied for photocatalytic H_2 production from water. Though the current efficiency of light conversion does not exceed 1%, the photocatalytic H_2 production remains from the viewpoint of practical application to be one of the most promising processes.

As far as the outlook for further developments is concerned, there are all reasons to believe that a slow but steady progress in the area would continue in the next decade. Nevertheless, who knows, new, unexpected bright ideas would emerge and ensure a vigorous success in global utilization of solar energy for mankind.

References

1 Energy Information Administration (2008) Office of Integrated Analysis and Forecasting, U.S. Department of Energy, Washington, DC. Annual Energy Outlook 2009. Available at http://www.eia.doe.gov/oiaf/ieo/index

2 Hoffmann, W.(2006) *Sol. Energy Mater. Sol. Cells*, **90**, 3285–3311.

Index

a

Ag/AgNO$_3$, photoelectrochemical cell   179
antennas   46
- bacterial antenna complex proteins   47–49
- light-harvesting antennas   46
- photosystems I and II harvesting antennas   49–53

Au(111) electrodes   141
AuNP/ITO electrodes   141
Az (An) electron exchange, logarithmic plots of   139

b

back electron transfer (BET) reaction   129
bacterial antenna complex proteins   47.
   See also Light-harvesting complex
BHJ solar cells   205
bisPC71BM, molecular structures of   228
bis(phenylethynyl)anthracene (BPEA)   99
blue copper protein   139, 140
Boltzmann constant   34
Born–Oppenheimer approximation (BOA)   33
borondipyrromethene (BDPY)   91
bovine serum albumin (BSA)
- with dual fluorophore–nitroxide probe   117
- FN1 incorporated into   116

bpyRu-C9-Ad bound to P450cam, structure of   115
Brunauer–Emmett–Teller (BET), surface area measurement   240
bulk heterojunction (BHJ)
- photocatalyst of interfacing CaFe$_2$O$_4$ and MgFe$_2$O$_4$ 238
- photovoltaics with   225
- polymer:fullerene   227, 228
- solar cells
- – consisting of P3OT and C$_{60}$, positive impact on   226
- – metal oxide nanocomposites, effect of   180, 212
- structures, development of   213

c

cadmium sulfide (CdS), nanoparticles of   238
C$_{60}$ anion radical   137
carbon nanodiamond (ND)   135
carbon nanotubes (CNTs)   103, 134, 180, 186, 220
carotenoid   99
cascade tunneling hypothesis   55
CdSe nanocrystals   145, 167, 222
CdSe nanoribbons, light irradiation   237
CdSe-sensitized electrodes, ZnS coating on   210
CdSe–TiO$_2$ nanocomposites, formation of   222
CdS nanocrystals   210
CdS quantum dots   210
CdS/TiO$_2$ composite system, photoinduced electron transfer   129
C101 dye   169
CEPO. See coherent electron-phonon oscillator (CEPO)
charge carrier   104, 154, 226
- systems   175–182
charge recombination (CR)   91
charge separation (CS)
- in donor–acceptor pairs   92
- – requirements for   92, 93
- electron transfer models   35
- for photogenerated radical pairs   28
- in photosynthetic reaction centers, level structure   60
- process   91

Solar Energy Conversion. Chemical Aspects, First Edition. Gertz Likhtenshtein.
© 2012 Wiley-VCH Verlag GmbH & Co. KGaA. Published 2012 by Wiley-VCH Verlag GmbH & Co. KGaA.

– spin effects on 28, 29
charge transfer
– modeling beyond Condon approximation 10
– NC effects in kinetics of 10, 15
– static and dynamic torsional disorder on kinetics of 10
charge transfer exciton (CTE) 224
– emission intensity 224
– photoluminescence intensity of 226
chemical bath deposition (CBD) 170, 210
chenodeoxycholic acid (DCA) 161, 223
Chlorobium tepidum 70
chlorophyll 45, 46, 50, 62, 70, 75, 120, 141, 241
cofacial porphyrin dimers (CPDs) 95
cofactor–protein complex 59
coherent electron–phonon oscillator (CEPO) 8, 9, 28
colloidal quantum dot photovoltaic devices 210
concerted, and multielectron processes 38–40
constrained density functional theory (CDFT) 8, 15
– electron transfer coupling elements from 8
– energies 8
– weak interactions encountered in 15
copper surface, photoelectron transfer 144
counterelectrodes (CEs) 182
CS. *See* charge separation (CS)
cubanes 243, 244
– cubanes [Mn$_4$O$_4$]$^{6+}$ 244
– PCET reactions of 245
– reaction pathways 247
CuPcF$_{16}$
– film deposited on a GaAs(100) wafer 132
– lowest unoccupied molecular level (LUMO) 133
– photoinduced electron transfer from 133
current density
– for hydrogen evolution at Pt electrode 34
– short-circuit 200
– of top and bottom cells 202
– *vs.* voltage 174
current–voltage characteristics 135
cyanobacteria 50, 52, 53, 68, 69, 72
cyclic tetrapyrroles 93–101
cyclic voltammetry 138, 139, 142
cyclopentadithiophene (CDT)-based low-bandgap copolymers 212
cytochrome P450scc (CYP11A1) 142

d

Dawson-type dirhenium(V)-oxido-bridge POM 237
D149 chemical structures of 167
dendric structures 134
density functional theory (DFT) 9, 127, 141, 159, 160, 175
deoxycholic acid (DCA) 167, 223
p-diaminobenzene 101
di-branched DSSC organic sensitizers, structures of 162
4,40-dimethyl-2,20-bipyridinium 240
direct adsorption (DA) 210
donor–acceptor systems 101–106, 134
donor F6T2, chemical structures of 206
DSSCs. *See* dye-sensitized solar cells (DSSCs)
dual flourophore–nitroxide molecules
– chemical structures 108
– factors affecting light energy conversion in 116–118
– photophysical and photochemical processes in 106, 107
– – system 1 107–109
– – system 2 109–113
dual FNO$^{\bullet}$ as model systems, for conversion of light energy 106
dye molecules, photoinduced processes on semiconductor nanoparticles 128
dye-sensitized heterojunctions (DSHs) 186
dye-sensitized solar cells (DSSCs) 91, 153, 157, 158, 161, 173, 183, 185, 199, 203, 208. *See also* individual cells
– carbon black-based counterelectrodes 184
– design and principle of 201
– efficiency 229
– electron transfer processes, schematic diagram of 154
– fabricated on 200
– fluorine-doped tin oxide (FTO) 183
– IPCE of 206
– IR 166
– quantum dot 210
– quasi-solid-state 188
– solid-state 185
– structure of 203
– three-dimensional 200
dye-sensitized solar cells I
– cathode 182–185
– charge carrier systems 175–182
– general description 153–155
– Grätzel design 155–157
– injection/recombination 171–175
– photoanode 168–171
– sensitizers

– – metalloporphyrins 159–161
– – organic dyes 161–167
– – ruthenium complexes 158, 159
– – semiconductor 167, 168
– on solar cells 151–153
– solid-state 185–189
dye-sensitized solar cells II
– fabrication of 219–223
– fullerene-based solar cells 223–229
– optical fiber 199–202
– polymers in 211–218
– quantum dot (QD) 208–211
– tandem approach 202–208
dye-sensitized titanium dioxide surfaces 216
dye solar cells 147, 153
dye structures 178

e

EIS. *See* electrochemical impedance spectroscopy (EIS)
electrochemical electron transfer
– adiabatic 37
– probability of elementary act in outer-sphere 37
– specificity of 33–38
electrochemical impedance spectroscopy (EIS) 138, 165, 175, 216
electrochemical proton-coupled electron transfer (EPCET) 34, 35
electron-deficient bithiazole (BT) 212
electron diffusion coefficient 175
electron diffusion length 171–173
electron donor–acceptor composite system 133
electron exchange processes 25
electron flow, through proteins 113–116
electronic excitations, in conjugated polyme/fullerene blends 227
electron–proton transfer coupling 29–33
electron transfer (ET) 1, 91, 172. *See also* charge transfer
– Born–Oppenheimer (BO) formulation 15
– cumulant expansion (CE) method to derive 12
– dependent of the Franck–Condon weighted density 13
– dynamics in low temperature 16
– effect of solvent fluctuations 17
– electronic and nuclear quantum mechanical effects 5–7
– energy *vs.* reaction coordinates 2
– Franck–Condon principle 1
– Gibbs energy for 12
– Marcus equation, validation 4

– methods to study, kinetics of 54
– role by electronic interaction 3
– schematic representation of 137
– temperature dependence 14
– theoretical models 1–7
– transition coefficient 3
– variation in energy of system and 2
electron transport 173. *See also* electron transfer
– blends suitable way to improve 226
– chain based on CdS/Cu_xS nanoheterostructures 238
– and chemically sensing properties 145
– with coherent electron–phonon oscillator 9
– dynamics of 173
– fullerene phase in 227
– intrinsic 169
– in molecular wire junctions 141
– pyrene-sensitized 118
– rate 169
– and recombination in dye-sensitized solar cells 173
– through dyad molecules 131
– voltammetric studies on 120
electron tunneling pathway 113, 114
ENDOR spectroscopy 98
energy conversion efficiency 157
energy-storable dye-sensitized solar cell (ES-DSSC) 181, 185
exchange integral 25

f

F-doped SnO_2 (FTO) 184
Fermi level 33–35, 37, 141, 151, 172
ferredoxin 45, 68, 69, 71
ferrocene–porphyrin–C_{60} triads 143
– self-assembled monolayers of 142
FE-SEM micrographs
– NW arrays, images 222
– of TiO_2 photoelectrode 163, 216
fill factor (FF) 163
flavodoxin 68, 71
fluorescence, electrochemical modulation of 142
fluorescence emission 138
fluorinated copper phthalocyanine. *See* $CuPcF_{16}$
fluorine-doped tin oxide (FTO) 183, 221
fluorophore–nitroxide compounds 107
Förster resonance energy transfer (FRET) 46
Franck–Condon factor 2, 23
free energy 2, 6, 12, 23, 31, 33, 59, 180
fullerene-based solar cells 223–229

g

GaN:ZnO catalysts, H_2 and O_2 evolution   247
Gertzel dye-sensitized solar cell   155
Gibbs energy   12, 18, 21–23, 72, 118
glassy carbon (GC) electrode   135
gold nanoclusters   141, 142
graphene nanoplatelets
– electrocatalytic properties of   184
graphene oxide (GO) suspension   256
graphene Pt nanoparticles, dispersion of   255
Grätzel design   155–157

h

hollow core-mesoporous shell carbon (HCMSC)   219
homodimeric PSII complex   73
hybrid polymer solar cells, morphology on   218
hydrogen, photogeneration of   240, 242
hydrophobic magnetic nanoparticles (NPs)   138
hydroxylamine derivatives   107
hyperfine sublevel correlation (HYSCORE) spectroscopy   81

i

idealized photovoltaic converter   152
impedance spectroscopy (IS)   171
incident photon-to-current efficiencies (IPCE)   159, 163, 180, 204, 206, 209, 211, 252
indium tin oxide (ITO)   131, 141
– electrodes   141
– layer   200
inducible nitric oxide synthase (iNOSoxy)   115
In_2S_3 SEM images of   220
integral encounter theory   6
intensity-modulated photocurrent spectroscopy   173
interfacial electron transfer (IET)   127
intramolecular electron transfer   103
ionic liquid electrolytes   181
IPCE. See incident photon-to-current efficiencies (IPCE)
IR absorption   133
IR bandgap   210
IR spectroscopy   130, 135, 216, 243
ITO nanoelectrodes   212

j

JK-64/JK-65/JK-64Hx, structure of   164
J–V curves, for devices constructed from compounds   131

k

$KCa_2Nb_3O_{10}$, structure   239
$K_4Nb_6O_{17}$, structure   239

l

lamination, schematic representation of   214
Landau–Zener model, of electron transfer   1–3
layered basic zinc acetate (LBZA) nanoparticle   173
layered hydroxide zinc carbonate (LHZC)   170
LC/SAM/Au electrode, electron transfer on   138
lead selenide (PbSe) nanocrystals   131
Lhca4 polypeptide   51
LHC II trimer   52
light energy conversion   45, 69, 91, 127
light-harvesting antennas   46
– requirements for   47
light-harvesting complex
– dynamic processes in   48, 49
– of plants   46
– structure of   47, 48
light-harvesting efficiency   206
long-range electron transfer (LRET)   24–28
– analysis of biological systems   28
– decay factor   25
– dependence of logarithm of relative parameters of   26
– rate constant of TTET   25
– triplet–singlet spin conversion of radical pair   26
– values of attenuation parameter of   27
low-bandgap polymer fullerene solar cell (pBEHTB)   224
lowest unoccupied molecular orbital (LUMO)   133, 166, 180, 215, 218
LRET. See long-range electron transfer (LRET)

m

manganese–oxo cube $[Mn_4O_4]^{n+}$ 243
Marcus equation   242
– analysis   139
– classical   242
– semiclassical   103
– validation   4
Marcus model, of electron transfer   3–7
– driving force and reorganization energy   9–17
– electron coupling   7–9
Marcus theory   4, 8, 9, 13, 15, 95, 140
– applied to interfacial ET kinetics   38
– potential dependence yielded an extremely low value for   140

MCM-48 silica 138
3-mercaptopropionic acid (MPA) 210
mesoporous carbon (MC) 171
mesoporous titanium dioxide 163
metal cations, doping of 250
metallic bonds 136
metalloporphyrins 160
– application of 159
MKZ-21 molecular structures of 164
Mn_3O_4
– GaN:ZnO modified
– – water splitting on 248
– Mn_3O_4 NPs as 246
Mn_4O_4-cubane water oxidation catalysts, development of 243
Mn–oxo cubane complex 245
molar extinction coefficient 158
MoS_2 ultrasonication of 238
multiwalled carbon nanotube (MWCNT)-free solar cells 216

n

Nafion 245
– coated polypyrrole 180, 185
– film 244
– schematic persantation 246
nanocarbon materials 133
nanostructured NiO film 202
nanowire (NW) arrays 199
naphthalenediimide (NDI) 98
naphthaleneimide (NI) 98
nickel oxide (NiO)
– nanopowders 206
– photocathodes 206
– thin films 204
– – erythrosin J-sensitized nanostructured 204
NK-4432, molecular structures of 167
NK-6037, molecular structures of 167
nonadiabatic electron transfer 11
– dependence of ET constants logarithm on 23
– electronic potential energy surfaces 22
nonadiabatic theory, for electron transfer 8
nonequilibrity, on driving force 21–23
nonlinear recombination kinetics, effect of 171
NP composite 173, 174

o

open-circuit voltage 160
organic sensitizers 163
– comprising donor 161

$OTE/SnO_2/ND$–TCPP, current–voltage characteristics 135
oxygen evolution complex (OEC) 72, 242

p

pBTTT, molecular structures of 228
PC71BM, molecular structures of 228
PCET. See proton-coupled electron transfer (PCET)
p-conducting Nafion membrane 244
perturbation molecular orbital (PMO) theory 9
Ph C_{61} butyric acid Me ester (PCBM) 131, 144, 205, 213, 217, 224, 226
(6,6)-phenyl-C61-butyric acid methyl ester (PCBM)) solar cells 205, 218, 224, 225
phospholipid-linked naphthoquinones 135, 136
photoanodes 168–171
– photovoltaic performance of 219
photocatalytic dihydrogen production 236. See also photocatalytic water splitting
– H_2 evolution
– – over TiO_2 236, 237
– – semiconductor 237–240
– – from water based on platinum and palladium 240, 241
– O_2 evolution
– – water splitting 241
photocatalytic reduction 235
photocatalytic water splitting 241, 255. See also water splitting
– heterogeneous process of 249
– H_2 evolution 241
– – Mn clusters 242–248
– – thermodynamics, and feasable mechanism of 241, 242
– O_2 evolution 241
– schematic energy diagrams of 249
– on semiconductor 250
photoconductive atomic force microscopy 224
photoconversion properties 218
photocurrent generation mechanism 143
photodetector 199
photoelectrochemical cell 235
– catalyzes 243
photoelectrochemical system 153
– charge transport, differerence 236
photoelectrochemical water splitting, using hematite 251
photoelectrochem system 250
photoelectron transfer, on copper surface 144
photoevolution dioxygen 241

photogenerated catalysis   236
photoinduced charge injection   132
photoinduced charge transfer   145
photoinduced electron injection   133
photoinduced electron transfer   7, 15, 91, 95, 101, 114, 133, 141, 143, 144
photoinduced hydrogen producing system   241
photoinduced interlayer electron transfer, in lipid films   118–121
photoinduced processes, occurring in (ZnP)2-Cu(I) (phen)2, 95
photooxidation, of water   46
photosynthesis   45, 46
– problems of   45
photosystem II (PS II) complex   77
photosystems I and II harvesting antennas   49–53
photovoltaic current   154
photovoltaic (PV) fiber   200
photovoltaic parameters   182
phycobilisome   52, 53
phycocyanobilin   52
phycoerythrobilin   52
picosecond laser flash photolysis experiments   130
plastic film electrodes   128
plastoquinol–plastoquinone exchange   73
platinum thermal cluster catalyst (PTCC)   180
poly(3,4-alkylenedioxythiophene) films   185
polyaniline (PANI)   183
poly[3,4-ethylenedioxythiophene: paratoluenesulfonate] (PEDOT:PTS)   184
polyethyleneterephthalate (PET) substrate   215
poly(3-hexylthiophene) (P3HT)   131, 144, 212, 213, 218, 226
– molecular structures and absorption spectra of   214
– P3HT:PCBM bulk heterojunction solar cells   218
polymer electrolyte   216
polymer/fullerene blend films   228
polymer:fullerene bulk heterojunction organic solar cells   227
polymer–fullerene solar cells
– charge transfer and transport   224
– open-circuit voltage of   225
polymer photovoltaic devices   211
polymer solar cells   211–218
polymer tandem solar cell, device structure of   205
poly(3,4-methylenedioxythiophene)   183

poly-(3-(2-methylhexyloxycarbonyl) dithiophene) (P3MHOCT)   213
polyphenol oxidase (PPO), adsorption of   136
poly (3,4-ethylenedioxythiophene):poly(styrene sulfonic acid) (PEDOT:PSS)   217
poly(p-phenylenevinylene) (PPV)   213
polypyrrole nanoparticles (NPs)   164
poly(4-styrenesulfonic acid) (PSS) nanowires   145
poly(6,6′,12,12′-tetraoctylindeno[1,2-b] fluorene-co-5,7-dithien-2-yl-thieno [3,4-b] pyrazine) (PIF-DTP)   216
polythiophene (PT)   213
polythiophene:fullerene solar cells, thermal stability of   225
porphyrin (Por)   98, 135, 159, 160, 161
– molecular orbitals (MOs) of   159
potassium niobate nanoscrolls   239
power conversion efficiency (PCE)   204
1-propyl-3-methylimidazolium iodide (PMII)   181
proton-coupled electron transfer (PCET)   29
– heterogeneous   35
– importance in redox reactions with   244
– magnetic properties of protonated semiquinone and   63
– reactions of cubane model complexes   245
– regimes of   31
– requires reorganization of solvent and   30
– role in chemical and biological processes   29
– theoretical description of   31
proton transfer pathway, in photosynthetic reaction centers   74
Pseudomonas aeruginosa   139
Pt nanoparticles   171
pulsed laser excitation   159
pyrene-sensitized electron transport   118

q
QE. *See* quantum efficiency (QE)
quantum-dot-sensitized solar cell (QDSC)   208–211
quantum dot solar cells   189, 208, 210
quantum efficiency (QE)   138, 153, 218, 220, 250, 251
quantum kramers theory   8
quantum theory, of electron transfer reactions   34

r
Raman scattering spectra   240
rapid-scan FTIR difference spectroscopy   75

reaction center of photosynthetic bacteria
  (RCPB)   53–56
– electron transfer, and molecular
    dynamics in   64–68
– kinetics and mechanism, of electron
    transfer in   58–64
– model of   55
– structure of   56–58
reaction centers of photosystems
    I and II   68, 69
– reaction center of photosystem II
    72–76
– reaction centers of PS I   69–72
red absorbing chromophores   163
ReI(CO)$_3$(dmp)-His124)|(Trp122)|
    AzCuI   113
– architecture of   114
– kinetic model, of photoinduced electron
    transfer in   113, 114
resonance integral   25
reversible hydrogen electrode (RHE)   252
Rhacophorus viridis   55, 56
Rhodobacter sphaeroides   48
– energy transfer
– – within B850 ring of light-harvesting
    complex   48
– – effect of Bchl redox potential of   59
– mutant reaction centers from   68
– pheophytin-modified reaction
    centers   60
– proton transfer pathways   65, 74
– reaction center from   56–59, 64, 75
– – crystallographic structures   63
– – light minus dark difference absorption
    spectra   61
– – structure and mechanism   57
– role of dynamic effects on   67
Rhodopseudomonas palustris   47
Rhodopseudomonus viridus   54
rotaxane molecules   94
Ru bipyridyl complexes   130, 244
Ru–I complex geometry of   175
Ru^{II}–Pd^{II} complex Pd colloid system,
    dihydrogen evolution   239
ruthenium complexes, ligands of   158

S

SAMgold electrodes   140
SBA-15/Co$_3$O$_4$, TEM images of   253
SBA-15/NiO, oxygen evolution   254
scanning electron microscopy (SEM)   173,
    205, 220
scanning Kelvin probe microscopy
    (SKPM)   223
scanning tunneling microscope (STM) break
    junction method   130
self-assembled monolayer (SAM)   244
semiconductor-sensitized analogue
    (SSSC)   167, 168
semiconductors, redox processes   127–147
– on carbon materials   133–136
– electron transfer
– – donor–bridge–nanoparticle acceptor
    complexes   130–132
– electron transfer in   144–147
– on metal surfaces   136–144
– sensitized TiO$_2$
– – interfacial electron transfer
    dynamics   127–130
silicon naphthalocyanine bis(trihexylsilyl
    oxide) (SiNc)   227
silicon phthalocyanine (SiPc) triads   99
single-walled carbon nanotubes
    (SWNTs)   103, 134, 135
– electron donor–acceptor hybrids   134
– immobilization strength   135
– polyphenol oxidase (PPO), adsorption
    of   136
solar cells
– components, fabrication of   219
– energy conversion efficiency   153
– equivalent circuit   151, 152
– photocatalysts/dye sensitizers in   127
solar hydrogen production systems   250
sol–gel-processed nanoparticles   189
solid-state dye-sensitized photovoltaic
    cell   187
solid-state/nanocrystalline solar cells   187
solvent-free liquid redox electrolytes   181
Soret band   161
spin-coated PEDOT:PSS   215
spin exchange (SE)   25
spin-selective charge transport pathways
    102
SRhB–silane–SnO$_2$, schematic structure
    of   133
stacked cup C nanotube (SCCNT)   209
Stark shift   131
structure-sensitizing dyes   203
substrate–ITO–PEDOT:PSS–(active layer)–
    aluminum   215
sulforhodamine B (SRhB)   130
superexchange   25
supported bilayer lipid membrane
    (s-BLM)   138
SWNTs. *See* single-walled carbon nanotubes
    (SWNTs)
Synechococcus elongatus   69

t

tandem dye-sensitized solar cells (TDSC)   203, 207
TC201-TC602, molecular structures of   166
TDDFT. *See* time-dependent density functional theory (TDDFT)
TDSC. *See* tandem dye-sensitized solar cells (TDSC)
tetra *n*-butylammonium (TBA)   158
tetrathiafulvalene (TTF) sensitizers, photooxidized   128
Thermochromatium tepidum   56
Thermosynechococcus elongatus   72
thiazolothiazole (TZ)   212
thiophene-substituted metalloporphyrins, DFT study of   160
thylakoid membranes   118
time-dependent density functional theory (TDDFT)   56, 63, 159
time-resolved fluorescence spectra of dansyl-TEMPO dual probe   117
titanium dioxide (TiO_2)
– bandgap excitation of   178
– CdSe nanoparticles, coupling of   146, 223
– chemical vapor deposition   128
– chromophores   156
– conduction band (CB)   208
– Cu_2O microgrid   251
– CuSCN layers   188
– different molecular layers   165
– distance dependence of   179
– DSSC composed of   202
– dye-sensitized solar cells   169
– dye/solution interface   129
– energy level diagram   146, 178
– fabrication protocols for   207
– FE-SEM micrographs   216, 217
– hollow fibers
– – DSSC, nanoribbon network electrode for   169
– – nanostructured   169
– mesoporous films   170
– nanobelt–ZnO nanowire array   168
– nanoparticles   175
– nanorod/nanotube (NR/NT) adjacent film   219
– nanotube(NT)array   177
– photoanode, nanocrystalline   221
– photovoltaic performance   177
– scheme of   211, 251
– semiconductor photocatalysts   239
– SEM images of   220
– transient kinetics of   172
titanium dioxide nanotubes (TiO_2 NTs)   170

transmission electron microscopy   173
transparent conducting oxide (TCO)   170, 186
transparent electrode   154
triethanolamine (TEOA)   240
trinitrodicyanomethylenefluorene SiPc-(TNDCF)   101
trinitrofluorenone SiPc-(TNF)   101
triphenylamine (TPA)   102
– derivatives   165
triphenylamine-based metal-free organic dyes (TC15)   188
triplet–triplet energy transfer (TTET)   25
TTexchange integral   25
tunneling-electron rate constants, plots of logarithms   116

u

ultrafast electron transfer (UFET)   8, 28, 139, 145, 209
ultrafast vibrational spectroscopy   131
ultraviolet photoemission spectroscopy (UPS)   132
UV–visible absorption spectra   170

v

valence band (VB)   127, 202, 207, 252
variational transition-state theory   7
vectorial multistep electron transfer   137

w

water oxidation system (WOS)   235. *See also* photosystem II (PS II) complex
– based on titanium oxides   250, 251
– heterogeneous catalysts for
– – general   248–250
– O_2 evolution   245
– photocatalysis   246, 250
– semiconductors for   251–256
water oxidizing complex (WOC)   82, 242, 243, 245, 246
water splitting   249. *See also* photocatalytic water splitting
– hydrogen generation   242
– O_2 evolution   241
– photochemical splitting of   235
– RuO_2 nanoparticles   254
– seawater   254
– by two-photon process   256
WOC. *See* water oxidizing complex (WOC)
WOS. *See* water oxidation system (WOS)

x

X-ray diffraction (XRD)   186, 240, 242

z

zeolite-based materials 250
zeolite membrane 238
zinc naphthalocyanine 225
zinc oxide (ZnO)
– Al transparent current collector layer 199
– nanowire arrays 170, 173, 174, 201, 222
– – schematics of 174
– porous electrode 189
zinc phthalocyanine 95, 225
zinc porphyrin 93, 96, 99
– arrays 97
– β-substituted 160
– and C_{60} fullerene moieties incorporated 93
– hexaphenylbenzene core bearing 99
ZINDO-1 parameter method 132
ZrO_2 surface, role of 144

Zusman model 17–21. *See also* Electron transfer (ET)
– equations generalization to 18
– equation to include quantum nuclear dynamics 19
– ET reaction rates calculation 18, 19
– features of ET of complex molecules 20, 21
– Gibbs energy activation 18
– influence of solvent polarization diffusion 19
– resonance splitting, lowers activation free energy 21
– treatment of experimental data for complex systems 20